NANOPARTICULATE DRUG DELIVERY SYSTEMS

NANOPARTICULATE DRUG DELIVERY SYSTEMS

Strategies, Technologies, and Applications

Edited by

YOON YEO
College of Pharmacy
Weldon School of Biomedical Engineering
Purdue University
West Lafayette, Indiana

Library of Congress Cataloging-in-Publication Data:

Nanoparticulate drug delivery systems; strategies, technologies, and applications / edited by Yoon Yeo

 Includes bibliographic references and index.

 ISBN 978-1-118-14887-7 (cloth)

Printed in the United States of America

10 9 8 7 6 5 4 3 2 1

CONTENTS

PREFACE

Enthusiasm for nanomedicine has grown exponentially over the years with an expectation that nanomedicine will enable the delivery of drugs or imaging agents to tissues and organs that they would otherwise not reach effectively. Early nanomedicine indeed addressed several critical problems inherent to some drugs, such as poor solubility and toxic side effects. Liposomal doxorubicin, micelle formulations of paclitaxel, and protein-bound paclitaxel are now used in clinical practice. At the same time, new approaches have continuously emerged with the view to develop ever more sophisticated, multifunctional nanoparticulate systems for more effective diagnosis and safe therapy of diseases.

On the other hand, this field has faced several challenges in translating novel ideas into clinical benefits. For example, the "active targeting" strategy, originally expected to increase tumor accumulation of nanoparticles by orders of magnitude, has fallen short of the expectations. Polymeric micelles or nanoparticles, designed to circulate in the blood for a prolonged period of time, are unstable in the presence of amphiphilic biological components, losing their ability to reach target tissues in intact form. Consequently, despite the increasing complexity, newer nanoparticle systems bring about only modest therapeutic benefits, failing to gain significant attention from commercial, clinical, and/or regulatory sectors. In order to further advance the field of nanomedicine and to develop clinically effective products, it is necessary to make a practical assessment of the potential and challenges of modern nanomedicines.

This book brings together a collection of recent nanomedicine technologies and discusses their promises and the remaining challenges. It does not intend to duplicate the existing compilations of established nanoparticle systems, nor does it attempt to broaden the topic to systems that are less relevant to drug delivery. Therefore, instead of covering established nanoparticulate systems that have been described in a number of recent books and review articles, we focus on the rationales and preclinical evaluation of relatively new nanoparticulate drug carriers.

The chapters of this book are organized with two goals in mind. The first chapter presents a general overview of targeted nanomedicine. Chapters 2–8 discuss nanoparticulate drug delivery systems that have gained increasing recognition in the recent literature. Chapters 9–12 discuss new opportunities and barriers in the biology relevant to drug delivery based on nanomedicine. Each chapter reviews a state of the art of the topic with extensive references and concludes with an open assessment of the remaining challenges. I hope that this arrangement helps readers to formulate innovative nanomedicine systems and to design evaluation strategies without reliving the existing experiences, be they positive or negative.

As the editor of this book, I am most appreciative of the insightful and comprehensive contributions of the chapter authors. Special thanks go to all the authors, who immediately agreed to contribute their time and effort, and have endured reminders and editing requests. I would also like to thank the staff members of the publisher, Wiley-Blackwell, especially Jonathan T. Rose, for their patience and support. It is my hope that this book will serve as a starting point for stimulating discussions and new experimentations toward the development of better nanomedicines that can translate in the near future into clinical benefits for patients.

YOON YEO

CONTRIBUTORS

Jin Woo Bae, Department of Biopharmaceutical Sciences, College of Pharmacy, University of Illinois, Chicago, IL, USA

You Han Bae, Department of Pharmaceutics, University of Utah, Salt Lake City, UT, USA

Gaurav Bajaj, Division of Clinical Pharmacology and Therapeutics, The Children's Hospital of Philadelphia, Philadelphia, PA, USA

Jianjun Cheng, Department of Materials Science and Engineering, University of Illinois at Urbana–Champaign, Urbana, IL, USA

Ji-Xin Cheng, Weldon School of Biomedical Engineering, Purdue University, West Lafayette, IN, USA

Kuiwon Choi, Center for Theragnosis, Biomedical Research Institute, Korea Institute of Science and Technology, Seongbuk-gu, Seoul, Republic of Korea

Nathan P. Gabrielson, Department of Materials Science and Engineering, University of Illinois at Urbana–Champaign, Urbana, IL, USA

Oleg V. Gerasimov, Department of Chemistry, Purdue University, West Lafayette, IN, USA

Khaled Greish, Department of Pharmacology and Toxicology, Otago School of Medical Sciences, University of Otago, Dunedin, New Zealand; Department of Oncology, Faculty of Medicine, Suez Canal University, Egypt

Christin P. Hollis, College of Pharmacy, University of Kentucky, Lexington, KY, USA

Seungpyo Hong, Department of Biopharmaceutical Sciences, College of Pharmacy, University of Illinois, Chicago, IL, USA

Leaf Huang, Division of Molecular Pharmaceutics and Center for Nanotechnology in Drug Delivery, Eshelman School of Pharmacy, University of North Carolina at Chapel Hill, Chapel Hill, NC, USA

Kwangmeyung Kim, Center for Theragnosis, Biomedical Research Institute, Korea Institute of Science and Technology, Seongbuk-gu, Seoul, Republic of Korea

Heebeom Koo, Center for Theragnosis, Biomedical Research Institute, Korea Institute of Science and Technology, Seongbuk-gu, Seoul, Republic of Korea

Ick Chan Kwon, Center for Theragnosis, Biomedical Research Institute, Korea Institute of Science and Technology, Seongbuk-gu, Seoul, Republic of Korea

Sangbin Lee, School of Life Sciences and Biotechnology, Korea University, Seoul, South Korea

Seung-Young Lee, Weldon School of Biomedical Engineering, Purdue University, West Lafayette, IN, USA

Tonglei Li, Department of Industrial and Physical Pharmacy, Purdue University, West Lafayette, IN, USA

Hayley Nehoff, Department of Pharmacology and Toxicology, Otago School of Medical Sciences, University of Otago, Dunedin, New Zealand

Joseph W. Nichols, Department of Bioengineering, University of Utah, Salt Lake City, UT, USA

Lin Niu, Department of Pharmaceutics, College of Pharmacy, University of Minnesota, Minneapolis, MN, USA

Yu-Kyoung Oh, College of Pharmacy, Seoul National University, Daehak-dong, Seoul, South Korea

Jayanth Panyam, Department of Pharmaceutics, College of Pharmacy, University of Minnesota, Minneapolis, MN, USA; Masonic Cancer Center, University of Minnesota, Minneapolis, MN, USA

Joo Yeon Park, College of Pharmacy, Seoul National University, Daehak-dong, Seoul, South Korea

Ryan M. Pearson, Department of Biopharmaceutical Sciences, College of Pharmacy, University of Illinois, Chicago, IL, USA

Marquita Qualls, Department of Chemistry, Purdue University, West Lafayette, IN, USA

Lee Dong Roh, College of Pharmacy, Seoul National University, Daehak-dong, Seoul, South Korea

Gayong Shim, College of Pharmacy, Seoul National University, Daehak-dong, Seoul, South Korea

Pochi Shum, Department of Chemistry, Purdue University, West Lafayette, IN, USA

Li Tang, Department of Materials Science and Engineering, University of Illinois at Urbana–Champaign, Urbana, IL, USA

Sebastien Taurin, Department of Pharmacology and Toxicology, Otago School of Medical Sciences, University of Otago, Dunedin, New Zealand

David H. Thompson, Department of Chemistry, Purdue University, West Lafayette, IN, USA

Rong Tong, Department of Chemical Engineering, Massachusetts Institute of Technology, Cambridge, MA, USA Laboratory for Biomaterials and Drug Delivery, Department of Anesthesiology, Division of Critical Care Medicine, Children's Hospital Boston, Harvard Medical School, Boston, MA, USA

Yuhua Wang, Division of Molecular Pharmaceutics and Center for Nanotechnology in Drug Delivery, Eshelman School of Pharmacy, University of North Carolina at Chapel Hill, Chapel Hill, NC, USA

Yoon Yeo, Department of Industrial and Physical Pharmacy, Purdue University, West Lafayette, IN, USA

Qian Yin, Department of Materials Science and Engineering, University of Illinois at Urbana–Champaign, Urbana, IL, USA

EDITOR

Yoon Yeo, Ph.D., is an Assistant Professor of Industrial and Physical Pharmacy at the College of Pharmacy, with a joint appointment as Assistant Professor at the Weldon School of Biomedical Engineering, at Purdue University. She earned her B.S. in Pharmacy and M.S. in Microbial Chemistry at Seoul National University in Korea, and her Ph.D. in Pharmaceutics at Purdue University in the United States. She completed postdoctoral training in Chemical Engineering at Massachusetts Institute of Technology and returned to Purdue as a faculty member. Her research focuses on nanoparticle surface engineering for drug delivery to solid tumors, inhalable drug/gene delivery for cystic fibrosis therapy, and functional biomaterials based on carbohydrates. Dr. Yeo has published 52 peer-reviewed papers and 7 book chapters, and received the NSF CAREER Award (2011) and New Investigator Awards from the American Association of Pharmaceutical Scientists (2009) and American Association of Colleges of Pharmacy (2008).

1

TUMOR-TARGETED NANOPARTICLES: STATE-OF-THE-ART AND REMAINING CHALLENGES

GAURAV BAJAJ

Division of Clinical Pharmacology and Therapeutics, The Children's Hospital of Philadelphia, Philadelphia, PA, USA

YOON YEO

Department of Industrial and Physical Pharmacy, Purdue University, West Lafayette, IN, USA

1.1 INTRODUCTION

Bringing a drug only to target tissues without spilling any molecule in unwanted places would be an ideal goal for any pharmacological therapies [1]. Owing to the dose-limiting side effects of chemotherapy, a number of drug delivery strategies have been developed in the context of cancer, where "targeted" therapy is most anticipated. Most tumor-targeted drug delivery systems are based on the fact that cancer cells express various molecular markers that are distinguished from those of normal cells. In particular, nanomedicines have received enormous attention in the past decades as a potential tool to increase the selectivity of chemotherapy and diagnosis, due to the small size conducive to circulation and the large surface area to volume ratio that facilitates surface functionalization. Several nanomedicine products have been launched in the market or in the clinical development stage as summarized in Table 1.1. Newer approaches are actively developed to increase target selectivity, although the number of targeted nanomedicines that reached the later phase of

Nanoparticulate Drug Delivery Systems: Strategies, Technologies, and Applications, First Edition.
Edited by Yoon Yeo.

TABLE 1.1 Nanomedicines on the Market or in Phase II/III Clinical Development

Type of Nanoparticles	Formulation	Drug	Brand Name (Company)	Indication(s)	Current Status and Reference
Lipid-based or liposomes	Lipid–drug complex	Amphotericin B	Abelcet (Sigma-Tau)	Antimicrobial	Market [3]
	PEGylated liposome	Doxorubicin	Doxil and Caelyx (Janssen)	Breast cancer, ovarian cancer, multiple myeloma, Kaposi's sarcoma	Market [4]
	Non-PEGylated liposome	Doxorubicin	Myocet (Enzon)	Breast cancer	Market [5]
	Non-PEGylated liposome	Daunorubicin	DaunoXome (Galen)	Kaposi's sarcoma	Market [6]
	Non-PEGylated liposome	Cytarabine	DepoCyt (Sigma-Tau)	Lymphomatus meningitis, leukemia, glioblastoma	Market [7]
	Non-PEGylated liposome	Amphotericin B	Ambisome (Gilead)	Fungal infection, cryptococcal meningitis	Market [8]
	PEGylated liposome	Cisplatin	Lipoplatin (Regulon)	Various malignancies	Phase III, investigational drug [9–13]
	Heat-activated PEGylated liposome	Doxorubucin	Thermodox (Celsion Corporation)	Hepatocellular carcinoma, recurrent chest wall breast cancer	Phase III [14]
	Sphingomyelin-based liposomes	Vincristine sulfate (targeted, Optisome™, details not provided)	Marqibo® (Talon Therapeutics)	Non-Hodgkin's lymphoma	NDA submitted [15,16]
Albumin-bound particles	Albumin bound nanoparticle	Paclitaxel (Albumin)	Abraxane® (Astellas)	Breast cancer	Market [17,18]
Polymer-drug/protein/nucleic acid conjugates	PEG conjugated	L-Asparaginase	Oncaspar® (Sigma-Tau)	Acute lymphoblastic leukemia	Market [19]
	Poliglumex (poly-L-glutamic acid) conjugated	Paclitaxel	Opaxio™ or previously Xyotax (Cell Therapeutics Inc.)	Nonsmall cell lung cancer, ovarian cancer	Phase III [20–23]

	Conjugated with uniquely engineered macromolecular polymer core using specialized linkers	Irinotecan	Etirinotecan Pegol (Nektar Therapeutics)	Breast cancer, ovarian cancer, colorectal cancer	Phase III [24]
	PEGylated anti-VEGF aptamer	Pegaptanib sodium (VEGF)	Macugen (Eyetech, Inc.)	Age-related macular degeneration	Market [25]
	PEG conjugated	Recombinant human granulocyte colony-stimulating factor	Neulasta (Amgen)	Prevention of chemotherapy-associated neutropenia	Market [26]
Polymeric micelles	Methoxy PEG-PLA copolymer micelle	Paclitaxel	Genexol-PM® (Samyang Co.)	Breast cancer, lung cancer, ovarian cancer	Phase II [27,28]
Nanocrystals	Nanocrystal drug particles	Aprepitant	Emend (Merck)	Antiemetic	Market [29]
	Nanocrystal drug particles	Sirolimus	Rapamune (Pfizer)	Immunosuppressive	Market [29]
	Nanocrystal drug particles	Fenofibrate	Tricor (Abbott)	Hypercholesterolemia or mixed dyslipidemia	Market [29]
	Nanocrystal drug particles	Megestrol acetate	Megace ES (BMS)	Synthetic progestin	Market [29]
Dendrimers	Carbopol® gel (cross-linked acrylic acid)	SPL7013 (antimicrobial agent)	VivaGel (Starpharma, Dendritic Nanotechnologies Inc., United States)	Bacterial vaginosis	Market, Phase III [3,30]
Cyclodextrin-based nanoparticles	Transferrin receptor-targeted cyclodextrin NPs, RONDEL™	siRNA targeting the M2 subunit of ribonucleotide reductase	CALAA-01 (Calando Pharmaceuticals)	Inhibition of tumor growth	Phase I trial, [31,32]
Inorganic nanoparticles	Streptavidin covalently attached to quantum nanocrystals (QDot)	—	Qdot® 800 Streptavidin Conjugate (Invitrogen)	Imaging, diagnostics, detection of proteins, nucleic acids	Market [33,34]
	Superparamagnetic iron oxide nanoparticles coated with carboxydextran	—	Resovist (Schering)	Hepatocellular carcinoma	Market [35]
	Iron nanoparticles	—	Feridex (AMAG Pharmaceuticals)	Detection of liver lesions	Market [36]

Notes: The table was created through internet search, product websites, and previously published articles [37,38]. Antibody–drug conjugates are omitted, but some publications list them under nanomedicine [38].

3

clinical trials is relatively low considering the prevalent excitement and the investment made till date [2].

This chapter discusses the rationale of targeted nanomedicines, different approaches to achieve this goal, current challenges, and considerations for their future development. Here, nanomedicines refer to particulate materials with a diameter in nanometer scale, prepared with natural or synthetic polymers, lipids, or inorganic solids, in the form of polymer–drug conjugates, liposomes, polymeric nanoparticles (NPs) or micelles, dendrimers, nanotubes, nanocrystals, nanorods, nanoshells, or nanocages [2]. Nanomedicines used as a carrier of a drug are often called nanocarriers. Nanoparticles and nanocarriers are also interchangeably used. We will use the term "nanoparticles" to represent various types of nanocarriers in this chapter.

1.2   FUNCTIONS OF NANOPARTICLES

Fates of a drug entering circulation largely depend on the properties of the drug. Small hydrophilic drug molecules do not bind to proteins in the blood and are quickly eliminated from the body. Hydrophobic drugs are eliminated relatively slowly through kidneys due to plasma protein binding, metabolized in the liver to a hydrophilic metabolite, and then eliminated from the body. On the other hand, the disposition and clearance of drugs encapsulated in NPs depend relatively less on the drug itself than on the properties of NPs during circulation [39].

Systemically administered NPs should be stable enough to retain a drug during circulation and able to extravasate at the target tissues and release the drug in the target tissues and/or inside the target cells to exert an action. Advancement in the nanotechnology has made it possible to incorporate therapeutic drugs in various types of NPs. NPs have contributed to the delivery of poorly water-soluble drugs and improving the bioavailability of drugs with poor stability [2,40,41]. One of the major concerns in cancer chemotherapy is multidrug resistance as most anticancer drugs are substrates of drug efflux transporters such as P-glycoprotein. NPs can be designed to overcome the resistance by bypassing efflux transporters [2,42]. Most significantly, NPs have gained enormous interest as a way of increasing drug delivery to target tissues, in particular, tumors. A detailed discussion of the use of NPs in drug delivery is available in other review articles [2,43].

NPs can be combined with contrast or imaging agents to visualize pathological tissues and monitor the progression of diseases [44]. Liposomes, micelles, and dendrimers are used to carry paramagnetic ions of indium-111 (^{111}In), technetium (^{99m}Tc), manganese (Mn), and gadolinium (Gd). Semiconductor quantum dots or gold NPs have been widely explored for target-specific imaging as combined with various functionalization strategies [45–47].

1.3   TUMOR-TARGETED NANOPARTICLES

Several strategies have been employed to achieve tumor-targeted drug delivery based on NPs. We briefly introduce these targeting strategies before we discuss challenges in translating NPs to commercial products.

1.3.1 Passive Targeting

Delivering anticancer agents to the tumors is a multistep process. Drugs delivered into the bloodstream should reach intratumoral region from systemic circulation by crossing tumor vasculature so that drug molecules have access to the tumor cells and stroma. Many solid tumors are characterized by leaky vasculature and poor lymphatic drainage [43]. New vasculature in the tumor is formed to sustain the tumor growth and provide nutrients; however, the structure is usually abnormal and defective with holes ranging from 400 to 600 nm in diameter [48–50]. The defective structure of blood vessels surrounding tumors allows NPs to extravasate and accumulate in the tumor tissues. The impaired lymphatic drainage further helps retain NPs in the tumors. Therefore, NPs with a size less than the cutoff of holes in the vasculature have a selective advantage in reaching tumors. This phenomenon is referred to as the enhanced permeability and retention (EPR) effect [51–53].

For NPs to take advantage of the EPR effect, it is important that the NPs circulate for a long period of time. It is widely known that NPs with hydrophobic or charged surfaces are readily opsonized and taken up by the reticuloendothelial system (RES), which is responsible for clearing macromolecules or particles present in the blood before they reach the target tissues [54,55]. Therefore, most NPs are surface-modified with a hydrophilic and electrically neutral polymer such as polyethylene glycol (PEG) (PEGylated) [54,56,57]. The PEGylated surface is resistant to protein binding and thus avoids the recognition by the RES [55,58,59]. For long-term circulation and effective extravasation via leaky vasculature of tumors, $\sim$50–300 nm is considered an optimal size [60,61].

1.3.2 Active Targeting

Active targeting strategy was developed to further enhance tumor accumulation of NPs beyond the level possible with long-circulating NPs. Here, NP surface is decorated with targeting ligands that can actively bind to cell receptors overexpressed on the tumor cells [2,49,62–65]. Frequently used targeting ligands include antibodies, peptides, nucleic acids, and small molecular weight ligands binding to specific receptors [2,65,66]. For example, NPs are conjugated with the monoclonal antibody teratuzumab or herceptin to target HER2 receptor on breast cancer cells [59,67]. A docetaxel-NP formulation containing a small molecule ligand which targets prostate-specific membrane antigen has recently entered a Phase I clinical trial [68].

The rationale of active targeting is that a specific interaction between targeting molecules and cell receptors will increase the amount of NPs retained in tumors. An example is shown in a biodistribution study of ^{111}In-labeled polymeric micelle NPs [69]. Nontargeted NPs (25 nm) were rapidly cleared from plasma as compared to nontargeted NPs with a size of 60 nm, resulting in twofold less total tumor accumulation. On the other hand, another 25 nm NPs functionalized with epidermal growth factor achieved a greater tumor accumulation than nontargeted ones in mice with tumors overexpressing EGF-receptors, reaching the same level as that of 60 nm nontargeted NPs [69]. However, it should be borne in mind that the contribution of a targeting ligand is predicated on the extravasation of NPs to the tumors; thus, the ability to survive circulation for a long period of time still remains one of the most important requirements in the active targeting strategy. For effective active targeting, the density of target receptors should be in range of 10^4 or 10^5 copies per cell [2,70]. In addition to the number of binding sites, binding affinity also plays an important role. Multivalency of ligands can further enhance the affinity for target cells [71]. Receptors in a cell are internalized via different pathways and go through a continuous process of recycling [72]. NPs with targeting ligands can be internalized via the receptor-mediated endocytosis pathways and release drugs inside the cells.

1.3.3 Target-Activated Systems

Target-activated systems refer to NPs that remain stable until they reach target tissues and are activated by the extracellular or intracellular cues [64,73]. The NPs are designed to circulate similar to passively targeted NPs and transform to a more cell-interactive form in response to conditions unique to tumor tissues or intracellular environments. These conditions include acidic pH or enzymes abundant in the tumor extracellular matrix (ECM).

1.3.3.1 pH-Activated Systems. NPs responsive to acidic pH can be classified into two categories: one activated in the tumor ECM, which may have a pH as low as ~6, and the other activated in the intracellular lysosomes, which acidifies to pH 4–5. Tumors exhibit a slightly acidic pH in the range 6.8–7.2 compared to the physiological pH of 7.4 [74]. This acidity is due to the accumulation of lactic acid, induced by the increased glycolysis of tumor cells [75]. The acidosis of tumor is further enhanced as tumors develop hypoxia, which induces or selects for hyperglycolic cells [76]. The pH-sensitive NPs are composed of polymers, which are protonated at acidic pHs. Upon protonation of the polymer, the NPs are disintegrated to release drugs or change the surface properties to enter cells [64,77].

Most NPs entering cells via receptor-mediated endocytosis pathways end up in endosomes, which undergo a maturation process into late endosomes and lysosomes [72]. The luminal pH in these vesicles is acidic, varying from pH 6 (endosomes), pH 5–6 (late endosomes), to pH 4.5 (lysosomes) [72]. NPs are designed to escape the vesicles and/or release the payload in response to the low pH. Polycationic polymers such as polyethyleneimine or fusogenic peptides can be used in the formulations [78]. On entering the cell, a drug should reach its final destination to produce an action, which can be a nucleus or other organelles such as mitochondria, lysosome, and endoplasmic reticulum. NP trafficking to the specific organelles may be achieved by the use of specific peptide sequences [47,79].

1.3.3.2 Enzymatically Activated Systems. Enzymes overexpressed in the tumor microenvironment are used for activating the NPs at tumor sites. An example of such enzymes is matrix metalloproteinases (MMPs) [80,81], which play a critical role in the invasion of tumor cells and angiogenesis [82]. Several studies have employed

MMPs for removing the PEG protection of NP surface at the tumor sites [83–85]. In these approaches, PEG is linked to the NP surface via a peptide linker sensitive to MMPs. Recently, cathepsin B has been utilized to activate a nanoprobe in a tumor-specific manner [86]. Cathepsin B is a lysosomal cysteine protease that plays a role in tumor progression [87]. The nanoprobe consists of glycol chitosan conjugated with a cathepsin B substrate peptide, a near infrared fluorescent dye, and a dark quencher, which self-assemble into 280 nm NPs. Upon the entrance into tumor cells, the cathepsin B substrate peptide is cleaved by intracellular cathepsin B, allowing for dequenching of the dye and emission of the fluorescence signals [86].

1.4 REMAINING CHALLENGES IN THE DEVELOPMENT OF TUMOR-TARGETED NANOPARTICLES

Several NP products, such as liposomes and polymer–drug conjugates, are currently in the market or in the late stage of development process, improving therapeutic efficacy and safety profiles of chemotherapeutics [36,40]. A large number of targeted NPs have been developed in the last three decades in an attempt to further improve the efficacy of NP-based therapy. However, several challenges remain to be overcome for developing NPs that can attract commercial interest and achieve significant clinical outcomes. Current limitations and challenges in the advancement of nanomedicine have been thoroughly discussed in recent review articles [36,60,88–92].

1.4.1 NP Stability

Stability of circulating NPs matters in at least two contexts. NPs carrying a drug must survive the body's natural ability to clear the NPs and should be able to retain a drug during circulation. The former requirement is largely addressed by surface modification with a "stealth" coating with polymers such as PEG. However, recent studies have identified potential disadvantages of PEG [73]. For example, PEG can interfere with NP–cell interactions [93] and endosomal escape of NPs [93] after the disposition of NPs. It is also suggested that PEGylated liposomes are implicated with immune responses, which lead to an accelerated blood clearance of the second dose liposomes [94,95]. Alternative polymers or surface protection strategies are actively explored [73].

Stable retention of a drug during circulation is another important challenge. Despite the thermodynamic stability in buffers, some polymeric micelles or matrix-type NPs have been shown to quickly disintegrate and/or leak the payload upon contact with various blood components [64,96–98]. Chen et al. demonstrated that polymeric micelles composed of PEG-poly(D,L-lactic acid) block-copolymer dissociated and released the encapsulated fluorescent dyes within 15 min after intravenous injection [96]. Another study showed that a drug and the carrier (liposomes) exhibited different pharmacokinetics in blood, indicating that the drug was escaping the carrier prematurely [99]. The instability of NPs in circulation is a significant (yet often neglected) problem in the development stage. Success of a new NP system depends on the availability of a method that can reliably predict the stability of NPs in a biological environment.

1.4.2 Heterogeneous Tumor Vasculature

Tumor accumulation of NPs mainly depends on their passive extravasation via the leaky vasculature feeding the tumors. While the defective architecture of tumor vasculature is widely exploited by the NP-based drug delivery, it is often ignored that the tumor blood vessels are heterogeneous [100,101] and do not exhibit the same degree of leakiness [102]. The heterogeneity of the EPR effect is even greater when different tumor models are compared. A biodistribution study of ^{111}In-labeled PEGylated liposomes in 17 patients with locally advanced cancers showed that the levels of tumor liposome uptake varied from 0.5% to 3.6% of injected dose (ID) at 72 h, which translated into 2.7–53% of ID/kg of tumor (ID/kg) [103]. This study also showed that the tumor uptake of liposomes varied significantly with the site of the primary tumors, from $5.3 \pm 2.6\%$ ID/kg in breast cancers, $18.3 \pm 5.7\%$ ID/kg in lung tumors, to $33.0 \pm 15.8\%$ ID/kg in head and neck cancers [103]. Therefore, NPs relying on the EPR effect alone are unlikely to achieve reliable and predictable clinical outcomes.

1.4.3 NP Distribution in Tumors

The lack of functional lymphatics and the high vascular permeability result in elevation of the pressure in the interstitial tissue of solid tumors [104]. Furthermore, tumors develop various conditions to increase the stiffness of tissue [105]. The high interstitial fluid pressure and the stiffness of the tissue interfere with migration of a drug in tumor, limiting its effect on the central region of the tumor, a potential source of tumor relapse and metastasis [106–108].

Drug delivery with NPs is more significantly influenced by this challenge because of the size and surface properties [105]. Goodman et al. predicted that movement of 100–200 nm NPs in tumor spheroids would be much restricted even with the treatment of an ECM-degrading enzyme [109]. It was also experimentally demonstrated that the intratumoral distribution of 60 nm polymeric micelles was limited and localized in proximity to the blood vessel [69]. The size restriction is particularly problematic in the treatment of tumors with hypovascular and hypopermeable properties such as pancreatic tumors [110]. In the case of NPs employing active targeting strategy, high affinity binding to the receptors can lead to the "binding site barrier effect" in solid tumors, where the NP–receptor interactions may prevent migration of NPs to the interior of tissues [2,69,111,112]. While an active targeting strategy has the potential to improve the delivery of drugs through the intracellular uptake of NPs, the effect may be limited to the periphery of tumors when the size and binding site barrier play a dominant role in intratumoral disposition of the NPs.

1.4.4 Regulatory Considerations

The unique properties of NPs enabled by the size and surface pose different safety issues than free drugs and larger dosage forms [113]. The U.S. Food and Drug

Administration's Nanotechnology Task Force, formed in August 2006, solicited the agency's action to develop a regulatory guidance for manufacturers and researchers [114]. However, much remains uncertain as to how their safety should be tested and how much time and cost will be required for the approval of NP products, especially those designed with multiple functionalities. Such uncertainties negatively influence the investors and development partners, when the new product brings about only mild improvement over existing products.

New drug products must pass the regulatory scrutiny that examines their safety, effectiveness, and potential hazards. Reproducibility and predictability of the product's performance are the most important quality criteria under any test methods and models. Accordingly, it is important to ensure that NPs have consistent properties, such as size, charge, shape, and density of surface functional groups. Increasing complexity of a formulation increases the difficulty of quality control as well as the development cost. Therefore, the current effort to develop new types of NPs must undergo a careful cost–benefit analysis prior to a significant investment.

1.5 FUTURE PERSPECTIVES

Several recent studies attempt to address the challenges discussed so far [105]. For example, Maeda et al. proposed the use of drugs that artificially augment the EPR effect [102]. Angiotensin II was used to elevate systemic blood pressure [115], and nitroglycerin or other agents inducing nitric oxide production were used to facilitate vascular blood flow and permeability [116]. These approaches have shown the potential to overcome the heterogeneity of the EPR effect. To overcome the limitations in NP distribution in tumors, ECM-degrading enzymes such as hyaluronidase and collagenase were proposed as a potential aid to reduce the stiffness of ECM and the interstitial fluid pressure [109,117]. While ongoing studies are likely to make these challenges under control, additional issues await future efforts. Here we discuss the remaining homework for the advancement of nanomedicine.

1.5.1 *In Vitro* Models

Preclinical studies of anticancer drugs and NPs are first carried out in cell culture models. The cell culture model is a convenient and inexpensive way of screening the effectiveness of a drug or an NP formulation. However, it is not unusual that these drugs and NPs showing great efficacy in cell culture models are often proven much less effective *in vivo*. The lack of predictability is due in part to the fact that a single cell population grown in a two-dimensional (2D) layer barely reflects the nature of human tumors, which involves an intricate cross talk between different cell populations and three-dimensional (3D) organization of cells and ECM [118]. There is an increasing understanding of the significance of 3D architecture of tumors, cell populations, matrix composition and mechanics, and the gradients of signaling materials in modeling tumors *in vitro*. Accordingly, active efforts are made to build more realistic *in vitro* tumor models [119–121]. Lessons learned from tissue engineering and regenerative medicine are urgently solicited in these efforts.

1.5.2 *In Vivo* Models

Owing to the relatively low cost and well-established protocols, mouse models with allograft or human xenograft tumors are widely used in the evaluation of *in vivo* performance of NPs. In these models, cancer cells are inoculated (typically subcutaneously) in immunodeficient mice, allowed to grow into visually identifiable tumors, and treated with a new formulation to examine the pharmacokinetics, biodistribution, and pharmacological effects. However, several cases show that the results of preclinical studies are poorly correlated with Phase III outcomes. For example, in a preclinical study with mice bearing C-26 colon carcinoma, PEGylated liposomes carrying doxorubicin achieved significant survival benefits than free doxorubicin, attributable to their accumulation in tumors [122]. On the other hand, the PEGylated liposomal doxorubin was at best equivalent to free doxorubin in clinical efficacy (progression-free survival and overall survival) in a randomized Phase III trial with metastatic breast cancer patients [123]. In another example, a macromolecular conjugate of PTX and poly(L-glutamic acid) (PTX poliglumex) demonstrated a prolonged circulation half-life and greater tumor uptake and antitumor activity as compared to Taxol (PTX solubilized with Cremophor EL) at the same dose in a mouse model with syngeneic ovarian carcinoma [124]. However, the developer officially withdrew its application for a marketing authorization of PTX poliglumex for the first-line treatment of lung cancer patients because of the lack of therapeutic advantages over the comparators and unexplained toxicity [125].

In explaining the gap between the results of xenograft models and clinical outcomes, several limitations of current animal models may be considered [118,126]. First, the size and growth rate of tumors in mice are not comparable to those of human patients. Human patients with tumors large enough to detect with visual inspection would be candidates for surgical debulking rather than chemotherapy. Metastatic or microscopic residual tumors would be more appropriate subjects of targeted chemotherapy, but not much is known about the vasculature structure of those tumors and the effectiveness of the EPR effect. Second, since many animal models employ allografts or human tumor xenografts, the use of immunodeficient mice is inevitable. The potential consequences of immune responses to NPs are underrated in these models. Third, when human xenografts are inoculated in mouse models, the tumors recruit substances to build ECM from mice as they grow. The potential impact of this artificial arrangement on the architecture of ECM, cell–ECM interactions, and tumor propagation is barely considered in the interpretation of preclinical studies. Fourth, even though a xenograft can represent important attributes of the original tumors, it is uncertain whether it captures the genetic and epigenetic variability of tumors in its entirety [118].

When preclinical studies in animal models are essential steps providing the rationale for clinical trials, successful development of NP products depends critically on the availability of reliable animal models. At this point, it is yet unclear what would be the best animal model for predicting clinical outcomes of a drug product. One potential alternative to the subcutaneous xenograft is an orthotopic model, where a xenograft grown in proximity to the tissues or organs from which the tumor cell line was derived [118]. This model has advantages over the subcutaneous model as it provides an environment closer to a normal milieu of the tumor. One may also envision that genetically modified animals will have a promise in recapitulating the progression of human tumors and increasing the predictability of clinical outcomes.

1.5.3 New Targets

Although the current state-of-the-art targeting strategies have largely relied on specific interactions between cell surface markers and ligands, a number of studies have also found that the efficacy demonstrated *in vitro* or *in vivo* model does not translate at the clinical level. This has partly to do with the fact that a tumor is not a collective mass of a single-cell population expressing a consistent level of markers at all times. To the contrary, a tumor is highly heterogeneous and dynamic both in composition and genetic makeup. This makes it difficult to achieve a consistent outcome with an NP targeting a single type of molecular target [60,91]. Targeting multiple markers simultaneously is a conceivable alternative to the current targeting strategy. In addition, quiescent cancer cells or tumor-initiating cells are suggested as a potential target of drug delivery [127]. The tumor-initiating cells or cancer stem cells (CSCs), a subpopulation of cancer cells similar to stem cells, are believed to differentiate into new tumors when specific conditions are met. CSCs are known to be resistant to current radio- and chemotherapy, and some believe that the surviving CSCs are partly responsible for metastasis and recurrence of the disease [127]. Potential targets for NP-based drug delivery to CSCs include cellular markers such as CD44 and CD133 and hypoxic conditions that lead to the formation of CSC niches [127].

1.5.4 New Therapeutic Agents

While NPs have contributed to reducing toxic side effects of traditional chemotherapeutic drugs [123,128], their accumulation (and side effects) in the RES and bystander organs is currently inevitable. NPs may be a more useful tool for the delivery of alternative therapeutic agents, which are inherently specific to tumor cells (thus benign to normal tissues) but unstable in circulation or do not have an effective means to access target cells. Nucleic acid therapeutics such as small-interfering RNA, microRNA, or antisense oligonucleotides have proven effective in suppressing intrinsic drug resistance and survival of cancer cells, but their delivery is challenging. Various NPs are currently developed for the systemic delivery of these gene therapeutics [129], and one of them has entered a Phase I clinical trial in 2008 [130].

Acknowledgments

This work was supported by NSF DMR-1056997 and NIH R21 CA135130.

REFERENCES

1. Strebhardt, K., Ullrich, A. (2008). Paul Ehrlich's magic bullet concept: 100 years of progress. *Nature Reviews Cancer 8*, 473–480.
2. Peer, D., Karp, J. M., Hong, S., Farokhzad, O. C., Margalit, R., Langer, R. (2007). Nanocarriers as an emerging platform for cancer therapy. *Nature Nanotechnology 2*, 751–760.
3. Diamond, M. and The Sepracoat Adhesion Study Group. (1998). Reduction of de novo postsurgical adhesions by intraoperative precoating with Sepracoat (HAL-C) solution: a prospective, randomized, blinded, placebo-controlled multicenter study. *Fertility and Sterility 69*, 1067–1074.
4. Doxil Available at http://www.doxil.com/assets/DOXIL_PI_Booklet.pdf Accessed June 30, 2012.
5. Myocet Available at http://www.ema.europa.eu/ema/index.jsp?curl=pages/medicines/human/medicines/000297/human_med_000916.jsp&mid=WC0b01ac058001d124 Accessed June 30, 2012.
6. Daunoxome Available at http://www.daunoxome.com/ Accessed June 30, 2012.
7. Depocyt Available at http://depocyt.com/patients/about.asp Accessed June 30, 2012.
8. Ambisome Available at http://www.ambisome.com/MOA.aspx Accessed June 30, 2012.
9. Lipoplatin Available at http://lipoplatin.com/info_drug.php Accessed June 30, 2012.
10. Boulikas, T. (2009). Clinical overview on Lipoplatin: a successful liposomal formulation of cisplatin. *Expert Opinion on Investigational Drugs 18*, 1197–1218.
11. Fantini, M., Gianni, L., Santelmo, C., Drudi, F., Castellani, C., Affatato, A., Nicolini, M., Ravaioli, A. (2011). Lipoplatin treatment in lung and breast cancer. *Chemotherapy Research and Practice 2011*, 125–192.
12. Mylonakis, N., Athanasiou, A., Ziras, N., Angel, J., Rapti, A., Lampaki, S., Politis, N., Karanikas, C., Kosmas, C. (2010). Phase II study of liposomal cisplatin (Lipoplatin) plus gemcitabine versus cisplatin plus gemcitabine as first line treatment in inoperable (stage IIIB/IV) non-small cell lung cancer. *Lung Cancer 68*, 240–247.
13. Stathopoulos, G. P., Boulikas, T. (2012). Lipoplatin formulation review article. *Journal of Drug Delivery 2012*, 581363.
14. Thermodox http://celsion.com/docs/technology_thermodox Accessed December 12, 2012.
15. Marqibo Available at http://www.talontx.com/pipeline.php Accessed June 30, 2012.
16. Rodriguez, M. A., Pytlik, R., Kozak, T., Chhanabhai, M., Gascoyne, R., Lu, B., Deitcher, S. R., Winter, J. N. (2009). Vincristine sulfate liposomes injection (Marqibo) in heavily pretreated patients with refractory aggressive non-Hodgkin lymphoma: report of the pivotal phase 2 study. *Cancer 115*, 3475–3482.
17. Green, M. R., Manikhas, G. M., Orlov, S., Afanasyev, B., Makhson, A. M., Bhar, P., Hawkins, M. J. (2006). Abraxane, a novel Cremophor-free, albumin-bound particle form

of paclitaxel for the treatment of advanced non-small-cell lung cancer. *Annals of Oncology 17,* 1263–1268.

18. Abraxane http://www.abraxane.com/mbc/hcp/download/Abraxane_Prescribing_Information .pdf Accessed December 12, 2012.

19. Oncaspar Available at http://www.oncaspar.com/providers-oncaspar.asp Accessed June 30, 2012.

20. Dipetrillo, T., Milas, L., Evans, D., Akerman, P., Ng, T., Miner, T., Cruff, D., Chauhan, B., Iannitti, D., Harrington, D., Safran, H. (2006). Paclitaxel poliglumex (PPX-Xyotax) and concurrent radiation for esophageal and gastric cancer: a phase I study. *American Journal of Clinical Oncology 29,* 376–379.

21. Singer, J. W. (2005). Paclitaxel poliglumex (XYOTAX, CT-2103): a macromolecular taxane. *Journal of Controlled Release 109,* 120–126.

22. Singer, J. W., Shaffer, S., Baker, B., Bernareggi, A., Stromatt, S., Nienstedt, D., Besman, M. (2005). Paclitaxel poliglumex (XYOTAX; CT-2103): an intracellularly targeted taxane. *Anticancer Drugs 16,* 243–254.

23. Opaxio http://www.celltherapeutics.com/opaxio Accessed December 12, 2012.

24. Erinotecan Available at http://www.nektar.com/product_pipeline/oncology_nktr-102. html Accessed June 30, 2012.

25. Ng, E. W., Shima, D. T., Calias, P., Cunningham, E. T., Jr., Guyer, D. R., Adamis, A. P. (2006). Pegaptanib, a targeted anti-VEGF aptamer for ocular vascular disease. *Nature Reviews Drug Discovery 5,* 123–132.

26. Molineux, G. (2004). The design and development of pegfilgrastim (PEG-rmetHuG-CSF, Neulasta). *Current Pharmaceutical Design 10,* 1235–1244.

27. Kim, D. W., Kim, S. Y., Kim, H. K., Kim, S. W., Shin, S. W., Kim, J. S., Park, K., Lee, M. Y., Heo, D. S. (2007). Multicenter phase II trial of Genexol-PM, a novel Cremophor-free, polymeric micelle formulation of paclitaxel, with cisplatin in patients with advanced non-small-cell lung cancer. *Annals of Oncology 18,* 2009–2014.

28. Lee, K. S., Chung, H. C., Im, S. A., Park, Y. H., Kim, C. S., Kim, S. B., Rha, S. Y., Lee, M. Y., Ro, J. (2008). Multicenter phase II trial of Genexol-PM, a Cremophor-free, polymeric micelle formulation of paclitaxel, in patients with metastatic breast cancer. *Breast Cancer Research and Treatment 108,* 241–250.

29. Junghanns, J. U., Muller, R. H. (2008). Nanocrystal technology, drug delivery and clinical applications. *International Journal of Nanomedicine 3,* 295–309.

30. Rupp, R., Rosenthal, S. L., Stanberry, L. R. (2007). VivaGel (SPL7013 Gel): a candidate dendrimer—microbicide for the prevention of HIV and HSV infection. *International Journal of Nanomedicine 2,* 561–566.

31. CALAA-01 (clinical-trial) Available at http://clinicaltrials.gov/ct2/show/NCT00689065?term=calando&rank=1 Accessed June 30, 2012.

32. CALAA-01 http://www.arrowheadresearch.com/programs/calaa-01 Accessed December 12, 2012.

33. Qdot Nanocrystal Available at http://www.invitrogen.com/site/us/en/home/brands/Molecular-Probes/Key-Molecular-Probes-Products/Qdot.html Accessed June 30, 2012.

34. Qdot 800 Available at http://products.invitrogen.com/ivgn/product/Q10171MP Accessed June 30, 2012.

35. Resovist Available at http://radiologie-uni-frankfurt.de/sites/radiologie-uni-frankfurt. de/content/e43/e2321/e2331/resofinal_eng.pdf Accessed June 30, 2012.

36. Feridex Available at http://www.amagpharma.com/products/feridex_iv.php Accessed June 30, 2012.

37. Wagner, V., Dullaart, A., Bock, A. K., Zweck, A. (2006). The emerging nanomedicine landscape. *Nature Biotechnology 24*, 1211–1217.

38. Davis, M. E., Chen, Z. G., Shin, D. M. (2008). Nanoparticle therapeutics: an emerging treatment modality for cancer. *Nature Reviews Drug Discovery 7*, 771–782.

39. Zamboni, W. C. (2008). Concept and clinical evaluation of carrier-mediated anticancer agents. *Oncologist 13*, 248–260.

40. Farokhzad, O. C., Langer, R. (2009). Impact of nanotechnology on drug delivery. *ACS Nano 3*, 16–20.

41. Li, S. D., Huang, L. (2008). Pharmacokinetics and biodistribution of nanoparticles. *Molecular Pharmaceutics 5*, 496–504.

42. Wong, H. L., Rauth, A. M., Bendayan, R., Manias, J. L., Ramaswamy, M., Liu, Z., Erhan, S. Z., Wu, X. Y. (2006). A new polymer-lipid hybrid nanoparticle system increases cytotoxicity of doxorubicin against multidrug-resistant human breast cancer cells. *Pharmaceutical Research 23*, 1574–1585.

43. Jain, R. K., Stylianopoulos, T. (2010). Delivering nanomedicine to solid tumors. *Nature Reviews Clinical Oncology 7*, 653–664.

44. Vladimir, T. (2009). Multifunctional and stimuli-sensitive pharmaceutical nanocarriers. *European Journal of Pharmaceutics and Biopharmaceutics 71*, 431–444.

45. Pericleous, P., Gazouli, M., Lyberopoulou, A., Rizos, S., Nikiteas, N., Efstathopoulos, E. P. (2012). Quantum dots hold promise for early cancer imaging and detection. *International Journal of Cancer 131*, 519–528.

46. Shi, D., Bedford, N. M., Cho, H.-S. (2011). Engineered multifunctional nanocarriers for cancer diagnosis and therapeutics. *Small 7*, 2549–2567.

47. Tkachenko, A. G., Xie, H., Coleman, D., Glomm, W., Ryan, J., Anderson, M. F., Franzen, S., Feldheim, D. L. (2003). Multifunctional gold nanoparticle−peptide complexes for nuclear targeting. *Journal of the American Chemical Society 125*, 4700–4701.

48. Allen, T. M., Cullis, P. R. (2004). Drug delivery systems: entering the mainstream. *Science 303*, 1818–1822.

49. Gabizon, A. A., Shmeeda, H., Zalipsky, S. (2006). Pros and cons of the liposome platform in cancer drug targeting. *Journal of Liposome Research 16*, 175–183.

50. Yuan, F., Dellian, M., Fukumura, D., Leunig, M., Berk, D. A., Torchilin, V. P., Jain, R. K. (1995). Vascular permeability in a human tumor xenograft: molecular size dependence and cutoff size. *Cancer Research 55*, 3752–3756.

51. Greish, K. (2007). Enhanced permeability and retention of macromolecular drugs in solid tumors: a royal gate for targeted anticancer nanomedicines. *Journal of Drug Targeting 15*, 457–464.

52. Maeda, H. (2001). The enhanced permeability and retention (EPR) effect in tumor vasculature: the key role of tumor-selective macromolecular drug targeting. *Advances in Enzyme Regulation 41*, 189–207.

53. Matsumura, Y., Maeda, H. (1986). A new concept for macromolecular therapeutics in cancer chemotherapy: mechanism of tumoritropic accumulation of proteins and the antitumor agent smancs. *Cancer Research 46*, 6387–6392.

54. Oku, N., Tokudome, Y., Asai, T., Tsukada, H. (2000). Evaluation of drug targeting strategies and liposomal trafficking. *Current Pharmaceutical Design 6*, 1669–1691.

55. Owens, D. E., 3rd, Peppas, N. A. (2006). Opsonization, biodistribution, and pharmacokinetics of polymeric nanoparticles. *International Journal of Pharmaceutics 307*, 93–102.

56. van Vlerken, L. E., Vyas, T. K., Amiji, M. M. (2007). Poly(ethylene glycol)-modified nanocarriers for tumor-targeted and intracellular delivery. *Pharmaceutical Research 24*, 1405–1414.

57. Allen, T. M., Chonn, A. (1987). Large unilamellar liposomes with low uptake into the reticuloendothelial system. *FEBS Letters 223*, 42–46.

58. Torchilin, V. P. (2007). Targeted pharmaceutical nanocarriers for cancer therapy and imaging. *AAPS Journal 9*, E128–E147.

59. Allen, T. M. (2002). Ligand-targeted therapeutics in anticancer therapy. *Nature Reviews Cancer 2*, 750–763.

60. Bae, Y. H., Park, K. (2011). Targeted drug delivery to tumors: myths, reality and possibility. *Journal of Controlled Release 153*, 198–205.

61. Moghimi, S. M., Hunter, A. C., Murray, J. C. (2001). Long-circulating and target-specific nanoparticles: theory to practice. *Pharmacology Reviews 53*, 283–318.

62. Yokoyama, M. (2005). Drug targeting with nano-sized carrier systems. *Journal of Artificial Organs 8*, 77–84.

63. Wang, M., Thanou, M. (2010). Targeting nanoparticles to cancer. *Pharmacological Research 62*, 90–99.

64. Gullotti, E., Yeo, Y. (2009). Extracellularly activated nanocarriers: a new paradigm of tumor targeted drug delivery. *Molecular Pharmaceutics 6*, 1041–1051.

65. Yu, B., Tai, H. C., Xue, W., Lee, L. J., Lee, R. J. (2010). Receptor-targeted nanocarriers for therapeutic delivery to cancer. *Molecular Membrane Biology 27*, 286–298.

66. Mann, A. P., Somasunderam, A., Nieves-Alicea, R., Li, X., Hu, A., Sood, A. K., Ferrari, M., Gorenstein, D. G., Tanaka, T. (2010). Identification of thioaptamer ligand against E-selectin: potential application for inflamed vasculature targeting. *PLoS One, 5*. e13050.

67. Albanell, J., Baselga, J. (1999). Trastuzumab, a humanized anti-HER2 monoclonal antibody, for the treatment of breast cancer. *Drugs Today (Barc) 35*, 931–946.

68. Hrkach, J., Von Hoff, D., Ali, M. M., Andrianova, E., Auer, J., Campbell, T., De Witt, D., Figa, M., Figueiredo, M., Horhota, A., Low, S., McDonnell, K., Peeke, E., Retnarajan, B., Sabnis, A., Schnipper, E., Song, J. J., Song, Y. H., Summa, J., Tompsett, D., Troiano, G., Van Geen Hoven, T., Wright, J., LoRusso, P., Kantoff, P. W., Bander, N. H., Sweeney, C., Farokhzad, O. C., Langer, R., Zale, S. (2012). Preclinical development and clinical translation of a PSMA-targeted docetaxel nanoparticle with a differentiated pharmacological profile. *Science Translational Medicine 4*, 128ra39.

69. Lee, H., Fonge, H., Hoang, B., Reilly, R. M., Allen, C. (2010). The effects of particle size and molecular targeting on the intratumoral and subcellular distribution of polymeric nanoparticles. *Molecular Pharmaceutics 7*, 1195–1208.

70. Park, J. W., Hong, K., Kirpotin, D. B., Colbern, G., Shalaby, R., Baselga, J., Shao, Y., Nielsen, U. B., Marks, J. D., Moore, D., Papahadjopoulos, D., Benz, C. C. (2002). Anti-HER2 immunoliposomes: enhanced efficacy attributable to targeted delivery. *Clinical Cancer Research 8*, 1172–1181.

71. Hong, S., Leroueil, P. R., Majoros, I. J., Orr, B. G., Baker, J. R., Jr., Banaszak Holl, M. M. (2007). The binding avidity of a nanoparticle-based multivalent targeted drug delivery platform. *Chemical Biology 14*, 107–115.

72. Alberts, B., Wilson, J. H., Hunt, T., *Molecular Biology of the Cell*, Garland Science, New York, 2008.

73. Amoozgar, Z., Yeo, Y. (2012). Recent advances in stealth coating of nanoparticle drug delivery systems. *Wiley Interdisciplinary Reviews: Nanomedicine and Nanobiotechnology 4*, 219–233.

74. Gerweck, L. E., Seetharaman, K. (1996). Cellular pH gradient in tumor versus normal tissue: potential exploitation for the treatment of cancer. *Cancer Research 56*, 1194–1198.

75. Vander Heiden, M. G., Cantley, L. C., Thompson, C. B. (2009). Understanding the Warburg effect: the metabolic requirements of cell proliferation. *Science 324*, 1029–1033.

76. Raghunand, N., Gatenby, R. A., Gillies, R. J. (2003). Microenvironmental and cellular consequences of altered blood flow in tumours. *British Journal of Radiology 76*, S11–S22.

77. Lee, E. S., Gao, Z., Bae, Y. H. (2008). Recent progress in tumor pH targeting nanotechnology. *Journal of Controlled Release 132*, 164–170.

78. Breunig, M., Bauer, S., Goepferich, A. (2008). Polymers and nanoparticles: intelligent tools for intracellular targeting? *European Journal of Pharmaceutics and Biopharmaceutics 68*, 112–128.

79. Szeto, H. (2006). Cell-permeable, mitochondrial-targeted, peptide antioxidants. *The AAPS Journal 8*, E277–E283.

80. Chau, Y., Dang, N. M., Tan, F. E., Langer, R. (2006). Investigation of targeting mechanism of new dextran-peptide-methotrexate conjugates using biodistribution study in matrix-metalloproteinase-overexpressing tumor xenograft model. *Journal of Pharmaceutical Sciences 95*, 542–551.

81. Hatakeyama, H., Akita, H., Kogure, K., Harashima, H. (2007). Development of a novel systemic gene delivery system for cancer therapy with a tumor-specific cleavable PEG-lipid. *Yakugaku Zasshi-Journal of the Pharmaceutical Society of Japan 127*, 1549–1556.

82. Egeblad, M., Werb, Z. (2002). New functions for the matrix metalloproteinases in cancer progression. *Nature Reviews Cancer 2*, 161–174.

83. Terada, T., Iwai, M., Kawakami, S., Yamashita, F., Hashida, M. (2006). Novel PEG-matrix metalloproteinase-2 cleavable peptide-lipid containing galactosylated liposomes for hepatocellular carcinoma-selective targeting. *Journal of Controlled Release 111*, 333–342.

84. Hatakeyama, H., Akita, H., Ito, E., Hayashi, Y., Oishi, M., Nagasaki, Y., Danev, R., Nagayama, K., Kaji, N., Kikuchi, H., Baba, Y., Harashima, H. (2011). Systemic delivery of siRNA to tumors using a lipid nanoparticle containing a tumor-specific cleavable PEG-lipid. *Biomaterials 32*, 4306–4316.

85. Niidome, T., Ohga, A., Akiyama, Y., Watanabe, K., Niidome, Y., Mori, T., Katayama, Y. (2010). Controlled release of PEG chain from gold nanorods: targeted delivery to tumor. *Bioorganic & Medicinal Chemistry 18*, 4453–4458.

86. Ryu, J. H., Kim, S. A., Koo, H., Yhee, J. Y., Lee, A., Na, J. H., Youn, I., Choi, K., Kwon, I. C., Kim, B.-S., Kim, K. (2011). Cathepsin B-sensitive nanoprobe for *in vivo* tumor diagnosis. *Journal of Materials Chemistry 21*, 17631–17634.

87. Lopez-Otin, C., Matrisian, L. M. (2007). Emerging roles of proteases in tumour suppression. *Nature Reviews Cancer 7*, 800–808.

88. Ruenraroengsak, P., Cook, J. M., Florence, A. T. (2010). Nanosystem drug targeting: facing up to complex realities. *Journal of Controlled Release 141*, 265–276.

89. Gaumet, M., Vargas, A., Gurny, R., Delie, F. (2008). Nanoparticles for drug delivery: the need for precision in reporting particle size parameters. *European Journal of Pharmaceutics and Biopharmaceutics 69*, 1–9.

90. Florence, A. T. (2007). Pharmaceutical nanotechnology: more than size. Ten topics for research. *International Journal of Pharmaceutics 339*, 1–2.

91. Bae, Y. H. (2009). Drug targeting and tumor heterogeneity. *Journal of Controlled Release 133*, 2–3.

92. Bawa, R. (2007). Patents and nanomedicine. *Nanomedicine 2*, 351–374.

93. Du, H., Chandaroy, P., Hui, S. W. (1997). Grafted poly-(ethylene glycol) on lipid surfaces inhibits protein adsorption and cell adhesion. *Biochimica et Biophysica Acta (BBA)— Biomembranes 1326*, 236–248.

94. Ishihara, T., Maeda, T., Sakamoto, H., Takasaki, N., Shigyo, M., Ishida, T., Kiwada, H., Mizushima, Y., Mizushima, T. (2010). Evasion of the accelerated blood clearance phenomenon by coating of nanoparticles with various hydrophilic polymers. *Biomacromolecules 11*, 2700–2706.

95. Ishihara, T., Takeda, M., Sakamoto, H., Kimoto, A., Kobayashi, C., Takasaki, N., Yuki, K., Tanaka, K.-i., Takenaga, M., Igarashi, R., Maeda, T., Yamakawa, N., Okamoto, Y., Otsuka, M., Ishida, T., Kiwada, H., Mizushima, Y., Mizushima, T. (2009). Accelerated blood clearance phenomenon upon repeated injection of PEG-modified PLA-nanoparticles. *Pharmaceutical Research 26*, 2270–2279.

96. Chen, H., Kim, S., Li, L., Wang, S., Park, K., Cheng, J. X. (2008). Release of hydrophobic molecules from polymer micelles into cell membranes revealed by Forster resonance energy transfer imaging. *Proceedings of National Academy of Sciences of the United States of America 105*, 6596–6601.

97. Chen, H., Kim, S., He, W., Wang, H., Low, P. S., Park, K., Cheng, J. X. (2008). Fast release of lipophilic agents from circulating PEG-PDLLA micelles revealed by *in vivo* forster resonance energy transfer imaging. *Langmuir 24*, 5213–5217.

98. Gullotti, E., Yeo, Y. (2012). Beyond the imaging: limitations of cellular uptake study in the evaluation of nanoparticles. *Journal of Controlled Release 164*, 172–176.

99. de Smet, M., Heijman, E., Langereis, S., Hijnen, N. M., Grüll, H. (2011). Magnetic resonance imaging of high intensity focused ultrasound mediated drug delivery from temperature-sensitive liposomes: an *in vivo* proof-of-concept study. *Journal of Controlled Release 150*, 102–110.

100. Dvorak, H. F., Weaver, V. M., Tlsty, T. D., Bergers, G. (2011). Tumor microenvironment and progression. *Journal of Surgical Oncology 103*, 468–474.

101. Sitohy, B., Nagy, J. A., Dvorak, H. F. (2012). Anti-VEGF/VEGFR therapy for cancer: reassessing the target. *Cancer Research 72*, 1909–1914.

102. Maeda, H. (2010). Tumor-selective delivery of macromolecular drugs via the EPR effect: background and future prospects. *Bioconjugate Chemistry 21*, 797–802.

103. Harrington, K. J., Mohammadtaghi, S., Uster, P. S., Glass, D., Peters, A. M., Vile, R. G., Stewart, J. S. W. (2001). Effective targeting of solid tumors in patients with locally

advanced cancers by radiolabeled pegylated liposomes. *Clinical Cancer Research 7*, 243–254.

104. Jain, R. K. (1994). Barriers to drug delivery in solid tumors. *Scientific American 271*, 58–65.

105. Holback, H., Yeo, Y. (2011). Intratumoral drug delivery with nanoparticulate carriers. *Pharmaceutical Research 28*, 1819–1830.

106. Heldin, C. H., Rubin, K., Pietras, K., Ostman, A. (2004). High interstitial fluid pressure—an obstacle in cancer therapy. *Nature Reviews Cancer 4*, 806–813.

107. Minchinton, A. I., Tannock, I. F. (2006). Drug penetration in solid tumours. *Nature Reviews Cancer 6*, 583–592.

108. Jang, S. H., Wientjes, M. G., Lu, D., Au, J. L. S. (2003). Drug delivery and transport to solid tumors. *Pharmaceutical Research 20*, 1337–1350.

109. Goodman, T. T., Chen, J. Y., Matveev, K., Pun, S. H. (2008). Spatio-temporal modeling of nanoparticle delivery to multicellular tumor spheroids. *Biotechnology and Bioengineering 101*, 388–399.

110. Cabral, H., Matsumoto, Y., Mizuno, K., Chen, Q., Murakami, M., Kimura, M., Terada, Y., Kano, M. R., Miyazono, K., Uesaka, M., Nishiyama, N., Kataoka, K. (2011). Accumulation of sub-100 nm polymeric micelles in poorly permeable tumours depends on size. *Nature Nanotechnology 6*, 815–823.

111. Juweid, M., Neumann, R., Paik, C., Perez-Bacete, M. J., Sato, J., van Osdol, W., Weinstein, J. N. (1992). Micropharmacology of monoclonal antibodies in solid tumors: direct experimental evidence for a binding site barrier. *Cancer Research 52*, 5144–5153.

112. Saga, T., Neumann, R. D., Heya, T., Sato, J., Kinuya, S., Le, N., Paik, C. H., Weinstein, J. N. (1995). Targeting cancer micrometastases with monoclonal antibodies: a binding-site barrier. *Proceedings of National Academy of Sciences of the United States of America 92*, 8999–9003.

113. Prescott, C. (2010). Regenerative nanomedicines: an emerging investment prospective? *Journal of The Royal Society Interface 7*, S783–S787.

114. US Food and Drug Administration. 2007. Nanotechnology Task Force Report 2007.

115. Nagamitsu, A., Greish, K., Maeda, H. (2009). Elevating blood pressure as a strategy to increase tumor-targeted delivery of macromolecular drug SMANCS: cases of advanced solid tumors. *Japanese Journal of Clinical Oncology 39*, 756–766.

116. Seki, T., Fang, J., Maeda, H. (2009). Enhanced delivery of macromolecular antitumor drugs to tumors by nitroglycerin application. *Cancer Science 100*, 2426–2430.

117. Eikenes, L., Tari, M., Tufto, I., Bruland, O. S., Davies, C. D. (2005). Hyaluronidase induces a transcapillary pressure gradient and improves the distribution and uptake of liposomal doxorubicin (Caelyx™) in human osteosarcoma xenografts. *British Journal of Cancer 93*, 81–88.

118. Kamb, A. (2005). What's wrong with our cancer models? *Nature Reviews Drug Discovery 4*, 161–165.

119. Hearnden, V., MacNeil, S., Battaglia, G. (2011). Tracking nanoparticles in three-dimensional tissue-engineered models using confocal laser scanning microscopy. *Methods in Molecular Biology 695*, 41–51.

120. Hosoya, H., Kadowaki, K., Matsusaki, M., Cabral, H., Nishihara, H., Ijichi, H., Koike, K., Kataoka, K., Miyazono, K., Akashi, M., Kano, M. R. (2012). Engineering fibrotic

tissue in pancreatic cancer: a novel three-dimensional model to investigate nanoparticle delivery. *Biochemical and Biophysical Research Communications 419*, 32–37.

121. Mitra, M., Mohanty, C., Harilal, A., Maheswari, U. K., Sahoo, S. K., Krishnakumar, S. (2012). A novel *in vitro* three-dimensional retinoblastoma model for evaluating chemo-therapeutic drugs. *Molecular Vision 18*, 1361–1378.

122. Huang, S. K., Mayhew, E., Gilani, S., Lasic, D. D., Martin, F. J., Papahadjopoulos, D. (1992). Pharmacokinetics and therapeutics of sterically stabilized liposomes in mice bearing C-26 colon carcinoma. *Cancer Research 52*, 6774–6781.

123. O'Brien, M. E. R., Wigler, N., Inbar, M., Rosso, R., Grischke, E., Santoro, A., Catane, R., Kieback, D. G., Tomczak, P., Ackland, S. P., Orlandi, F., Mellars, L., Alland, L., Tendler, C., Grp, C. B. C. S. (2004). Reduced cardiotoxicity and comparable efficacy in a phase III trial of pegylated liposomal doxorubicin HCl (CAELYX™/Doxil®) versus conventional doxorubicin for first-line treatment of metastatic breast cancer. *Annals of Oncology 15*, 440–449.

124. Li, C., Yu, D.-F., Newman, R. A., Cabral, F., Stephens, L. C., Hunter, N., Milas, L., Wallace, S. (1998). Complete regression of well-established tumors using a novel water-soluble poly(L-glutamic acid)-paclitaxel conjugate. *Cancer Research 58*, 2404–2409.

125. European Medicines Agency. 2009. Questions and answers on the withdrawal of the marketing authorisation application for Opaxio (Paclitaxel poliglumex).

126. Damia, G., D'Incalci, M. (2009). Contemporary pre-clinical development of anticancer agents—what are the optimal preclinical models? *European Journal of Cancer 45*, 2768–2781.

127. Vinogradov, S., Wei, X. (2012). Cancer stem cells and drug resistance: the potential of nanomedicine. *Nanomedicine 7*, 597–615.

128. Hawkins, M. J., Soon-Shiong, P., Desai, N. (2008). Protein nanoparticles as drug carriers in clinical medicine. *Advanced Drug Delivery Reviews 60*, 876–885.

129. David, S., Pitard, B., Benoît, J.-P., Passirani, C. (2010). Non-viral nanosystems for systemic siRNA delivery. *Pharmacological Research 62*, 100–114.

130. Davis, M. E., Zuckerman, J. E., Choi, C. H. J., Seligson, D., Tolcher, A., Alabi, C. A., Yen, Y., Heidel, J. D., Ribas, A. (2010). Evidence of RNAi in humans from systemically administered siRNA via targeted nanoparticles. *Nature 464*, 1067–1070.

2

APPLICATIONS OF LIGAND-ENGINEERED NANOMEDICINES

GAYONG SHIM, JOO YEON PARK, LEE DONG ROH, AND YU-KYOUNG OH
College of Pharmacy, Seoul National University, Daehak-dong, Seoul, South Korea

SANGBIN LEE
School of Life Sciences and Biotechnology, Korea University, Seoul, South Korea

2.1 INTRODUCTION

Over the past two decades, a variety of therapeutic modalities have been investigated for the treatment of cancer. Among the most promising new modalities in the field of medical applications is nanomedicine, which has proven to be an outstanding approach for improving therapeutic index [1,2]. Nanomedicines can achieve remarkable improvements in anticancer efficacy not only by increasing bioavailability and half-life, but also by promoting tumor-accumulation of therapeutic entities through the enhanced permeability and retention (EPR) effect [3]. Moreover, through modifications of the surface of nanoparticles, nanomedicines have showed the ability to improve target-specificity and further enhance the efficacy [4].

2.1.1 Nanoparticulate Drug Delivery

Nanoparticles are a proven modality for altering the *in vivo* fates of various therapeutic entities, such as small molecules, nucleic acids, and peptides [5]. Nanoparticulate drugs can achieve prolonged retention *in vivo*, increased accumulation in tumor tissues, and enhanced pharmacodynamics [1]. Depending on the materials used to construct them, nanoparticles are mainly classified as lipid- or

Nanoparticulate Drug Delivery Systems: Strategies, Technologies, and Applications, First Edition.
Edited by Yoon Yeo.
© 2013 John Wiley & Sons, Inc. Published 2013 by John Wiley & Sons, Inc.

21

polymer-based. The series of nanomedicines on the market and in clinical trials substantiates the potential of nanoparticles as drug-delivery platforms [5,6]. Among nanomedicines, lipid-based nanomedicines, particularly liposomes, have a long history of commercialization. Following *in vivo* administration, the surface of liposomes easily interacts with serum proteins, promoting their elimination by the reticuloendothelial system. More prolonged circulation in the blood has been achieved by introducing poly(ethylene glycol) (PEG) moieties on liposomal surfaces, thereby reducing the interaction of liposomes with blood components. Such surface coating with PEG has been applied not only to liposomes but also to other nanoparticles. The current status of nanomedicines on the market is summarized in Table 2.1.

TABLE 2.1 Current Status of Nanomedicine on the Market

Product Name	Active Drug	Company	Indications
Lipid-based			
Daunoxome	Daunorubicin	Nexstar	Kaposi's sarcoma
Myocet	Doxorubicin	Zeneus	Combinational therapy of recurrent breast cancer
Doxil/Caelyx	Doxorubicin	Tibotec	Refractory Kaposi's sarcoma
	Doxorubicin	ALZA	Ovarian cancer/recurrent breast cancer
DepoCyt	Cytarabine	Enzon	Lymphomatous meningitis
Onco TCS	Vincristine	Enzon/Inex	Non-Hodgkin's lymphoma
Ambisome	Amphotericin	Nexstar	Fungal infection in immunocompromised patients
Visudyne	Verteforfin	QLT/Norvatis	Wet age-related macular degeneration
Depo Dur	Morphine sulfate	Pacira Pharm	Postsurgical pain
Polymer-based			
Neulasta	PEG–GCSF	Amgen	Neutropenia associated with cancer chemotherapy
Pegasis	PEG–IFN2a	Genentech	Melanoma/Chronic myeloid leukemia/renal-cell carcinoma
Oncaspar	PEG–L-asparaginase	Enzon	Acute lymphoblastic leukemia
Zinostatin stimalamer	SMANCS	Yamanouchi	Hepatocellular carcinoma
Genexol-PM	Paclitaxel	Samyang	Metastatic breast cancer
Others			
Abaxane	Albumin-bound paclitaxel	Abraxis bioScience/ AstraZeneca	Metastatic breast cancer

GCSF, granulocyte colony-stimulating factor; IFNα, interferon-α; SMANCS, styrene maleic anhydride neocarzinostatin.

2.1.2 Engineered Nanoparticulate Drug Delivery

Although nanoparticulate drugs have been equipped with attractive features such as improvements in pharmacokinetic and pharmacodynamic profiles, they still suffer from the lack of an active driving force for tissue or cell-specific delivery, relying instead on the EPR effect. Under *in vivo* conditions, nanoparticles confront formidable biological barriers that impede their delivery to target tissues, including the reticuloendothelial system, vascular endothelium, and blood enzymes [7]. Coating nanoparticle surfaces with PEG increases their stability in the bloodstream, but it does not enhance their interactions with specific types of tissues or cells. To the contrary, there have been reports that higher PEG content in nanomedicines can even reduce interactions of nanoparticles with cells [8]. Numerous approaches have been developed to overcome such limitations through modification of nanoparticles with various ligands, including antibodies, peptides and proteins, sugars and polysaccharides, and chemicals.

2.2 LIGAND-MEDIATED DELIVERY

2.2.1 Antibodies

Because of their specific antigen-recognition properties, antibodies have long been studied as a way of increasing the specificity of nanomedicine delivery [9,10]. The types of antibody variants considered for this purpose include single-chain variable fragments (scFv), F(ab′)$_2$ fragments, and whole IgG antibodies. In cancer-related applications, targeting is typically achieved by tagging nanoparticles with antibodies against proteins known to be overexpressed on the surface of tumor cells relative to normal cells. Commonly studied cancer target proteins include HER2, selectins, and the epidermal growth factor receptor (EGFR). The primary methods for tagging antibodies to nanoparticles are covalent chemical conjugation and specific biomimetic interaction methods. *In vivo* applications of antibody-tagged nanomedicines are summarized in Table 2.2.

2.2.1.1 In Vivo *Applications of Antibody-Tagged Nanomedicines*

Nanoparticles Targeted to Tumors. Taking advantage of the fact that HER2 is overexpressed on several cancer cells, researchers have tagged liposomes entrapping vincristine [11] or doxorubicin [12] with HER2-specific antibodies. Liposomes entrapping the anticancer drug vincristine were tagged with anti-HER2 antibodies by coupling to maleimide-modified PEG lipids [11]. Modification of vincristine liposomes with anti-HER2 antibody was shown to provide greater efficacy than free vincristine or unmodified liposomal vincristine against the HER2-overexpressing breast cancer cell lines, BT474-M2 and SKBR3. For *in vivo* applications, anti-HER2 antibody-modified immunoliposomes, unmodified liposomes, and free vincristine were intravenously administered to BT474-M2-xenografted mice at a dose of 1 mg/kg once a week for 3 weeks. Compared to other groups, immunoliposomes

TABLE 2.2 *In Vivo* Applications of Antibody-Tagged Nanomedicines

Target	Disease or Target Tissue	Nanoparticles	Active Substances	*In Vivo* Model	References
HER2	Breast cancer	Liposome	Vincristine	BT474-M2-bearing mice	11
HER2	Breast cancer	Liposome	Doxorubicin	MCF-7/HER2-bearing mice	12
C-met	Lung cancer	Liposome	Doxorubicin	H1993-bearing mice	13
BRCAA1	Gastric cancer	Fe_3O_4 nanoparticles	Fluorescent magnetic nanoprobes	MGC803-bearing mice	14
EGFR	Glioblastoma multiforme	Liposome	Sodium borocaptate	EGFR-overexpressing U87-bearing mice	16
EGFR	Glioblastoma multiforme	Liposome	Fluorescent dye (8-hydroxypyrene–1,3,6-trisulfonic acid trisodium salt)	EGFR-overexpressing U87-bearing mice	25
EGFR	Hepatocellular carcinoma	Poly(lactic acid-*co*-L-lysine) nanoparticle	Fluorescent dye (rhodamine B)	SMMC7721-bearing mice	17
E-selectin	MCa-4 mammary tumor	Liposome	Combretastatin	MCa-4-bearming mice	18
THY 1.1	Mesangial cells in kidney	Liposome	Doxorubicin	Wistar rats	19
P-selectin	Myocardial infarction	Liposome	VEGF	Myocardial infarction-induced rats	20
LFA-1	Human immunodeficiency virus-infected cells	Liposome	Anti-CCR5 siRNA	Humanized mice	21

produced a greater reduction in tumor volumes and were well tolerated without causing substantial weight loss.

In addition to anti-HER2 antibody, quantum-dot (Q-dot) nanoparticles have been further conjugated to doxorubicin-loaded liposomes to create multifunctional imaging and therapeutic nanomedicines [12]. In this study, anti-HER2/ErbB2 scFv fragments were conjugated to one end of liposomal surface PEG, and carboxyl Q-dots were chemically tethered to the other end of liposomal surface PEG. *In vivo* imaging revealed a greater accumulation of anti-HER2 scFv- and Q-dot-conjugated doxorubicin immunoliposomes than Q-dot-conjugated liposomes in tumor tissues following intravenous administration in MCF-7/HER2-xenografted mice.

Anti-cMet scFv has been used to modify doxorubicin-loaded liposomes and Q-dot nanoparticles [13]. Doxorubicin-loaded liposomes were modified with anti-cMet scFv by coupling to maleimide-modified PEG chains on the external surface of liposomes. Anti-cMet scFv was similarly linked to maleimide-activated Q-dots. In H460-derived lung cancer xenografted mice, intravenous administration of 2 mg/kg of doxorubicin in an anti-cMet scFv-modified liposome formulation was shown to be almost twice as effective in reducing tumor volume as unmodified liposomes. Molecular imaging in mice revealed a greater distribution of anti-cMet scFv-modified Q-dots in tumors than unmodified Q-dots. However, a substantial distribution of Q-dots to tissues other than tumors was noted in the upper tail region, regardless of scFv modification. Such a distribution of Q-dots to the normal tail region may limit the targeted imaging of tumor tissues.

In addition to Q-dots, fluorescent magnetic nanoparticles have been conjugated with anti-BRCAA1 antibody to allow *in vivo* fluorescence and magnetic resonance imaging of gastric cancers [14]. In these applications, fluorescent magnetic nanoparticles were first functionalized with a carboxyl group and then conjugated with anti-BRCAA1 antibody using 1-ethyl-3-(3-dimethylaminopropyl)-carbodiimide and *N*-hydroxylsuccinimide as coupling agents. Treatment of the MGC803 gastric cancer cell line with anti-BRCAA1 antibody-tethered fluorescent magnetic nanoparticles provided a twofold increase in cellular-uptake efficiency compared to plain fluorescent magnetic nanoparticles. *In vivo* fluorescence imaging showed that the tumor tissue distribution of anti-BRCAA1 antibody-coupled nanoparticles in MGC803-xenografted mice peaked 6 h after intravenous administration. Magnetic resonance imaging obtained within 12 h after administration revealed a higher tumor distribution of anti-BRCAA1 antibody-coated magnetic nanoparticles compared to plain nanoparticles.

Antibodies against EGFR, which is overexpressed in malignant glioblastoma but weakly expressed or undetectable in normal brain [15], have been used for ligand modification of liposomes and polymeric nanoparticles. In particular, sodium borocaptate-loaded liposomes tagged with anti-EGFR antibody have been used for boron neutron capture therapy [16]. After intravenous administration of immunoliposomes to U87-EGFR cell-xenografted mice, the distribution of sodium borocaptate-loaded liposomes to the tumor tissues was tested by quantitation of boron compounds. Compared to unmodified liposomes, anti-EGFR antibody-tagged liposomes delivered higher amounts of boron compounds to tumor tissues, but not to the normal brain, liver, or blood.

Poly(lactic acid-*co*-L-lysine) nanoparticles have been modified with anti-EGFR-antibodies [17]. For chemical tethering, these antibodies were linked to the surface of poly(lactic acid-*co*-L-lysine) nanoparticles using the coupling agents, 1-ethyl-3-(3-dimethylaminopropyl)-carbodiimide and *N*-hydroxylsuccinimide, and rhodamine B was loaded into the nanoparticles as a tracer of cellular uptake. In the SMMC-7721 hepatocellular carcinoma cell line, the internalization of anti-EGFR-antibody-conjugated poly(lactic acid-*co*-L-lysine) nanoparticles was decreased in the presence of excess amounts of free anti-EGFR antibodies, indicating that nanoparticles were endocytosed via EGFR receptors. In SMMC-7721-bearing mice, polymeric nanoparticles with or without anti-EGFR antibody were distributed to tumor tissues 4 h after intravenous administration, with anti-EGFR antibody-conjugated nanoparticles showing more prolonged retention within tumor tissues during the 48 h after dosing.

Modification of PEGylated liposomes with an anti-E-selectin antibody has been studied as a means to improve distribution to mammary tumor tissues [18]. In this application, an anti-E-selectin antibody was coupled to the distal end of maleimide (2 mol%)-functionalized PEG formulated in liposomes entrapping the antivascular drug, combretastatin A4 disodium phosphate (CA4DP). At equivalent doses (15 mg/kg), anti-E-selectin antibody-tagged liposomes incorporating CA4DP showed greater anticancer efficacy in mice bearing MCa-4 mammary tumor cell xenografts than plain liposomes incorporating CA4DP and was more efficacious than a higher dose (81 mg/kg) of free CA4DP. Moreover, when used in combination therapy, CA4DP chemotherapy using E-selectin antibody-tagged liposomes synergized with radiation therapy to produce greater anticancer effects than chemotherapy or radiation therapy alone.

Nanoparticles Targeted to Other Diseases. Although many antibody-modified nanomedicines have been studied for targeting tumor tissues, recent studies have reported applications of immunoliposomes for kidney diseases [19], vascular diseases [20], and antiviral therapy [21]. The therapeutic entities carried by these immunoliposomes include doxorubicin, cytokine proteins, and small interfering RNA (siRNA).

The F(ab')$_2$ fragment of an anti-OX7 antibody, which has an affinity for Thy 1.1 antigen on mesangial cells in the kidney, has been used to enhance the delivery of doxorubicin-loaded liposomes to rat kidney [19]. In this application, anti-OX7 F(ab')$_2$ was conjugated to the surfaces of liposomes via maleimide functionalized PEG lipids. Intravenously administered anti-OX7 F(ab')$_2$-conjugated liposomes were shown to specifically enhance delivery to mesangial cells in both kidneys. In contrast, free OX7 F(ab)$_2$ fragments coadministered with liposomes did not enhance renal delivery. Treatment of rats with low-dose doxorubicin (0.24 mg/kg) resulted in extensive glomerular damage in the case of immunoliposomes, whereas such treatment with unmodified liposomes or free drug did not. This study indicates the utility of immunoliposomes for treatment of renal diseases.

In myocardial infarction, ischemic myocardium sites are reported to exhibit upregulation of cell-adhesion molecules, such as P-selectin. On the basis of this pathology, an anti-P-selectin antibody was designed to target altered vascular endothelial surfaces [20]. In this study, anti-P-selectin antibody was conjugated to the maleimide group on the 1,2-distearoyl-*sn*-glycero-3-phosphoethanolamine-*N*-[(polyethylene glycol)2000] (DSPE–PEG2000) component of the liposome. Next, vascular endothelial growth factor was loaded into anti-P-selectin-conjugated liposomes by rehydrating freeze-dried immunoliposomes with vascular endothelial growth factor (VEGF) solutions. In the rat myocardial infarction model, treatment with VEGF-loaded immunoliposomes was reported to significantly improve cardiac function and the number of perfused vessels compared to controls [20].

For siRNA-based antiviral therapy, liposomes have been modified with an antibody against lymphocyte function-associated antigen-1 (LFA-1) [21]. In this approach, rather than using PEG as a stabilizing moiety, hyaluronan 751 kDa was used as a liposome-surface modifier to increase stability in the bloodstream. Hyaluronan-modified liposomes were then conjugated with anti-LFA-1 antibody using an amine-coupling method. CCR5-specific siRNA was entrapped by rehydrating lyophilized liposomes with CCR5-specific siRNA and a protamine complex solution. The resulting immunoliposomal siRNA was injected into humanized mice via the tail vein at a dose of 50 µg/mouse. Subsequent challenge of humanized mice with human immunodeficiency virus was shown to provide enhanced resistance to infection by reducing viral load.

2.2.1.2 Methods for Tagging Nanomedicines with Antibodies

Covalent Conjugation. Because antibodies must retain their three-dimensional conformation to specifically interact with antigen epitopes, methods used to tag nanomedicines with antibodies should be carefully controlled. Although new methods of antibody tagging have been developed, the most widely used method has been the chemical conjugation method, where the antibody is linked to activated groups on the surfaces of nanoparticles via covalent bonding [22].

Maleimide groups have been used to provide activated ends to nanoparticles, allowing conjugation to sulfhydryl groups in cysteine residues of the antibody. This method has been used to modify drug-entrapped liposomes with various antibodies, including the anti-cMet scFv [13], and anti-OX7 F(ab′)₂ fragments [19], and anti-HER2 [11], anti-E-selectin [18], and anti-P-selectin [20] antibodies, described earlier.

Alternatively, the surface of nanoparticles can be chemically modified using 1-ethyl-3-(3-dimethylaminopropyl)-carbodiimide and *N*-hydroxysuccinimide as coupling agents. Examples of this include the modification of poly(lactic acid-*co*-L-lysine) nanoparticles with anti-EGFR-antibody [17] and modification of fluorescent magnetic nanoparticles with anti-BRCAA1 antibody [14].

Chemical coupling methods, however, suffer from several potential limitations. First, chemical coupling may affect the three-dimensional conformation of antibodies. Second, antibodies bound by nonspecific coupling are randomly oriented.

Both altered conformation and random orientation may result in the loss of antibody activity. Moreover, tagging antibodies to nanoparticles through complex reaction procedures requires time and effort.

Noncovalent Binding. To overcome the limitations inherent in chemical coupling methods, researchers have developed noncovalent biomimetic linking methods based on specific interactions. The prototypical example of this is based on protein A, a protein produced by *Staphylococcus aureus* that binds to the Fc domain of antibodies through its ZZ motif. Protein A has been exploited to prevent opsonization in the bloodstream [23] and has been applied to biotechnology products for the separation of antibodies from protein mixtures and the design of protein chips [24]. In the current context, the antibody-affinity ZZ motif of protein A has been used to tag antibodies to the surface of liposomes without chemical coupling. In a previous study by Feng and colleagues, a ZZ-His peptide was first bound to the nickel component of Ni lipid-containing liposomes through Ni–His interactions [16]. The resulting liposomal surface ZZ was used to tag an anti-EGFR antibody by simply mixing at room temperature. This approach, however, is limited by the potential cytotoxicity of the liposomal Ni used for His binding. Moreover, directly tagging an antibody to the surface of liposomes with PEG lipid chains may reduce the probability of antibody interaction because of steric hindrance of neighboring PEG chains.

In a more recent study, Feng and colleagues sought to improve the ability of antibodies to recognize and bind to target cells by designing liposomes that positioned the antibody further out from the liposome surface than did the Ni liposomes used previously. To accomplish this, they first modified maleimide-functionalized PEG chains of liposomes with a thiolated form of Gaussia luciferase-ZZ-His [25]. Gaussia luciferase is a 19.9 kDa bioluminescent reporter protein that provides an imaging functionality. Anti-EGFR antibodies were then bound to the ZZ domain at the distal end of PEG chains on the liposomal surface. The presence of Gaussia luciferase enabled monitoring of the *in vivo* fate of immunoliposomes by bioluminescence imaging.

In addition to the ZZ motif, a cyclic 13-mer peptide has been reported to be a crucial sequence for the Fc-binding function of protein A [26]. Taking advantage of the Fc-binding 13-mer peptide (FcBP), Oh and colleagues linked the peptide to functionalized PEG lipid derivatives [27], and then used the FcBP-linked PEG lipids to formulate liposomes to which antibodies could be bound via the specific affinity of FcBP. Subsequent attachment of anti-CXCR4 antibody to FcBP-modified liposomes resulted in enhanced cellular uptake by a CXCR4-positive Ramos cell line compared to plain liposomes.

Compared with the conventional chemical conjugation methods (Figure 2.1a), protein A-based affinity methods for noncovalently tagging antibodies (Figure 2.1b) may have advantages in retaining the three-dimensional conformation of liposomes and controlling the orientation of antibodies. However, this Fc moiety affinity-based approach is also limited in that it cannot be used to coat liposomes with scFv or $F(ab')_2$ fragment type antibodies, which lack in the Fc portion.

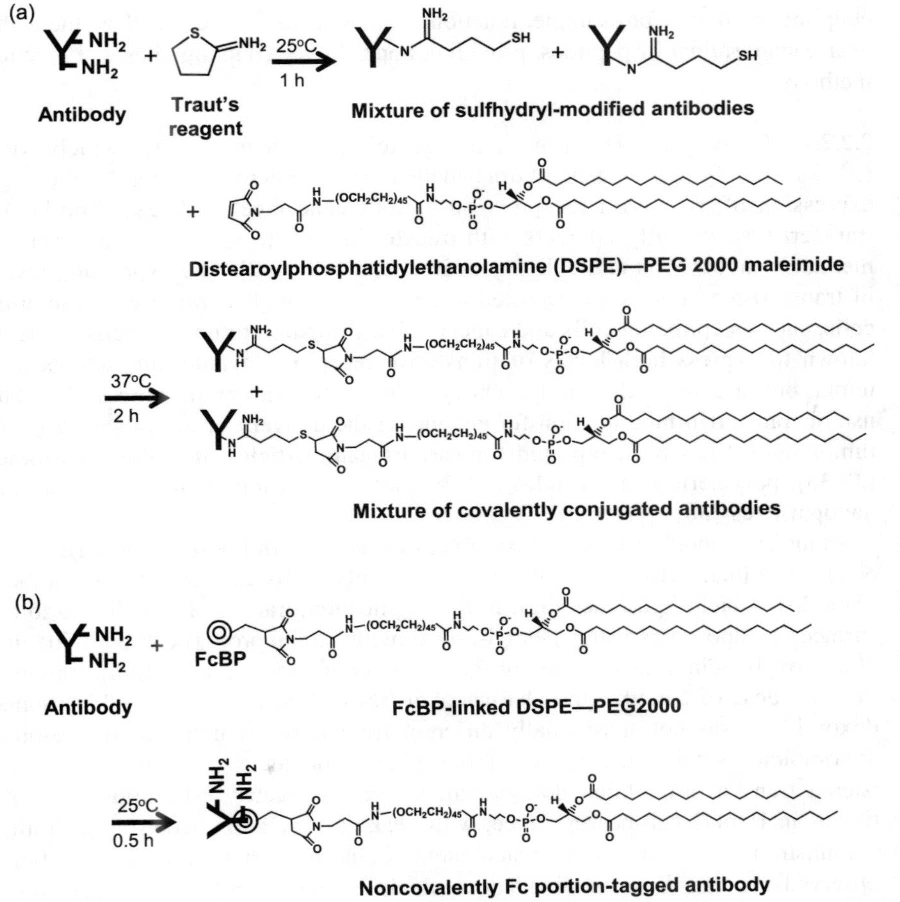

FIGURE 2.1 Covalent and noncovalent antibody-tagging methods. (a) Covalent tagging of antibody to the distal end of maleimide-functionalized PEG derivative of lipid. (b) FcBP-based noncovalent and specific tagging of antibody to a PEG derivative of lipid.

2.2.2 Peptides and Proteins

Compared to the variety of antibodies that can confer specific binding to overexpressed antigens on pathogenic cells, the menu of proteins and peptide ligands available for modification of nanoparticles is limited. With the exception of the widely studied transferrin, few proteins have been used for modifying nanoparticles. Peptide ligands have been more actively studied owing to the recent development of phage-display technologies, which allow for discovery of new peptides with high affinity for specific targets. Compared to protein ligands, peptide ligands are advantageous in that they can be easily synthesized and are thus commercially available. Moreover, their relatively small size makes it easy to handle in chemical coupling methods. For example, functional amino acids can be added synthetically near each terminal of peptides, providing control over coupling sites and reaction methods.

2.2.2.1 Transferrin. The iron-binding protein transferrin has been widely studied for modifying anticancer drug-loaded nanomedicines owing to the high expression of transferrin receptors on various cancer cells [28,29]. Iron-bound transferrin specifically interacts with transferrin receptors, resulting in receptor-mediated endocytosis and delivery of iron into cells [30]. The expression levels of transferrin receptors are reported to be relatively high in rapidly proliferating cells, such as epithelial cells and cancer cells, compared to normal cells. Cancers known to express high levels of transferrin receptors include lung adenocarcinoma, breast cancer, chronic lymphocytic leukemia, and brain tumors [31]. The use of transferrin modification for enhancing the delivery of anticancer drugs to tumor tissues has been reported for various nanoparticles, including liposomes [32,33], polymeric nanoparticles [34–36], albumin nanoparticles [37], and gold nanoparticles [38].

Transferrin modification of doxorubicin-loaded stealth liposomes was shown to enhance cellular delivery of doxorubicin to HepG2 cells compared to unmodified, plain doxorubicin liposomes [32]. In this application, transferrin was linked to the surfaces of liposomes using 1-ethyl-3-(3-dimethylaminopropyl)-carbodiimide and N-hydroxylsuccinimide as coupling agents. Following intravenous administration in rats at a dose of 5 mg/kg, the pharmacokinetics of transferrin-modified liposomal doxorubicin was not substantially different from those of unmodified liposomal doxorubicin. Notably, despite the similar pharmacokinetics, intravenously administered transferrin-modified doxorubicin showed a greater distribution to tumor tissues in HepG2 cell-bearing mice, with peak accumulation occurring 24 h after administration. The similar pharmacokinetics in rats but enhanced anticancer effects observed for transferrin-modified doxorubicin liposomes reflect enhanced tumor tissue distribution and cellular uptake.

Transferrin-modified liposomes have been studied as a delivery system for an antisense oligonucleotide [33]. In this application, G3139, an antisense oligonucleotide targeting Bcl-2, was complexed with a mixture of liposomes and protamines to yield lipoplexes, after which transferrin-conjugated PEG lipids were inserted into the lipoplexes. Thirty-six hours after intravenous administration of G3139 (5 mg/kg) in K562-bearing xenografted mice, tumor uptake of transferrin-modified lipoplexes was higher than that of the free form of G3139. Because there was no comparison between transferrin-modified liposomes and plain liposomes in this study, it is difficult to conclude that the observed increase in tumor uptake was due to transferrin, given the EPR effect noted for nanoparticles.

In addition to liposomes, various polymeric nanoparticles have been modified with transferrin. For example, biodegradable poly(DL-lactide-co-glycolide) (PLGA) nanoparticles have been loaded with paclitaxel and modified with transferrin [35]. The surface of paclitaxel-loaded PLGA nanoparticles was epoxy activated and reacted with transferrin. A pharmacokinetic study of paclitaxel (20 mg/kg) in Sprague–Dawley rats showed that the mean residence time of paclitaxel delivered by plain PLGA nanoparticles and transferrin-modified PLGA increased by 2.68 and 3.76 folds, respectively, as compared to a free paclitaxel solution. The distribution of paclitaxel to tumor tissues was highest with transferrin-modified PLGA, reaching a peak accumulation 2 h after intravenous administration.

Polycyanoacrylate nanoparticles carrying PEG-hydroxycamptothecin have also been modified with transferrin [39]. In normal Wistar rats, the pharmacokinetic profile of hydroxycamptothecin delivered in transferrin-modified PEG nanoparticles did not substantially differ from that in unmodified PEG nanoparticles. A fit to a two-compartment model showed that the elimination half-life of free PEG-hydroxycamptothecin was 0.68 h, but unmodified PEG nanoparticles and transferrin-modified PEG nanoparticles increased the half-life to 23.21 and 20.92 h, respectively. The tumor distribution of transferrin-modified PEG nanoparticles in S180 tumor-bearing mice, measured as area under the curve, was 2.12-fold higher than that of unmodified PEG nanoparticles, and reached a peak accumulation 4 h after intravenous administration.

Doxorubicin-loaded human serum albumin nanoparticles have recently been dual-modified with transferrin and tumor necrosis factor-related apoptosis-inducing ligand (TRAIL) [37]. Dual labeling with transferrin and TRAIL was reported to enable the delivery of nanoparticles to multiple tumor types, including HCT 116 and CAPAN-1 cells, and drug-resistant MCF-7/ADR cells. Moreover, dual-ligand modification with transferrin and TRAIL was shown to provide synergistic tumor cell-killing activity, reflecting the cytotoxicity of doxorubicin and the induction of apoptosis via the TRAIL ligand. Transferrin/TRAIL-modified doxorubicin nanoparticles intravenously administered in HCT116-xenografted mice showed accumulation in tumor tissues, reaching a peak 16 h after injection.

One concern regarding transferrin-mediated delivery is rapid intracellular trafficking kinetics [30]. After receptor-mediated endocytosis, transferrin receptor/ transferrin complexes release iron in the endosome and then rapidly recycle back to the cell surface. The short residence time of transferrin receptor-containing endosomes within cells may reduce the ability of transferrin to deliver the drugs entrapped in nanomedicines. One approach to improving the efficacy of transferrin-modified nanomedicines is genetically engineering variants of transferrin with decreased iron release rates [40]. This study tested genetic variants of transferrin in the context of a transferrin-conjugated cytotoxin, but this approach may be applicable to other transferrin-modified nanomedicines.

2.2.2.2 *Arginine–Glycine–Aspartic Acid (RGD).* Integrins are among the most widely studied receptors for the tumor-targeted delivery of nanomedicines. Integrins, composed of a heterodimer of α- and β-subunits, play a role in such cellular functions as cellular adhesion, migration, and proliferation [41]. Of several types of integrins, integrin isoforms $\alpha\upsilon\beta3$, $\alpha5\beta1$, and $\alpha\upsilon\beta6$ are overexpressed in various tumor tissues [42] and are thus attractive targets for cancer therapy. Integrins $\alpha\upsilon\beta3$ and $\alpha\upsilon\beta5$ have been well studied for cancer targeting since they are overexpressed not only in tumor cells but also in angiogenic endothelial cells [43].

The tripeptide, arginine–glycine–aspartic acid (RGD), is a part of certain extracellular proteins such as fibronectin and vitronectin that interact with integrins. This integrin-binding ligand has been used to modify the surfaces of nanoparticles carrying anticancer drugs. Since Pierschbacher and Ruoslahti first reported that RGD was involved in cell attachment [44], many groups have studied RGD for integrin-directed delivery of nanomedicines. Cyclic RGD is preferred to linear RGD because linear RGD has poor binding affinity for integrins [45]. Cyclic RGD has been linked to polymeric nanoparticles [46], dendrimers [47,48], and liposomes [49,50].

The cyclic pentapeptide RGDyK has been co-conjugated to PEGylated poly-amidoamine dendrimers with the anticancer drug doxorubicin for targeting to tumor neovascular endothelial cells and tumor tissues [47]. In this application, doxorubicin was conjugated via an acid-sensitive *cis*-aconityl linkage to allow the release of free doxorubicin in the acidic endosomal compartment after endocytosis. The cyclic RGDyK pentapeptide was then linked using bifunctional *N*-hydroxysulfosuccini-mide-polyoxyethylene-maleimide. Although no quantitative biodistribution data were provided, tumor sections were immunofluorescence-stained using an antivas-cular endothelial cadherin antibody to monitor the extent of tumor growth. Immu-nofluorescence staining revealed that vascular endothelial cadherin levels and, thus, tumor size were lowest in tumor tissues treated with cyclic RGDyK pentapeptide-linked conjugates.

In a recent study by the same group, cyclic RGD-linked PEGylated polyamido-amine dendrimers were conjugated with doxorubicin using an acid-insensitive succinic linkage or acid-sensitive aconityl linkage [48]. Quantitative biodistribu-tion data showed that tumor accumulation of cyclic RGD–polymer conjugates with doxorubicin attached through an acid-sensitive aconityl linkage was twofold lower than that with doxorubicin attached through an acid-insensitive succinic linkage. Despite the lower tumor accumulation, cyclic RGD–polymer conjugates with doxorubicin attached via an acid-sensitive aconityl linkage promoted the longest survival times in mice. The prolonged survival effect is attributed to the acid-labile release of doxorubicin in endolysosomes following uptake of the conjugates by the tumor cells. Unlike acid-insensitive linkage, acid-sensitive linkage may facilitate the liberation of free doxorubicin from the conjugates and increase the diffusion of the liberated doxorubicin from acidic lysosomes to the cytoplasm and subse-quently to the nucleus in an actively intercalating form. This study highlights the importance of free drug release at tumor sites, especially within intracellular environments.

Polycationic liposomes have also been modified with a cyclic RGD–PEG lipid derivative for systemic delivery of siRNA [49]. In this application the cationic charges of liposomes for electrostatic complexation with negatively charged siRNA were provided by dicetylphosphate-tetraethylenepentamine. The cyclic RGD modi-fication was introduced into liposomes using a PEG lipid derivative of cyclic RGD in the liposome preparation. After intravenous administration into C57BL/6 mice bearing luciferase-expressing B16F10-luc2 cells, cyclic RGD–PEG polycationic liposomes containing luciferase siRNA produced the greatest degree of luciferase knockdown in B16F10 metastatic lung tissues [49].

Although most RGD-modified nanomedicines have been developed for the purpose of tumor-targeted delivery, other *in vivo* applications have been reported based on the high expression of αvβ3 integrins on angiogenic vascular endothelial cells. For example, instead of anticancer drugs, the anti-inflammatory drug dexamethasone phosphate has been loaded into RGD–PEGylated liposomes and tested in a rat arthritis model [50]. In this study, the authors observed an antiarthritic effect of the dexamethasone phosphate loaded RGD–PEGylated liposomes.

2.2.2.3 Bombesin. Bombesin is a 14-amino-acid peptide with a high affinity for the gastrin-releasing peptide receptor, also called the bombesin receptor, which is highly expressed on the surface of prostate, breast, and small-cell lung carcinoma cells. Exploiting this overexpression of bombesin receptors, researchers have modified nanoparticles with bombesin peptide for tumor targeting and demonstrated enhanced delivery via bombesin receptors *in vitro* [51,52]. Bombesin-functionalized gold nanoparticles have also been designed as a delivery system for X-ray contrast agents for molecular computed tomography imaging of prostate tumors in mice [53].

2.2.2.4 Other Peptides. Peptides that can distinguish between tumor and normal tissues are attractive in the development of cancer diagnostics and targeted-delivery systems for anticancer agents. The *in vivo* applications of other peptide-modified nanomedicines are summarized in Table 2.3. Peptides can be used as antibody replacements for modification of nanomedicines if the corresponding peptide-binding receptors are clearly identified [54]. The phage display technique has contributed greatly to the discovery of numerous potent peptides that bind to specific tumor cells with high affinity. However, phage display-derived peptides are limited by the lack of understanding of their cognate receptors. The development of techniques that allow the identification of specific receptors for peptides is needed for the advancement of this field.

2.2.3 Sugars and Polysaccharides

2.2.3.1 Galactose. Galactose, a simple type of sugar, is a primary metabolic energy source for humans. Apart from the oral absorption of galactose in nutrients such as sugar beets, gum, and milk products, this monosaccharide is biochemically synthesized in the body as glucoproteins and glycolipids. In the nanomedicine field, galactose has been utilized for liver-targeted delivery because galactose-bearing asialoglycoproteins can bind the asialoglycoprotein receptors localized to paren-chymal liver cells (i.e., hepatocytes) [64]. Thus, it is possible to target nanoparticles to liver cells and liver-derived cells by tethering galactose molecules to the nanoparticle surface.

Galactosylated multimodular lipoplexes have been studied as a means to deliver plasmid DNA into primary hepatocytes [65]. Galactosylated cationic vectors, including lipids and polyethyleneimines, have been shown to increase the efficiency of nucleic acid transfection into hepatocytes compared with the vectors lacking galactose. On the other hand, it is possible that the galactosylated cationic

TABLE 2.3 *In Vivo* Applications of Peptide-Modified Nanomedicines

Peptide	Target	Nanoparticles	Active Substances	In Vivo Model	References
SFSIIHTPILPL	Hepatocellular carcinoma cells	Liposome	Doxorubicin	Mahlavu-derived hepatocellular carcinoma-bearing mice	55
GGGGYSAYPDSVPMMSK	Ephrin type-A receptor 2	Magnetic nanoparticle	Fluorescent dye (rhodamine)	BG-1-bearing mice	56
YHWYGYTPQNVI/LARLLT	Epidermal growth factor receptor	Liposome	Fluorescent dye (Cy5.5)	H1299-bearing mice	57,58
HVGGSSV	Irradiated tumors	Liposome	Doxorubicin	H460-bearing mice	59,60
CDPGYIGSR	Laminin receptor	Liposome	Fluorouracil	B16F10-bearing mice	61
HAIYPRH	Transferrin receptor	Polyamidoamine dedrimer nanoparticle	Doxorubicin	Bel-7402-bearing mice	62,63

nanoparticles might also have promoted nonspecific cellular uptake and cytotoxicity due to their positive charges. To reduce nonspecific cellular delivery, Letrou-Bonneval and colleagues developed multimodular assemblies with a guanidinium-cholesterol-based cationic liposomes core, which contained internally condensed plasmid DNA and external PEG with galactose at distal ends as a steric stabilizer. These colloidally stabilized, galactosylated lipoplexes were shown to increase the transfection efficiency of plasmid DNA in primary hepatocytes of Sprague–Dawley rats, whereas ungalactosylated lipoplexes did not.

N-Acetylgalactosamine, an amino sugar derivative of galactose, has also been used to target asialoglycoprotein receptors, specifically for receptor-targeted delivery of siRNA-carrying lipid nanoparticles [66]. To ascertain the role of N-acetylgalactosamine and asialoglycoprotein receptor interaction in the liver delivery of siRNA, these researchers intravenously administered factor VII-specific siRNA-loaded ionizable lipid nanoparticles containing various molar amounts of N-acetylgalactosamine-modified lipids to apolipoprotein $E^{-/-}$ mice at an siRNA dose of 0.2 mg/kg. A maximal gene-silencing effect in serum was observed at 0.15 mol% N-acetylgalactosamine moiety, indicating the importance of exogenous galactose modification of nanoparticles [66].

2.2.3.2 Mannose and Its Derivatives.

Mannose, a sugar monomer of a hexose, has several important biological roles, including the glycosylation of proteins. Unlike galactose interactions with asialoglycoprotein receptors on parenchymal liver cells, D-mannose is recognized by a mannose receptor generally present on nonparenchymal liver cells, such as Kupffer cells, macrophages resident in the liver.

Mannosylated liposomes have been studied for targeted delivery to alveolar macrophages [67]. Liposomes containing 9.1 mol% of mannosylation exhibited the most efficient uptake by alveolar macrophages in vitro. Tests showed that [³H] cholesteryl hexadecyl ether-labeled mannosylated liposomes administered by pulmonary aerosolization in Sprague–Dawley rats accumulated in alveolar macrophages, with a peak level 2.5-fold greater than that observed with plain liposomes occurring 2 h after administration.

Mannose as a targeting moiety has also been used to modify ultrasound-responsive bubble liposomes for gene delivery [68]. Complexes of such mannose-modified liposomes with plasmid DNA have demonstrated mannose receptor-expressing cell-specific gene transfer in vivo. In the conventional lipoplex form of plasmid DNA delivery, endosomal trafficking of lipoplexes induced the production of tumor necrosis factor-α through recognition of plasmid DNA by endosomal toll-like receptor 9. It was expected that ultrasound stimulation would allow the delivery of lipoplexes directly into the cytoplasm, reducing the likelihood of DNA recognition by tumor necrosis factor-α. As predicted, exposure to ultrasound following intravenous administration of mannose-modified bubble liposomes to mice was shown to significantly reduce toll-like receptor 9 production in the exposed tissue [69].

Mannan, a mannose-containing polysaccharide recognized by the mannose receptor of antigen-presenting cells, has been used to coat superparamagnetic iron oxide nanoparticles and applied to imaging of lymph nodes [70]. Mannan-coated superparamagnetic iron oxide nanoparticles injected into mice were shown to distribute to the lymph node to a greater degree than dextran-coated or poly(vinyl alcohol)-coated iron oxide nanoparticles. The accumulation of mannan-coated superparamagnetic iron oxide nanoparticles in the lymph node peaked after 1 h and gradually decreased until 24 h. Unlike mannan-coated nanoparticles, poly(vinyl alcohol)-coated iron oxide nanoparticles were mainly distributed in the spleen.

Mannose-6-phosphate has been used to modify human serum albumin for targeted delivery to cells expressing the mannose-6-phosphate/insulin-like growth factor-II receptor [71]. Such conjugates demonstrated an affinity for insulin-like growth factor-II receptor expressing B16 melanoma cells and were internalized by them. Two hour after intravenous injection into B16 bearing C57BL/6 mice, ^{125}I-doxorubin-linked conjugates were mainly distributed in the liver (50% of injected dose), followed by the lung, kidney, stomach, small intestine, and tumor. A total of 3% of the injected dose of ^{125}I-doxorubin-linked conjugates was distributed in tumor tissues, an amount less than 1/10 that in the liver. A particularly valuable aspect of this study is its quantitative presentation of biodistribution data as percentage of injected dose in 20 organs and tissues. Although the terminology "tumor targeting" is commonly used, the absolute percentage of distribution to the tumors as reported in this study suggests that the nanoparticles distribute in other tissues of the body to substantial extents.

2.2.3.3 *Hyaluronic Acid.*

Hyaluronic acid, an anionic glycosaminoglycan, is a carbohydrate abundant in the body. Notably, it is a major component of the extracellular matrix, which participates in cell proliferation, migration, and interaction with cell surface receptors, including intracellular adhesion molecule-1 and CD44, the receptor for hyaluronic acid-mediated motility [72,73].

Reducible polyethyleneimine-grafted hyaluronic acid has been synthesized and used for systemic delivery of siRNA against liver cirrhosis [74]. Systemic administration of apolipoprotein B-specific siRNA using reducible polyethyleneimine-grafted hyaluronic acid complexes was shown to reduce target gene expression and increase therapeutic efficacy in a mouse model of liver cirrhosis.

In liver fibrosis, the expression of matrix metalloproteinases is imbalanced; thus, one therapeutic strategy is to restore metalloproteinase activity by delivering plasmid DNA encoding matrix metalloproteinases. To this end, hyaluronic acid, which is negatively charged, can be used to form electrostatic triple complexes with preformed dual complexes of plasmid DNA encoding matrix metalloproteinase 13 and polyethyleneimine [75]. The resulting DNA–polyethyleneimine is shielded with hyaluronic acid by charge interactions. Intravenous administration of the hyaluronic acid-shielded plasmid DNA complexes into mice was shown to ameliorate the progression of liver fibrosis as evidenced by a reduction in collagen deposition. In this study, the shielding with hyaluronic acid was shown to reduce the toxicity of the polyethyleneimine vector. Without hyaluronic acid shielding, the maximum dose of plasmid DNA that could be systemically administered was 1 mg/kg, owing to the toxicity of polyethyleneimine. However, the shielding of hyaluronic acid reduced the toxicity of polyethyleneimine, allowing 100% survival of mice at a plasmid DNA dose of at least 3 mg/kg.

In a recent study, hyaluronic acid itself was used as nanoparticle material for anticancer drug delivery [76]. Hyaluronic acid is biodegraded by the endogenous enzyme hyaluronidase, which is known to be particularly abundant in malignant tissues and metastatic cancers. To exploit this abundance of hyaluronidase and facilitate rapid release of drug at tumor sites, researchers have formulated PEG-conjugated hyaluronic acid nanoparticles loaded with an anticancer drug, campto-thecin. Studies of the biodistribution of camptothecin-loaded PEGylated hyaluronic acid nanoparticles showed tumor localization in mice bearing CD44-expressing tumors. Camptothecin-loaded PEGylated hyaluronic acid nanoparticles prevented significant increases in tumor size for at least 35 days.

2.2.4 Small Molecule Chemicals

2.2.4.1 Folate. Folate—water soluble Vitamin B$_9$—plays several roles in bio-synthesis, including nucleic acid metabolism (e.g., DNA and RNA synthesis), prevention of DNA changes, and amino acid production. Notably, folate is required for the replication of DNA during cell division and proliferation. Cancer cells, which grow very rapidly, express high surface levels of folate receptors, which mediate cellular uptake of folate. Given the higher relative expression of folate receptors on tumor cells, it is not surprising that folate has been widely studied as a ligand molecule for nanomedicines carrying anticancer drugs or imaging agents [77,78]. *In vivo* applications of folate-modified nanomedicines are summarized in Table 2.4.

In the study of Yamada and colleagues [79], different *in vitro* and *in vivo* results were observed between doxorubicin in free form and folate-modified liposomes. *In vitro*, free doxorubicin showed higher antitumor activity than doxorubicin in folate-modified liposomes. In folate receptor-expressing KB cells, the concentration leading to 50% of cell death was 0.65 and 1.27 μg/ml for doxorubicin in free form and in liposomes with folate linked to the distal end of PEG lipid chain (mw 5000), respectively. In contrast, *in vivo* antitumor activity of doxorubicin was found to be higher in folate-liposomes as compared to free form. This difference was explained by the different *in vivo* distribution pattern of doxorubicin. The distribution of doxorubicin to tumor tissues was found only in the mice intravenously injected with folate-modified liposomes but not in the mice treated with free doxorubicin.

The importance of drug release at the target site has motivated development of an approach that controls the release of drug in a pH-responsive manner. Here, folate has been linked to pH-sensitive polymeric micelles composed of PEG–poly (aspartate–hydrazone–doxorubicin) for tumor tissue delivery of doxorubicin [80]. In this application, the anticancer drug doxorubicin was conjugated to pH-sensitive hydrazone to facilitate the drug release in acidic intracellular compartments, such as lysosomes and endosomes, where pH ranges from 5 to 6. Conjugation of folate to pH-sensitive micelles dramatically increased cellular uptake and cytotoxicity *in vitro* compared to unmodified micelles. However, an *in vivo* distribution study revealed doxorubicin accumulation in tumors after delivery using micelles was similar with or

TABLE 2.4 *In Vivo* Applications of Folate-Modified Nanomedicines

Nanomedicines	Active Substances	*In Vivo* Application	References
Liposome	Doxorubicin	M109-bearing mice	79
Liposome	Doxorubicin	M109-bearing mice, J6456-bearing mice, KB-bearing mice	82
Liposome	Docetaxel	Sarcoma-180-bearing mice	83
Liposome	Plasmid DNA	M109-bearing mice	84
Liposome	Plasmid DNA	SCC7-bearing mice	85
Liposome	Antisense Oligodeoxynucleotide	KB-bearing mice	86
Polymeric micelle	Doxorubicin	A2780/doxorubicin-resistant tumor-bearing mice	87
Polymeric micelle	Adriamycin	KB-bearing mice	80
Heparin-based nanoparticle	Paclitaxel	KB-3-1-bearing mice	81
Polyamidoamine dendrimer	EGFR-antisense oligonucleotides	C6-bearing rats	88
Nanohydrogel	5-Fluorouracil	Wistar rats	89
Albumin nanoparticles	^{188}Re, cisplatin	SKOV3-bearing mice	90
Chelator conjugate	^{99m}Tc	KB-bearing mice	91

without folate modification. Despite the similar tumor accumulation, folate-conjugated micelles showed lower *in vivo* toxicity and a reduced effective dose, increasing the therapeutic window by fivefold compared to the free drug. The absence of a difference in tumor accumulation between micelles with and without folate conjugation was explained by the dominance of the EPR effect in determining the tumor accumulation of drug carriers. The higher therapeutic window of folate-modified micelles indicates that folate conjugation may affect the tumor distribution of micelles after extravasation, rather than the migration of micelles from the bloodstream into the tumor tissues.

Consistent with the results observed in this micelle study, another group reported no difference in the tumor tissue accumulation of a folate-modified nanomedicine [81]. Specifically, molecular imaging revealed that the tumor tissue distribution of heparin-paclitaxel nanoparticles modified with folate ligands was similar to that of unmodified particles. To determine whether nanoparticles simply accumulated outside tumor cells or actually entered them, these researchers stained tumor cells in the tumor tissues with an antihuman CD326 antibody. This tumor cell-identification strategy revealed that folate-modified nanoparticles were predominantly taken up by tumor cells, whereas unmodified nanoparticles were present in the extracellular space. Moreover, tumor volume was found to be significantly smaller in the group treated with folate-modified nanoparticles compared to groups treated with unmodified nanoparticles or free drug. This increase in the *in vivo* anticancer efficacy of folate-modified nanoparticles suggests tumor-cell targeting after extravasation.

2.2.4.2 Alendronate. Although small molecule chemical ligands have many advantages such as lower immunogenicity, ease of handling in modification processes, and possibly lower costs, only a few small molecule chemical ligands other than folate have been studied.

Alendronate, a drug used to prevent and treat osteoporosis, has been studied as a bone-targeting moiety in a mouse model of bone metastasis [92]. In this application, alendronate was linked to an *N*-(2-hydroxypropyl) methacrylamide-paclitaxel conjugate. An *in vivo* study showed that the alendronate-linked conjugate-treated group achieved greater tumor regression than did groups treated with free paclitaxel or a mixture of free paclitaxel and alendronate. Moreover, the alendronate-linked conjugate showed less hemolytic activity than Cremophor EL, a commercial vehicle for paclitaxel. Although the authors speculated that the higher efficacy and safety of alendronate-linked conjugate was due to a more favorable pharmacokinetic profile of the conjugate, quantitative data on tumor tissue distribution were not provided in this study.

2.2.4.3 Anisamide. Sigma receptors are membrane-bound proteins with a high affinity for neuroleptics. Although sigma receptors are expressed on normal tissues, including liver, lungs, kidneys, and central nervous system, they are known to be overexpressed in diverse tumors. Huang and colleagues tested liposomes modified with the sigma receptor-binding ligand, anisamide, first studying anisamide-modified liposomes as delivery vehicles for doxorubicin [93]. A PEG-conjugated phospholipid was derivatized with an anisamide ligand, which was then incorporated into the doxorubicin-loaded liposome. *In vivo* imaging revealed that anisamide-conjugated liposomal doxorubicin showed significantly improved accumulation in the tumor.

In a recent study by the same group, anisamide-modified cationic liposomes were studied for systemic delivery of c-Myc-specific siRNA [94]. Four hours after intravenous administration to B16F10 melanoma tumor-bearing mice, anisamide-modified liposomes showed increased accumulation in tumors compared to unmodified liposomes. Western blots of tumor tissues showed that the c-Myc protein was silenced in the group treated with anisamide-modified liposomes, but not in other groups.

2.3 CURRENT CHALLENGES IN CLINICAL DEVELOPMENT OF LIGAND-MODIFIED NANOMEDICINE

2.3.1 Nonclinical and Clinical Studies

In addition to the *in vivo* case studies highlighted earlier, numerous animal studies are ongoing. Despite these efforts, very few powerful ligand-modified nanomedicine candidates have entered clinical trials. One such case is the transferrin-conjugated immunotoxin, TransMid (Xenova Biomedix Co.), which has entered clinical trials as a locally administered treatment for brain tumors [95]. In this conjugate, transferrin is bound to diphtheria toxin through a lysine linker and a thioether. After binding to glioma cells, the conjugate enters the cells via receptor-mediated endocytosis and releases toxin, which kills the cells. This immunotoxin completed Phase I and II clinical trials, but was stopped during Phase III trials in late 2006 by the sponsor, Xenova Biomedix, according to a conditional power analysis showing that the conjugate would not significantly improve upon standard treatments for high-grade brain tumors.

A second case, EC-145, is under development by Endocyte Inc. [96], although putting this chemical drug conjugate in the category of nanomedicine may be arguable. EC-145 is a conjugate of folate and vinka alkaloid, in which folate is linked to the anticancer agent desacetylvinblastine monohydrazide through a peptide spacer. After promising preclinical trial results [97], the conjugate is currently in Phase II clinical trials in patients with advanced ovarian cancer.

One other nanomedicine undergoing clinical testing is Calando Pharmaceutical's nanomaterial, denoted CALAA-01, which is now in Phase I clinical trials [5,98] after completion of multi-dosing tests in nonhuman primates [99,100]. CALAA-01 consists of a human transferrin-tagged cyclodextrin-based polymer for systemic administration of siRNA against the M2 subunit of ribonucleotide reductase. Researchers at Calando have reported that systemic delivery of siRNA-loaded nanoparticles significantly reduced target gene expression in patients. Because CALAA-01 is not only a targeted nanoparticle but also an siRNA therapeutic, this first clinical trial for the treatment of solid cancer represents a significant step for the field of nanomedicine. It is worthwhile to note that the focus of the study in question was on the systemic administration of RNAi therapeutics rather than on the targeting ligand, and comparative studies of siRNA-loaded nanoparticles with and without transferrin are lacking. It remains to be seen whether the transferrin ligand played a vital role in the results of the clinical study.

2.3.2 Challenges and Future Perspectives

Numerous ligand-modified nanomedicine candidates have shown promise *in vitro*, but there are multiple *in vivo* barriers that can nullify the targeting capabilities of nanoparticles. One of the biggest barriers against the delivery to target tissues is the massive uptake of nanoparticles in the reticuloendothelial systems such as liver. As quantitatively reported in the study using mannose-6-phosphate as a targeting moiety [71], described earlier, 50% of the injected dose was found in the liver, with less than 5% of the dose reaching the tumor tissue. To increase the delivery to tumors or other target sites, it is essential that the substantial distribution of nanoparticles to the liver be decreased.

One approach for reducing such off-target delivery is to increase the selectivity of the nanomedicine. To date, most nanomedicines have been modified with a single ligand whose receptors are overexpressed on target cells. However, it is practically not easy to identify suitable receptors that exist exclusively on target cells. As a result, off-target binding of nanomedicines to other normal cells that also express the same receptor is unavoidable. To circumvent this limitation and increase the selectivity for tumor cells, which actually overexpress multiple types of surface receptors, Saul and coworkers have recently suggested a dual-ligand approach [101]. In their *in vitro* study, liposomes were labeled with folate and an anti-EGFR antibody to further improve the specificity of tumor targeting. Although additional proof-of-concept experiments are needed for *in vivo* applications, a dual-ligand approach might be a future direction for enhancing the specificity of delivery.

One general expectation in ligand-modified nanomedicine is that the ligands actively improve the migration of nanoparticles from the bloodstream to target sites, such as tumor cells. This idea of "smart" ligands that guide nanoparticles from the rapidly flowing bloodstream to specific tumor sites is not fully realized in practice. Although several recent studies have shown dramatic increase in tumor accumulation of ligand-modified nanoparticles, others have consistently reported the lack of enhanced tumor tissue accumulation, even after high-density ligand modification. As discussed by the authors of these recent studies [80], most ligand-grafted nanoparticles arrive at tumor sites, not through the active guidance of ligands but through the predominant EPR effect.

Several approaches have been suggested for improving the penetration of nano-particles through vascular endothelial barriers. One such approach is to increase the action of the EPR effect for both passive and targeted delivery of nanomedicines. Several groups have attempted to enhance EPR effects by administration of nitric oxide donor chemicals [102]. Additional approaches include the use of components for vascular targeting [98] and the design of multifunctional nanoparticles with bio-inspired penetration enhancers [103].

Although conjugated ligands do not appear to have a profound effect on the tumor accumulation of nanomedicines, they may contribute to enhancing tumor cell uptake of nanoparticles after extravasation, as suggested by the case of folate-modified micelles [80], described earlier. These and similar studies suggest that the greater anticancer efficacy of ligand-modified nanoparticles, despite similar accumulation at tumor sites, may be due to enhanced intracellular delivery.

After the arrival of ligand-modified nanomedicines at tumor tissues, the remaining challenge is penetration through the tumor mass. Owing to their size, nano-particles at tumor sites have been shown to accumulate mainly in the peripheral area of the vasculature, rather than penetrating deeply into the tumor mass [78]. Reducing the sizes of nanoparticles in the tumor microenvironment has been suggested as a way of overcoming the size-dependent hindrance of nanoparticle penetration into the tumor matrix [104]. In this study, 100 nm nanoparticles were shrunk to a size of 10 nm by the action of proteases that are highly expressed in the tumor micro-environment. Such an approach might provide a framework for subsequent efforts to overcome the penetration limitation.

The composition of engineered nanoparticles is an important topic in the commercialization of nanomedicine. Currently, the most widely used method—chemical tethering of ligands onto preformed nanoparticles—has several limitations with respect to manufacturing and quality control [105]. The underlying chemistry requires optimization to prevent batch-to-batch variability and variations according to the types of ligands and vehicles [106]. Complicated synthesis procedures also require complex purification procedures, which typically lead to expensive and time-consuming scaling-up efforts. With this manufacturing issue in mind, recent studies have demonstrated the advantages of self-assembly, prefunctionalization, and genetic engineering strategies in eliminating or minimizing complex manufacturing processes.

Ligand-engineered nanomedicines need to be used in conjunction with patient-tailored diagnostic tools to achieve the best therapeutic outcome. Depending on the cancer type, the expression levels of the receptors for targeting ligands may vary. Moreover, only patients that overexpress the specific receptors would be responsive to the ligand-modified nanomedicine. Thus, for optimal therapeutic efficacy, patients may need to be diagnosed in advance to determine whether they are likely to be responsive to a specific ligand-modified nanomedicine.

2.4 CONCLUSIONS

A variety of *in vivo* applications of ligand-engineered nanoparticles have been described. Despite the merits of engineered nanoparticles, remaining *in vivo* barriers can severely limit the activity of these particles after reaching tumor sites. At present, an ideal nanoparticle that satisfies all the requirements for successful clinical trials does not exist. However, the first and difficult steps toward developing a concrete understanding of the drawbacks have been taken, and related research has already been started to conquer them. As these obstacles are surmounted, the list of clinically successful nanomedicines is expected to grow in the near future.

Acknowledgments

This work was supported by research grants from the Ministry of Education, Science, and Technology (2012007005), from the Bio & Medical Technology Development Program of the National Research Foundation funded by the Korean government (MEST) (No. 20120006123), and from the Korean Health Technology R&D Project (No. A092010), Ministry for Health, Welfare & Family Affairs, Republic of Korea.

REFERENCES

1. Davis, M. E., Chen, Z., Shin, D. M. (2008). Nanoparticle therapeutics: an emerging treatment modality for cancer. *Nature Reviews Drug Discovery 7*, 771–782.
2. Petros, R. A., DeSimone, J. M. (2010). Strategies in the design of nanoparticles for therapeutic applications. *Nature Reviews Drug Discovery 9*, 615–627.

3. Matsumura, Y., Maeda, H. (1986). A new concept for macromolecular therapeutics in cancer chemotherapy: mechanism of tumoritropic accumulation of proteins and the antitumor agent smancs. *Cancer Research 46*, 6387–6392.

4. Mohanty, C., Das, M., Kanwar, J. R., Sahoo, S. K. (2011). Receptor mediated tumor targeting: an emerging approach for cancer therapy. *Current Drug Delivery 8*, 45–56.

5. Zhang, L., Gu, F. X., Chan, J. M., Wang, A. Z., Langer, R. S., Farokhzad, O. C. (2008). Nanoparticles in medicine: therapeutic applications and developments. *Clinical Pharmacology and Therapeutics 83*, 761–769.

6. Irache, J. M., Esparza, I., Gamazo, C., Agüeros, M., Espuelas, S. (2011). Nanomedicine: novel approaches in human and veterinary therapeutics. *Veterinary Parasitology 180*, 47–71.

7. Jain, R. K., Stylianopoulos, T. (2010). Delivering nanomedicine to solid tumors. *Nature Reviews Clinical Oncology 7*, 653–664.

8. Hatakeyama, H., Akita, H., Harashima, H. (2011). A multifunctional envelope type nano device (MEND) for gene delivery to tumours based on the EPR effect: a strategy for overcoming the PEG dilemma. *Advances in Drug Delivery Review 63*, 152–160.

9. Yu, B., Tai, H. C., Xue, W., Lee, L. J., Lee, R. J. (2010). Receptor-targeted nanocarriers for therapeutic delivery to cancer. *Molecular Membrane Biology 27*, 286–298.

10. Fay, F., Scott, C. J. (2011). Antibody-targeted nanoparticles for cancer therapy. *Immunotherapy 3*, 381–394.

11. Noble, C. O., Guo, Z., Hayes, M. E., Marks, J. D., Park, J. W., Benz, C. C., Kirpotin, D. B., Drummond, D. C. (2009). Characterization of highly stable liposomal and immunoliposomal formulations of vincristine and vinblastin. *Cancer Chemotherapy and Pharmacology 64*, 741–751.

12. Weng, K. C., Noble, C. O., Papahadjopoulos-Sternberg, B., Chen, F. F., Drummond, D. C., Kirpotin, D. B., Wang, D., Hom, Y. K., Hann, B., Park, J. W. (2008). Targeted tumor cell internalization and imaging of multifunctional quantum dot-conjugated immunoliposomes *in vitro* and *in vivo*. *Nano Letters 8*, 2851–2857.

13. Lu, R. M., Chang, Y. L., Chen, M. S., Wu, H. C. (2011). Single chain anti-c-Met antibody conjugated nanoparticles for *in vivo* tumor-targeted imaging and drug delivery. *Biomaterials 32*, 3265–3274.

14. Wang, K., Ruan, J., Qian, Q., Song, H., Bao, C., Zhang, X., Kong, Y., Zhang, C., Hu, G., Ni, J., Cui, D. (2011). BRCAA1 monoclonal antibody conjugated fluorescent magnetic nanoparticles for *in vivo* targeted magnetofluorescent imaging of gastric cancer. *Nanobiotechnology 9*, 23.

15. Sauter, G., Maeda, T., Waldman, F. M., Davis, R. L., Feuerstein, B. G. (1996). Patterns of epidermal growth factor receptor amplification in malignant gliomas. *The American Journal of Pathology 148*, 1047–1053.

16. Feng, B., Tomizawa, K., Michiue, H., Miyatake, S., Han, X. J., Fujimura, A., Seno, M., Kirihata, M., Matsui, H. (2009). Delivery of sodium borocaptate to glioma cells using immunoliposome conjugated with anti-EGFR antibodies by ZZ-His. *Biomaterials 30*, 1746–1755.

17. Liu, P., Li, Z., Zhu, M., Sun, Y., Li, Y., Wang, H., Duan, Y. (2010). Preparation of EGFR monoclonal antibody conjugated nanoparticles and targeting to hepatocellular carcinoma. *Journal of Materials Science: Materials in Medicine 21*, 551–556.

18. Pattillo, C. B., Venegas, B., Donelson, F. J., Del Valle, L., Knight, L. C., Chong, P. L., Kiani, M. F. (2009). Radiation-guided targeting of combretastatin encapsulated immunoliposomes to mammary tumors. *Pharmaceutical Research 26*, 1093–1100.

19. Tuffin, G., Waelti, E., Huwyler, J., Hammer, C., Marti, H. P. (2005). Immunoliposome targeting to mesangial cells: a promising strategy for specific drug delivery to the kidney. *Journal of the American Society of Nephrology 16*, 3295–3305.

20. Scott, R. C., Rosano, J. M., Ivanov, Z., Wang, B., Chong, P. L., Issekutz, A. C., Crabbe, D. L., Kiani, M. F. (2009). Targeting VEGF-encapsulated immunoliposomes to MI heart improves vascularity and cardiac function. *The FASEB Journal 23*, 3361–3367.

21. Kim, S. S., Peer, D., Kumar, P., Subramanya, S., Wu, H., Asthana, D., Habiro, K., Yang, Y. G., Manjunath, N., Shimaoka, M., Shankar, P. (2010). RNAi-mediated CCR5 silencing by LFA-1-targeted nanoparticles prevents HIV infection in BLT mice. *Molecular Therapy 18*, 370–376.

22. Manjappa, A. S., Chaudhari, K. R., Venkataraju, M. P., Dantuluri, P., Nanda, B., Sidda, C., Sawant, K. K., Murthy, R. S. (2011). Antibody derivatization and conjugation strategies: application in preparation of stealth immunoliposome to target chemotherapeutics to tumor. *Journal of Controlled Release 150*, 2–22.

23. Schalén, C., Truedsson, L., Christensen, K. K., Christensen, P. (1985). Blocking of antibody complement-dependent effector functions by streptococcal IgG Fc-receptor and staphylococcal protein A. *Acta Pathologica, Microbiologica, et Immunologica Scandinavica. Section B, Microbiology 93*, 395–400.

24. Jung, Y., Kang, H. J., Lee, J. M., Jung, S. O., Yun, W. S., Chung, S. J., Chung, B. H. (2008). Controlled antibody immobilization onto immunoanalytical platforms by synthetic peptide. *Analytical Biochemistry 374*, 99–105.

25. Feng, B., Tomizawa, K., Michiue, H., Han, X. J., Miyatake, S., Matsui, H. (2010). Development of a bifunctional immunoliposome system for combined drug delivery and imaging *in vivo*. *Biomaterials 31*, 4139–4145.

26. De Lano, W. L., Ultsch, M. H., de Vos, A. M., Wells, J. A. (2000). Convergent solutions to binding at a protein–protein interface. *Science 287*, 1279–1283.

27. Oh, Y. K., Chang, R. S., Yu, Y. H., Kim, W. K. Lipopeptides with specific affinity to Fc region of antibodies and antigen-recognizing lipid nanoparticles comprising the same. 10-2010-0119486 (Korea, Patent).

28. Daniels, T. R., Delgado, T., Helguera, G., Penichet, M. L. (2006). The transferrin receptor part II: targeted delivery of therapeutic agents into cancer cells. *Clinical Immunology 121*, 159–176.

29. Daniels, T. R., Bernabeu, E., Rodríguez, J. A., Patel, S., Kozman, M., Chiappetta, D. A., Holler, E., Ljubimova, J. Y., Helguera, G., Penichet, M. L. (2012). Transferrin receptors and the targeted delivery of therapeutic agents against cancer. *Biochimica et Biophysica Acta 1820*, 291–317.

30. Yoon, D. J., Liu, C. T., Quinlan, D. S., Nafisi, P. M., Kamei, D. T. (2011). Intracellular trafficking considerations in the development of natural ligand-drug molecular conjugates for cancer. *Annals of Biomedical Engineering 39*, 1235–1251.

31. Daniels, T. R., Delgado, T., Rodriguez, J. A., Helguera, G., Penichet, M. L. (2006). The transferrin receptor part I: biology and targeting with cytotoxic antibodies for the treatment of cancer. *Clinical Immunology 121*, 144–158.

32. Li, X., Ding, L., Xu, Y., Wang, Y., Ping, Q. (2009). Targeted delivery of doxorubicin using stealth liposomes modified with transferrin. *International Journal of Pharmaceutics 373*, 116–123.

33. Zhang, X., Koh, C. G., Yu, B., Liu, S., Piao, L., Marcucci, G., Lee, R. J., Lee, L. J. (2009). Transferrin receptor targeted lipopolyplexes for delivery of antisense oligonucleotide G3139 in a murine K562 xenograft model. *Pharmaceutical Research 26*, 1516–1524.

34. Xu, Z., Gu, W., Huang, J., Sui, H., Zhou, Z., Yang, Y., Yan, Z., Li, Y. (2005). *In vitro* and *in vivo* evaluation of actively targetable nanoparticles for paclitaxel delivery. *International Journal of Pharmaceutics 288*, 361–368.

35. Shah, N., Chaudhari, K., Dantuluri, P., Murthy, R. S. R., Das, S. (2009). Paclitaxel-loaded PLGA nanoparticles surface modified with transferrin and Pluronic®P85, an *in vitro* cell line and *in vivo* biodistribution studies on rat model. *Journal of Drug Targeting 17*, 533–542.

36. Ren, W., Chang, J., Yan, C., Qian, X., Long, L., He, B., Yuan, X., Kang, C., Betbeder, D., Sheng, J., Pu, P. (2010). Development of transferrin functionalized poly(ethylene glycol)/poly(lactic acid) amphiphilic block copolymeric micelles as a potential delivery system targeting brain glioma. *Journal of Materials Science: Materials in Medicine 21*, 2673–2681.

37. Bae, S., Ma, K., Kim, T. H., Lee, E. S., Oh, K. T., Park, E-S., Lee, K. C., Youn, Y. S. (2012). Doxorubicin-loaded human serum albumin nanoparticles surface-modified with TNF-related apoptosis-inducing ligand and transferrin for targeting multiple. *Biomaterials 33*, 1536–1546.

38. Choi, C. H. J., Alabi, C. A., Webster, P., Davis, M. E. (2010). Mechanism of active targeting in solid tumors with transferrin-containing gold nanoparticles. *Proceedings of the National Academy of Sciences of the United States of America 107*, 1235–1240.

39. Hong, M., Zhu, S., Jiang, Y., Tang, G., Sun, C., Fang, C., Shi, B., Pei, Y. (2010). Novel anti-tumor strategy: PEG–hydroxycamptothecin conjugate loaded transferrin-PEG-nanoparticles. *Journal of Controlled Release 141*, 22–29.

40. Yoon, D. J., Chu, D. S., Ng, C. W., Pharm, E. A., Mason, A. B., Hudson, D. M., Smith, V. C., MacGillivray, R. T., Kamei, D. T. (2009). Genetically engineering transferrin to improve its *in vitro* ability to deliver cytotoxins. *Journal of Controlled Release 133*, 178–184.

41. Hsu, A. R., Veeravagu, A., Cai, W., Hou, L. C., Tse, V., Chen, X. (2007). Integrin $\alpha\upsilon\beta3$ antagonists for anti-angiogenic cancer treatment. *Recent Patents on Anti-Cancer Drug Discovery 2*, 143–160.

42. McCabe, N. P., De, S., Vasanji, A. (2007). Prostate cancer specific integrin $\alpha\upsilon\beta3$ modulates bone metastatic growth and tissue remodeling. *Oncogene 26*, 6238–6243.

43. Shim, G., Lee, S., Choi, H., Lee, J., Kim, C-W., Byun, Y., Oh, Y-K. (2011). Nanomedicines for receptor-mediated tumor targeting. *Recent Patents on Nanomedicine 1*, 138–148.

44. Pierschbacher, M. D., Ruoslahti, E. (1984). Cell attachment activity of fibronectin can be duplicated by small fragments of the molecule. *Nature 309*, 30–33.

45. Juliano, R. L., Alam, R., Dixit, V., Kang, H. M. (2009). Cell-targeting and cell-penetrating peptides for delivery of therapeutic and imaging agents. *Wires Nanomedicine and Nanobiotechnology 1*, 324–335.

46. Danhier, F., Vroman, B., Lecouturier, N., Crokart, N., Pourcelle, V., Freichels, H., Jérôme, C., Marchand-Brynaert, J., Feron, O., Préat, V. (2009). Targeting of tumor endothelium by RGD-grafted PLGA-nanoparticles loaded with paclitaxel. *Journal of Controlled Release 140*, 166–173.

47. Zhu, S., Qian, L., Hong, M., Zhang, L., Pei, Y., Jiang, Y. (2011). RGD-modified PEG–PAMAM–DOX conjugate: *in vitro* and *in vivo* targeting to both tumor neovascular endothelial cells and tumor cells. *Advanced Materials 23*, 84–89.

48. Zhang, L., Zhu, S., Qian, L., Pei, Y., Qiu, Y., Jiang, Y. (2011). RGD-modified PEG–PAMAM–DOX conjugates: *in vitro* and *in vivo* studies for glioma. *European Journal of Pharmaceutics and Biopharmaceutics 79*, 232–240.

49. Yonenaga, N., Kenjo, E., Asai, T., Tsuruta, A., Shimizu, K., Dewa, T., Nango, M., Oku, N. (2012). RGD-based active targeting of novel polycation liposomes bearing siRNA for cancer treatment. *Journal of Controlled Release 160*, 177–181.

50. Koning, G. A., Schiffelers, R. M., Wauben, M. H. M., Kok, R. J., Mastrobattista, E., Molema, G., ten Hagen, T. L. M., Storm, G. (2006). Targeting of angiogenic endothelial cells at sites of inflammation by dexamethasone phosphate–containing RGD peptide liposomes inhibits experimental arthritis. *Arthritis and Rheumatism 54*, 1198–1208.

51. Martin, A. L., Hickey, J. L., Ablack, A. L., Lewis, J. D., Luyt, L. G., Gillies, E. R. (2010). Synthesis of bombesin-functionalized iron oxide nanoparticles and their specific uptake in prostate cancer cells. *Journal of Nanoparticle Research 12*, 1599–1608.

52. Honer, M., Mu, L., Stellfeld, T., Graham, K., Martic, M., Fischer, C. R., Lehmann, L., Schubiger, P. A., Ametamey, S. M., Dinkelborg, L., Srinivasan, A., Borkowski, S. (2011). ^{18}F-labeled bombesin analog for specific and effective targeting of prostate tumors expressing gastrin-releasing peptide receptors. *The Journal of Nuclear Medicine 52*, 270–278.

53. Chanda, N., Kattumuri, V., Shukla, R., Zambre, A., Katti, K., Upendran, A., Kulkarni, R. R., Kan, P., Fent, G. M., Casteel, S. W., Smith, C. J., Boote, E., Robertson, J. D., Cutler, C., Lever, J. R., Katti, K. V., Kannan, R. (2010). Bombesin functionalized gold nanoparticles show *in vitro* and *in vivo* cancer receptor specificity. *Proceedings of the National Academy of Sciences of the United States of America 107*, 8760–8765.

54. Brown, K. C. (2010). Peptidic tumor targeting agents: the road from phage display peptide selections to clinical applications. *Current Pharmaceutical Design 16*, 1040–1054.

55. Lo, A., Lin, C-T., Wu, H-C. (2008). Hepatocellular carcinoma cell-specific peptide ligand for targeted drug delivery. *Molecular Cancer Therapeutics 7*, 579–589.

56. Scarberry, K. E., Dickerson, E. B., McDonald, J. F., Zhang, Z. J. (2008). Magnetic nanoparticle-peptide conjugates for *in vitro* and *in vivo* targeting and extraction of cancer cells. *Journal of the American Chemical Society 130*, 10258–10262.

57. Song, S., Liu, D., Peng, J., Sun, Y., Li, Z., Gu, J-R., Xu, Y. (2008). Peptide ligand-mediated liposome distribution and targeting to EGFR expressing tumor *in vivo*. *International Journal of Pharmaceutics 363*, 155–161.

58. Song, S., Liu, D., Peng, J., Deng, H., Guo, Y., Xu, L. X., Miller, A. D., Xu, Y. (2009). Novel peptide ligand directs liposomes toward EGF-R high-expressing cancer cells *in vitro* and *in vivo*. *The FASEB Journal 23*, 1396–1404.

59. Han, Z., Fu, A., Wang, H., Diaz, R., Geng, L., Onishko, H., Hallahan, D. E. (2008). Noninvasive assessment of cancer response to therapy. *Nature Medicine 14*, 343–349.

60. Lowery, A., Onishko, H., Hallahan, D. E., Han, Z. (2011). Tumor-targeted delivery of liposome-encapsulated doxorubicin by use of a peptide that selectively binds to irradiated tumors. *Journal of Controlled Release 150*, 117–124.

61. Dubey, P. K., Singodia, D., Vyas, S. P. (2010). Liposomes modified with YIGSR peptide for tumor targeting. *Journal of Drug Targeting 18*, 373–380.

62. Han, L., Huang, R., Liu, S., Huang, S., Jiang, C. (2010). Peptide-conjugated PAMAM for targeted doxorubicin delivery to transferrin receptor overexpressed tumors. *Molecular Pharmaceutics 7*, 2156–2165.

63. Han, L., Huang, R., Li, J., Liu, S., Huang, S., Jiang, C. (2011). Plasmid pORF-hTRAIL and doxorubicin co-delivery targeting to tumor using peptide-conjugated polyamido-amine dendrimer. *Biomaterials 32*, 1242–1252.

64. Zhu, L., Ye, Z., Cheng, K., Miller, D. D., Mahato, R. I. (2008). Site-specific delivery of oligonucleotides to hepatocytes after systemic administration. *Bioconjugate Chemistry 19*, 290–298.

65. Letrou-Bonneval, E., Chèvre, R., Lambert, O., Costet, P., André, C., Tellier, C., Pitard, B. (2008). Galactosylated multimodular lipoplexes for specific gene transfer into primary hepatocytes. *The Journal of Gene Medicine 10*, 1198–1209.

66. Akinc, A., Querbes, W., De, S., Qin, J., Frank-Kamenetsky, M., Jayaprakash, K. N., Jayaraman, M., Rajeev, K. G., Cantley, W. L., Dorkin, J. R., Butler, J. S., Qin, L. L., Racie, T., Sprague, A., Fava, E., Zeigerer, A., Hope, M. J., Zerial, M., Sah, D. W. Y., Fitzgerald, K., Tracy, M. A., Manoharan, M., Koteliansky, V., Fougerolles, A., Maier, M. A. (2010). Targeted delivery of RNAi therapeutics with endogenous and exogenous ligand-based mechanisms. *Molecular Therapy 18*, 1357–1364.

67. Chono, S., Kaneko, K., Yamamoto, E., Togami, K., Morimoto, K. (2010). Effect of surface-mannose modification on aerosolized liposomal delivery to alveolar macro-phages. *Drug Development and Industrial Pharmacy 36*, 102–107.

68. Un, K., Kawakami, S., Suzuki, R., Maruyama, K., Yamashita, F., Hashida, M. (2010). Development of an ultrasound-responsive and mannose-modified gene carrier for DNA vaccine therapy. *Biomaterials 31*, 7813–7826.

69. Un, K., Kawakami, S., Yoshida, M., Higuchi, Y., Suzuki, R., Maruyama, K., Yamashita, F., Hashida, M. (2011). The elucidation of gene transferring mechanism by ultrasound-responsive unmodified and mannose-modified lipoplexes. *Biomaterials 32*, 4659–4669.

70. Vu-Quang, H., Yoo, M-K., Jeong, H-J., Lee, H-J., Muthiah, M., Rhee, J. H., Lee, J-H., Cho, C-S., Jeong, Y. Y., Park, I-K. (2011). Targeted delivery of mannan-coated super-paramagnetic iron oxide nanoparticles to antigen-presenting cells for magnetic resonance-based diagnosis of metastatic lymph nodes *in vivo*. *Acta Biomaterialia 7*, 3935–3945.

71. Prakash, J., Beljaars, L., Harapanahalli, A. K., Zeinstra-Smith, M., Jager-Krikken, A., Hessing, M., Steen, H., Poelstra, K. (2010). Tumor-targeted intracellular delivery of anticancer drugs through the mannose-6-phosphate/insulin-like growth factor II recep-tor. *International Journal of Cancer 126*, 1966–1981.

72. Koo, H., Huh, M. S., Sun, I-C., Yuk, S. H., Choi, K., Kim, K., Kwon, I. C. (2011). *In vivo* targeted delivery of nanoparticles for theranosis. *Accounts of Chemical Research 44*, 1018–1028.

73. Kouvidi, K., Berdiaki, A., Nikitovic, D., Katonis, P., Afratis, N., Hascall, V. C., Karamanos, N. K., Tzanakakis. G. N. (2011). Role of receptor for hyaluronic acid-

mediated motility (RHAMM) in low molecular weight hyaluronan (LMWHA)-mediated fibrosarcoma cell adhesion. *Journal of Biological Chemistry 286*, 38509–38520.

74. Park, K., Hong, S. W., Hur, W., Lee, M-Y., Yang, J-A., Kim, S. W., Yoon, S. K., Hahn, S. K. (2011). Target specific systemic delivery of TGF-β siRNA/(PEI-SS)-*g*-HA complex for the treatment of liver cirrhosis. *Biomaterials 32*, 4951–4958.

75. Kim, E. J., Cho, H. J., Park, D., Kim, J. Y., Kim, Y. B., Park, T. G., Shim, C. K., Oh, Y. K. (2011). Antifibrotic effect of MMP13-encoding plasmid DNA delivered using poly-ethylenimine shielded with hyaluronic acid. *Molecular Therapy 19*, 355–361.

76. Choi, K. Y., Yoon, H. Y., Kim, J-H., Bae, S. M., Park, R-W., Kang, Y. M., Kim, I-S., Kwon, I. C., Choi, K., Jeong, S. Y., Kim, K., Park, J. H. (2011). Smart nanocarrier based on PEGylated hyaluronic acid for cancer therapy. *ACS Nano 5*, 8591–8599.

77. Low, P. S., Kularatne, S. A. (2009). Folate-targeted therapeutic and imaging agents for cancer. *Current Opinion in Chemical Biology 13*, 256–262.

78. Kolhatkar, R., Lote, A., Khambati, H. (2011). Active tumor targeting of nanomaterials using folic acid, transferrin and integrin receptors. *Current Drug Discovery Technology 8*, 197–206.

79. Yamada, A., Taniguchi, Y., Kawano, K., Honda, T., Hattori, Y., Maitani, Y. (2008). Design of folate-linked liposomal doxorubicin to its antitumor effect in mice. *Clinical Cancer Research 14*, 8161–8168.

80. Bae, Y., Nishiyama, N., Kataoka, K. (2007). *In vivo* antitumor activity of the folate-conjugated pH-sensitive polymeric micelle selectively releasing adriamycin in the intracellular acidic compartments. *Bioconjugate Chemistry 18*, 1131–1139.

81. Wang, X., Li, J., Wang, Y., Cho, K. J., Kim, G., Gjyrezi, A., Koenig, L., Giannakakou, P., Shin, H. J., Tighiouart, M., Nie, S., Chen, Z. G., Shin, D. M. (2009). HFT-T, a targeting nanoparticle, enhances specific delivery of paclitaxel to folate receptor-positive tumors. *ACS Nano 3*, 3165–3174.

82. Gabizon, A., Horowitz, A. T., Goren, D., Tzemach, D., Shmeeda, H., Zalipsky, S. (2003). *In vivo* fate of folate-targeted polyethylene-glycol liposomes in tumor-bearing mice. *Clinical Cancer Research 9*, 6551–6559.

83. Li, X., Tian, X., Zhang, J., Zhao, X., Chen, X., Jiang, Y., Wang, D., Pan, W. (2011). *In vitro* and *in vivo* evaluation of folate receptor-targeting amphiphilic copolymermodified liposomes loaded with docetaxel. *International Journal of Nanomedicine 6*, 1167–1184.

84. Hofland, H. E. J., Masson, C., Iginla, S., Osetinsky, I., Reddy, J. A., Leamon, C. P., Scherman, D., Bessodes, M., Wils, P. (2002). Folate-targeted gene transfer *in vivo*. *Molecular Therapy 5*, 739–744.

85. Duarte, S., Faneca, H., De Lima, M. C. P. (2012). Folate-associated lipoplexes mediate efficient gene delivery and potent antitumoral activity *in vitro* and *in vivo*. *International Journal of Pharmaceutics 423*, 365–377.

86. Leamon, C. P., Cooper, S. R., Hardee, D. E. (2003). Folate-liposome-mediated antisense oligodeoxynucleotide targeting to cancer cells: evaluation *in vitro* and *in vivo*. *Bioconjugate Chemistry 14*, 738–747.

87. Kim, D., Gao, Z., Lee, E. S., Bae, Y. H. (2009). *In vivo* evaluation of doxorubicin-loaded polymeric micelles targeting folate receptors and early endosomal pH in drug-resistant ovarian cancer. *Molecular Pharmaceutics 6*, 1353–1362.

88. Kang, C., Yuan, X., Li, F., Pu, P., Yu, S., Shen, C., Zhang, Z., Zhang, Y. (2009). Evaluation of folate-PAMAM for the delivery of antisense oligonucleotides to rat C6

glioma cells *in vitro* and *in vivo*. *Journal of Biomedical Materials Research A 93*, 585–594.

89. Blanco, M. D., Guerrero, S., Benito, M., Fernández, A., Teijón C., Olmo, R., Katime, I., Teijón, J. M. (2011). *In vitro* and *in vivo* evaluation of a folate-targeted copolymeric submicrohydrogel based on *N*-isopropylacrylamide as 5-fluorouracil delivery system. *Polymers 3*, 1107–1125.

90. Tang, Q-S., Chen, D-Z., Xue, W-Q., Xiang, J-Y., Gong, Y-C., Zhang, L., Guo, C-Q. (2011). Preparation and biodistribution of [188]Re-labeled folate conjugated human serum albumin magnetic cisplatin nanoparticles ([188]Re-folate-CDDP/HSA MNPs) *in vivo*. *International Journal of Nanomedicine 6*, 3077–3085.

91. Mindt, T. L., Müller, C., Melis, M., De Jong, M., Schibli, R. (2008). "Click-to-chelate": *in vitro* and *in vivo* comparison of a [99m]Tc(CO)$_3$-labeled N(τ)-histidine folate derivative with its isostructural, clicked 1,2,3-triazole analogue. *Bioconjugate Chemistry 19*, 1689–1695.

92. Miller, K., Eldar-Boock, A., Polyak, D., Segal, E., Benayoun, L., Shaked, Y., Satchi-Fainaro, R. (2011). Antiangiogenic antitumor activity of HPMA copolymer-paclitaxel-alendronate conjugate on breast cancer bone metastasis mouse model. *Molecular Pharmaceutics 8*, 1052–1062.

93. Banerjee, R., Tyagi, P., Li, S., Huang, L. (2004). Anisamide-targeted stealth liposomes: a potent carrier for targeting doxorubicin to human prostate cancer cells. *International Journal of Cancer 112*, 693–700.

94. Chen, Y., Bathula, S. R., Yang, Q., Huang, L. (2010). Targeted nanoparticles deliver siRNA to melanoma. *Journal of Investigational Dermatology 130*, 2790–2798.

95. Kratz, F., Elsadek, B. (2012). Clinical impact of serum proteins on drug delivery. *Journal of Controlled Release 161*, 429–445.

96. Dosio, F., Milla, P., Cattel, L. (2010). EC-145, a folate-targeted vinca alkaloid conjugate for the potential treatment of folate receptor-expressing cancers. *Current Opinion on Investigational Drugs 11*, 1424–1433.

97. Reddy, J. A., Dorton, R., Westrick, E., Dawson, A., Smith, T., Xu, L. C., Vetzel, M., Kleindl, P., Vlahov, I. R., Leamon, C. P. (2007). Preclinical evaluation of EC145, a folate-vinca alkaloid conjugate. *Cancer Research 67*, 4434–4442.

98. Timko, B. P., Whitehead, K., Gao, W., Kohane, D. S., Farokhza, O., Anderson, D., Langer, R. (2011). Advances in drug delivery. *Annual Review of Materials Research 41*, 1–20.

99. Davis, M. E. (2009). The first targeted delivery of siRNA in humans via a self-assembling, cyclodextrin polymer-based nanoparticle: from concept to clinic. *Molecular Pharmaceuticals 6*, 659–668.

100. Davis, M. E., Zuckerman, J. E., Choi, C. H. J., Seligson, D., Tolcher, A., Alabi, C. A., Yen, Y., Heidel, J. D., Ribas, A. (2010). Evidence of RNAi in humans from systemically administered siRNA via targeted nanoparticles. *Nature 464*, 1067–1070.

101. Saul, J. M., Annapragada, A. V., Bellamkonda, R. V. (2006). A dual-ligand approach for enhancing targeting selectivity of therapeutic nanocarriers. *Journal of Controlled Release 114*, 277–287.

102. Chrastina, A., Massey, K. A., Schnitzer, J. E. (2011). Overcoming *in vivo* barriers to targeted nanodelivery. *Wiley Interdisciplinary Reviews: Nanomedicine and Nanobiotechnology 3*, 421–437.

103. Yoo, J-W., Irvine, D. J., Discher, D. E., Mitragotri, S. (2011). Bio-inspired, bioengineered and biomimetic drug delivery carriers. *Nature Reviews Drug Discovery 10*, 521–535.

104. Wong, C., Stylianopoulos, T., Cui, J., Martin, J., Chauhan, V. P., Jiang, W., Popovic, Z., Jain, R. K., Bawendi, M. G., Fukumura, D. (2011). Multistage nanoparticle delivery system for deep penetration into tumor tissue. *Proceedings of the National Academy of Sciences of the United States of America 108*, 2426–2431.

105. Hirata, K., Maruyama, T., Watanabe, H., Maeda, H., Nakajou, K., Iwao, Y., Ishima, Y., Katsumi, H., Hashida, M., Otagiri, M. (2010). Genetically engineered mannosylated-human serum albumin as a versatile carrier for liver-selective therapeutics. *Journal of Controlled Release 145*, 9–16.

106. Patil, Y., Toti, U., Khdair, A., Ma, L., Panyam, J. (2009). Single-step surface functionalization of polymeric nanoparticles for targeted drug delivery. *Biomaterials 30*, 859–866.

3

LIPID NANOPARTICLES FOR THE DELIVERY OF NUCLEIC ACIDS

YUHUA WANG AND LEAF HUANG

Division of Molecular Pharmaceutics and Center for Nanotechnology in Drug Delivery, Eshelman School of Pharmacy, University of North Carolina at Chapel Hill, Chapel Hill, NC, USA

3.1 INTRODUCTION

As with the development of recombinant DNA technology, gene therapy emerged as a promising technology that would drastically improve the practice of medicine for treating inherited and acquired diseases. The idea underlying gene therapy is to treat human disease by the transfer of genetic materials into specific cells of the patient [1]. Nucleic acids, the fundamental elements that precisely controlled the expression of proteins, are very appealing therapeutic candidates due to the simplicity of the drug development strategy and minimal side effects compared with conventional drugs. Moreover, the discovery of RNA interference (RNAi) by Fire and Mello in the late 1990s has opened up an entirely new field of "gene therapy," using small RNA fragments that potently knockdown the target gene expression with high specificity and efficiency [2]. Gene therapy has hardly made its mark in medicine. Successful implementation of gene transfer in the clinic requires drug delivery techniques that can efficiently and specifically deliver nucleic acids to the site of action after systemic administration.

The physicochemical properties of nucleic acid, such as vulnerability to nucleases, high anionic charge content, and high molecular weight (MW), preclude naked unformulated molecules from performing their functions after systemic delivery. Therefore, a delivery vector should facilitate nucleic acid therapeutics to access the

Nanoparticulate Drug Delivery Systems: Strategies, Technologies, and Applications, First Edition. Edited by Yoon Yeo.
© 2013 John Wiley & Sons, Inc. Published 2013 by John Wiley & Sons, Inc.

target cell, cross the cell membrane, escape the endosome compartments, and reach the sites of action, which is the cytoplasm for small interfering RNA (siRNA) and microRNA, and the nucleus for plasmid DNA (pDNA). Apart from physical methods such as electroporation [3] and hydrodynamic injection [4], there are two major classes of gene delivery vehicles that have been extensively investigated and developed to deliver nucleic acids to the target cells *in vivo*: viral and nonviral vectors. The former class harnesses the extremely high transduction efficiency of virus particles to infect target cells. For this purpose, the viral genome is genetically modified to deplete virulent and proliferation elements, leaving only the expression machinery in most cases. However, this class of vectors could stimulate serious immune responses [5] or result in unwanted host genome integration, which in rare cases would induce cancer development [6,7]. The latter vectors are composed of liposomes, polymers, inorganic materials, and nucleic acid conjugates. The synthetic vectors, compared with their counterparts, are easier for scale-up production, lower in immunogenicity, and cheaper in manufacture cost. Although favorable in many aspects, nonviral vectors suffer from relatively low delivery efficiency due to multiple barriers between the site of administration and the site of action. Ever since the first landmark publications in 1987 [8], cationic liposome based lipid nanoparticles (LNPs) turn out to be the most widely used and efficient nanometric device for *in vitro* delivery of nucleic acid. Cationic liposomes condense DNA into an ill-defined particulate structure (lipoplex) with excess positive charges, which allow efficient, but nonspecific, binding with the negatively charged cell surface and subsequent internalization [9]. The condensation of DNA also results in protection from enzymatic degradation. Effective as they appear *in vitro*, most of the LNPs failed or behaved far from satisfactory *in vivo*, especially for systemic delivery to solid tumor.

In this chapter, we first elucidate the extracellular and the intracellular barriers the LNPs encounter after systemic administration. Understanding the challenges in the delivery will lead to a rational design of novel lipid molecules or nanometric devices that could address these challenges. Next, we use a bottom-up strategy to identify the latest progression in lipid component development in the LNPs, and then compare various formulations that were exploited to encapsulate nucleic acid for successful *in vivo* delivery. The nucleic acids in this review include pDNA, siRNA, and antisense oligonucleotide (ODN), which either serve for gain of function or loss of function to treat various diseases.

3.2 BARRIERS TO SYSTEMIC GENE DELIVERY

Except for some directly accessible tumor sites after surgery ablation, tumor sites are typically only reachable through systemic administration of therapeutics in the bloodstream. Unfortunately, for the LNP-based nucleic acid delivery systems, multiple barriers exist extracellularly and intracellularly. These barriers substantially impede the successful delivery and therefore undermine the efficacy of delivered therapeutics.

3.2.1 Extracellular Barriers

Extracellular barriers refer to the hurdles imposed on the LNPs in the blood circulation before tumor cells take them up. These barriers include the liposome instability in serum, reticuloendothelial system (RES) uptake, vascular endothelium, and extracellular matrix (Figure 3.1).

Positively charged surface, found in most LNPs, results in strong protein adsorption when they are exposed to the blood serum where a high concentration of proteins with countercharges are present. The protein adsorption leads to agglomeration of the LNPs, followed by entrapment in the lung capillary and in the spleen macrophage enriched mesh tissue [10]. Some of the adsorbed blood proteins, for example, IgM and C4, are the known opsonins that facilitate the uptake of the LNPs by the RES cells, mainly the liver Kupffer cells and the splenic macrophages [10]. Conventional LNPs that encapsulate nucleic acid usually have a short blood circulation half-life, which could be substantially prolonged by coating the particles with polysaccharides or poly(ethylene glycol) (PEG) residues [10,11]. The hydrophilic coating endows the LNP with resistance to RES and hence the coated formulation is named *stealth liposomes*. Prolonged circulation provides more opportunity for LNPs to access the tumor site.

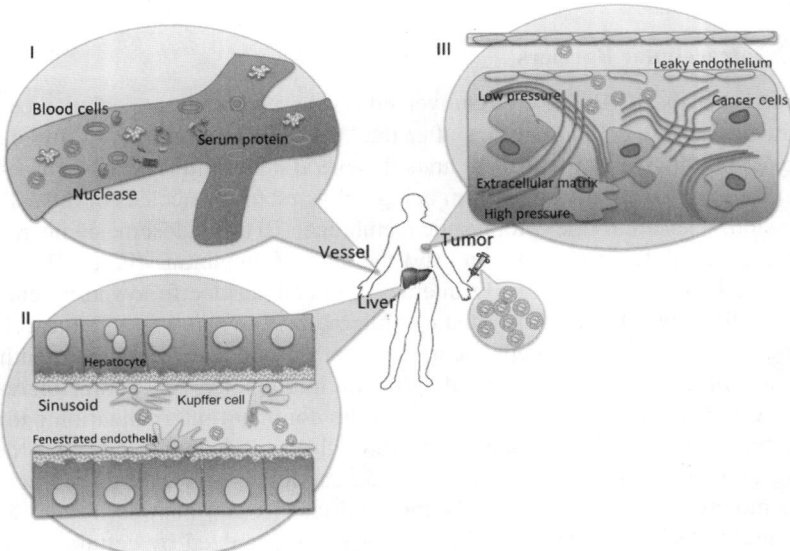

FIGURE 3.1 Illustration of barriers to intravenous delivery of gene therapeutics. After LNP is intravenously injected into human body, serum protein tends to be adsorbed to the surface of LNP, destabilizing colloidal stability of LNP and causing the release of cargo (Inset I). The RES in the liver may take up LNPs, which reduces half-life of LNPs in the blood circulation (Inset II). Circulating LNPs extravasate into the interstitial space tumor tissue through the leaky endothelial lining and result in high accumulation in the tumor tissue due to the lack of lymphatic drainage (Inset III). However, LNPs have to traverse through the crowded extracellular matrix against the increasing fluid pressure to access the proximity of cancer cell.

Egress from the bloodstream and across the endothelial barriers also poses an insurmountable obstacle for liposome-mediated nucleic acid delivery. Molecules larger than 5–10 nm cannot readily traverse the capillary endothelium of healthy tissue [12] but keep circulating until they are cleared from the blood by renal clearance or other mechanisms. Most solid tumors, liver, and spleen allow the extravasation of large molecules or particles due to different anatomical character-istics of the endothelium in these organs [13]. It was reported by Maeda that tumor vascular has elevated permeability, which accounted for higher extravasation of macromolecules into interstitial tissue spaces [14] and a prolonged retention due to deficient lymphatic drainage [15]. This tumor-specific feature was named *enhanced permeability and retention* (EPR) effect, which has been extensively exploited as the major passive tumor-targeting strategy to achieve a higher accumulation in the tumor than plasma or other tissues. After extravasation from the blood vessel to the tumor site by the EPR effect, the LNPs are required to penetrate into the extracellular matrix filled with polysaccharide and protein network to gain access to the tumor cell membrane. This matrix permits only diffusion of small molecules and proteins yet not the conventional LNPs which are usually 100 nm or larger. Furthermore, the defective lymphatic elevates the interstitial pressure in tumor mass, which leads to radially outward convection opposing the inward diffusion [16].

3.2.2 Intracellular Barriers

Provided that the LNPs have circumvented all the potential extracellular barriers and traveled to the cellular surface after the "Odyssey Journey," with physical in-tegrity and biological activity, there are still several major intracellular barriers that hinder foreign genetic materials from being delivered to the nucleus of the cells for expression. They are (i) the cytoplasmic membrane, (ii) the endosome compartment, (iii) the intracellular trafficking, and (iv) the nuclear membrane (Figure 3.2).

The LNPs may nonspecifically interact with cell surface followed by endocy-tosis. Studies have also demonstrated that internalization pathway and intracellular routing of particles is size dependent. The clathrin-mediated endocytic pathway has a size limit for internalization of ~200 nm in diameter. As the size increases, caveolae-mediated internalization becomes the dominant internalization pathway [17], and the LNPs are less likely to go through the endosome route. In order to realize targeted gene delivery, the LNPs are modified with various biologically active moieties such as arginine-glycine-aspartate (RGD) peptide, folate, anisa-mide, and transferrin, which can be specifically recognized by certain receptors that are overexpressed on the cancer cell surface and facilitate the internalization of the LNPs. The receptor-mediated internalization of LNPs is primarily via endocytotic pathway.

After cellular uptake via endocytosis, the LNPs are encapsulated in the membrane vesicles derived by invagination of clathrin-coated pits, namely, endosomes. The pH in the lumen of early endosomes gradually experiences a drop from neutral to pH 6.0 as the vacuolar ATPase proton pump is activated. As endosomes mature, the pH value continues to drop to 5.0 when late endosomes fuse with lysosomes [18]. Nucleic

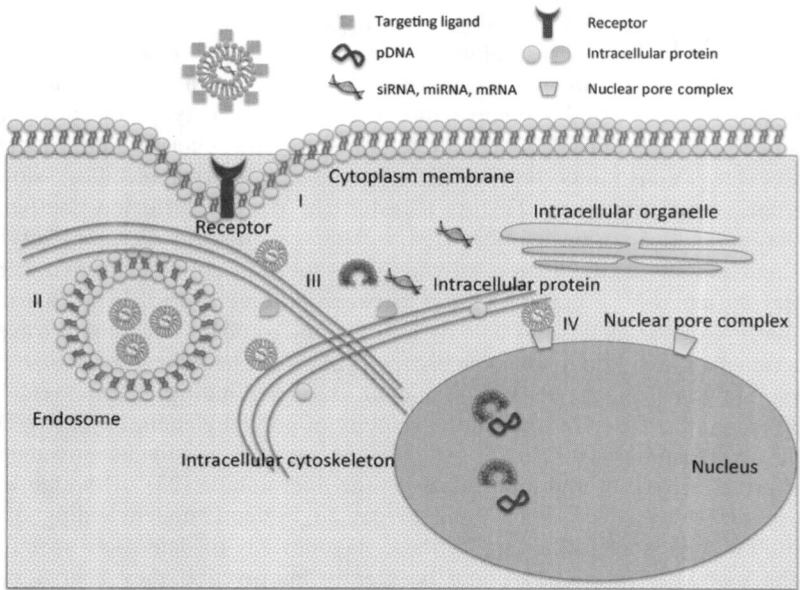

FIGURE 3.2 Schematic illustration of proposed pathway of gene delivery mediated by the LNPs. The targeting motif on the surface of the LNPs binds to the receptor overexpressed on the surface of cancer cells, allowing the internalization of the complexes via receptor-mediated endocytosis (Step I). The LNPs will be encapsulated in the endosomal compartment. The cargo needs to be released into the cytoplasm before the endosomes fuse with lysosomes (Step II). The cargos such as siRNA, mRNA, or miRNA will take effect in the cytoplasm as released. For pDNA, the carrier has to escort the cargo through the intracellular matrix until it reaches the periphery of the nucleus (Step III). If the size of the LNP complexes is more than 40 nm, the complex can enter the nucleus only at the mitosis phase of the cell cycle when the nuclear membrane dissolves. If the LNP complex is small enough (<40 nm), the complex may pass through the nuclear pore complex (NPC) (Step IV). Inside the nucleus, the pDNA will be decomplexed from the carrier for transcription.

acids that are entrapped in the endosome are subject to degradation if they have not escaped from the endosome before endosome–lysosome fusion occurs. This usually leads to poor efficacy because nucleic acids would not ultimately reach the cytoplasm or nucleus. Therefore, endosome entrapment is always considered one of the major intracellular barriers to nonviral gene delivery. In general, when the LNPs are endocytosed, the cationic lipids can interact with anionic membrane lipids and form ion pairs. Hence, the endosome membrane is destabilized by the formation of hexagonal phase (H_{II}), and nucleic acids are released into the cytoplasm subsequently [19].

Once successfully released into the cytoplasm, cargos such as siRNA, antisense RNA, miRNA, and mRNA could directly take effect without further transportation. However, pDNA must traverse through cytoplasm to reach the nucleus, which is a poorly characterized event so far and also considered a rate-limiting step in the pDNA-mediated gene transfer. According to Goodsell's depiction [20], cytoplasm is a milieu that is crowded with high concentration of soluble proteins, subcellular organisms and compartments, cytoskeleton structure, and all other small molecules. All of these cellular components, especially three types of cytoskeletal filaments, namely, microtubules, microfilaments, and intermediate filaments, impose steric constraints on the diffusion of the macromolecules, slowing down long-range transport [21]. As a result, the passive diffusion coefficients of macromolecules are inversely related to their size. It seems very unlikely that Brownian motion is the major driving force for this colloidal nonviral gene delivery system whose particle size is between 80 and 150 nm [21]. In nature, adenovirus overcomes this obstacle by utilizing microtubule structure to travel to the perinuclear area after endosome escape [22]. This phenomenon has also been observed in nonviral gene delivery. Suh et al. employed multiple particle tracking (MPT) technique to determine that the velocity of polyethyleneimine (PEI)–DNA complexes transport was 0.2 μm/s, which is the same magnitude as movement of motor proteins dyneins and kinesins along the microtubule [23]. Although a few mechanistic studies have been done to get an in-depth understanding of the microtubule-mediated trafficking of these nanodevices to date, it is speculated that the motion of the LNPs could be greatly improved if they bear a signal molecule that can bind specifically to the microtubule [24,25].

The nuclear envelope is always considered a crucial barrier to efficient nonviral gene delivery. Over the past 10 years, mechanisms for the nuclear transport of pDNA have been investigated and described. Many approaches have been developed based on exploitation of these mechanisms. In general, when carrier–DNA complexes reach perinuclear membrane, there are three scenarios for the LNPs to finally reach nucleoplasm, which are categorized according to the particle size. When the particle size is below 10 nm, it can passively diffuse through the nuclear pore complex (NPC). However, this does not apply to most of the LNPs. When the particle size is above 40 nm, it tends to accumulate around the nuclear membrane and wait for the mitosis phase to enter the nuclear compartment when the nuclear envelope is dissolved. This is partly why growth-arrested cells and other non-dividing cells are considered difficult to transfect. Also, this accounts for the observed cell-cycle dependent expression pattern of the cells transfected by nonviral vectors. When the particle size is between 10 and 40 nm and the particle carries a nuclear localization signal (NLS), it could be actively transported into the nucleus through the nuclear pore complex (NPC) via the NLS-mediated nuclear importing machinery. NLS is a short basic amino acid sequence that can be recognized by importin-α, an NLS receptor, which then dimerizes with importin-β. The importin-αβ heterodimer complex then carries the NLS bearing cargo through NPC for nuclear transport. NLS has been incorporated into cationic lipids [26,27] and proved to facilitate pDNA nuclear entry. The location of NLS in the vector is of significant importance regarding the active functionality. The signal should be readily accessible to the importin protein. If NLS overlaps with the DNA-binding domain, there is a good chance that the NLS-directed nuclear transport would be compromised.

3.3   COMPONENTS OF LNPs

The components in the LNPs usually include various lipids, polymer (polypeptide or synthetic polymer) and lipid-PEG conjugates, and so on. There are extensive investigations on a rational design basis, to ameliorate the components of the LNPs for improved delivery efficiency. Here we review the most representative ones.

3.3.1   Cationic Lipid

The encapsulation of nucleic acids of various molecular weights, ranging from 20 base pairs to several thousand base pairs, essentially depends on the cationic charges of carrier molecules. A multivalent cation is the key component because the electrostatic interaction between the cationic lipid and the anionic cargos results in condensation of nucleic acid into nanometric particles through an entropy-driven process [28].

Commonly used cationic lipids are usually classified into three categories [29] based on the charges on their headgroups: (1) monovalent aliphatic lipids such as N-(1-(2,3-dioleyloxy) propyl)-N,N,N-trimethylammonium chloride (DOTMA) [8], 1,2-dioleyl-3-trimethylammonium propane (DOTAP) [30]; (2) multivalent aliphatic lipids with headgroups containing several amine functions such as dioctadecylami-doglycylspermine (DOGS) [31], 2,3-dioleyloxy-N-(2-[sperminecarboxyamido] ethyl-N,N-dimethyl-1-propaniminium bromide (DOSPA) [32]; (3) cationic lipid derivatives such as 3β-(N-(N',N'-dimethylaminoethane)-carbamoyl)cholesterol (DC-Chol) [33] and bis(guanidinium)-tren-cholesterol (BGTC) [34].

The cationic lipids not only are responsible for DNA condensation but also play a critical role in interacting with the negatively charged cellular membrane and triggering an efficient endocytosis for uptake. This advantage is well exploited in nonspecific tissue culture transfection *in vitro*. Nevertheless, the advantage *in vitro* turns out to be the disadvantage *in vivo*. The positive charge is detrimental to *in vivo* transfection because the charges on the particle surface tend to attract negatively charged serum proteins, subsequently leading to fast blood clearance and poor tissue distribution for the aforementioned reasons.

This problem is addressed by either coating liposomes with neutrally charged PEG polymer, which not only screen the positive charge but also provide a hydrophilic steric barrier that significantly reduces protein adsorption. Alternatively, cationic lipids with a pH-sensitive ionizable headgroup emerge. With pK_a slightly lower than 7.0, ionizable cationic lipid can efficiently condense nucleic acids at acidic pH and maintain a neutral or low cationic surface charge at physiological pH (7.4) after *in vivo* administration. This strategy provides sophisticated control over surface charge of the LNPs in the blood circulation and prevents nonspecific cellular uptake, representing a recent trend in the development of lipid materials for nucleic acid delivery, especially for siRNA.

The idea of incorporating ionizable lipids in nucleic acid delivery was first reported by Semple et al. [35]. In order to overcome the unfavorable pharmaco-kinetics (PK) and toxicity profiles for cationic lipid based LNPs, a novel ionizable aminolipid (1,2-dioleoyl-3-dimethylammonium propane, DODAP) was used to efficiently encapsulate antisense ODN in the LNPs. Later, another ionizable lipid DLinDMA (1,2-dilinoleyloxy-N,N-dimethyl-3-aminopropane) has proven to be highly effective regarding *in vivo* transfection efficiency [35]. The liposome made of DLinDMA showed close to neutral surface charge in the blood circulation at physiological pH. After endocytosis, the pH drops in the endosomal compartment, and the lipid is protonated. The cationic lipid is supposed to interact with naturally occurring anionic phospholipids in the endosomal membrane by electrostatic interaction, forming ion pairs that induce conformational transition from bilayer to hexagonal H_{II} phase, which destabilizes endosomal membrane for nucleic acid release. The LNPs formed with ionizable lipids have been tested in rodents, nonhuman primates [36], and is now being evaluated in human clinical trials.

As a benchmark for the design of novel ionizable cationic lipid, a pool of ~1200 lipid molecules with various modifications on the hydrocarbon chain, linker, and headgroup have been implicated [37]. A sensitive mouse Factor VII model was used as the *in vivo* screening system to evaluate liposome-based hepatocyte siRNA delivery efficiency [38]. A series of modifications have been implemented in the headgroups, linkers, and hydrocarbon chains [37]. The strategy for the headgroup modifications was to change the size, acid-dissociation constant, and number of ionizable groups and to evaluate the corresponding delivery efficiency. Piperrazino, morpholino, trimethylamino, or bis-dimethylamino cations were tested, with none showing better performance than dimethylamino headgroup of 2,2-dilinoleyl-4-dimethylaminomethyl-[1,3]-dioxolane (DLin-K-DMA). In addition, the distance between the dimethylamino group and the dioxlane linker was varied by introducing various numbers of the methylene group. The results showed that insertion of a single additional methylene group (2,2-dilinoleyl-4-(2-dimethylaminoethyl)-[1,3]-dioxolane) produced a dramatic increase in potency with an *in vivo* ED50 as low as 0.1 mg/kg, which turned out to be the lowest dose for silencing Factor VII *in vivo* [37].

Linker region was optimized by testing different rates of chemical or enzymatic stability and hydrophobicity according to modifications. The results indicated that rapidly degradable linker did not necessarily correlate with high *in vivo* activity. Highly reactive ester-containing lipid showed reduced *in vivo* activity compared to less reactive alkoxy-containing lipid. The lipids were speculated to undergo an unwanted degradation before the vector successfully delivered a cargo to the cells.

For hydrocarbon chains, a correlation between number of *cis* double bonds in the chains and delivery activity was studied. It was determined that linoleyl lipid containing two double bonds per hydrocarbon chain is optimal in a transfection model, mediated by fusion with endosomal membrane for cytoplasmic release of the nucleic acid [39].

3.3.2 Neutral Lipid

In order to avoid unfavorable PK caused by positive surface charge after systemic delivery, Sood and colleagues have developed a method to formulate siRNA in a neutral lipid dioleoylphosphatidylcholine (DOPC)-based-liposome for systemic delivery [40]. Since DOPC has an amphiphilic choline headgroup, which displays neutral charge at physiological pH, a conventional approach to encapsulate siRNA with a charge–charge interaction is not applicable here. Therefore, a *tert*-butanol/ Tween 20 system was employed to encapsulate siRNA through a series of process including evaporation, lyophilization, and rehydration [40]. The loading efficiency of siRNA was estimated to be about 65%. The DOPC liposome was able to deliver anticancer siRNA in metastatic liver carcinoma [41,42] and orthotopic ovarian cancer model [40,43–45] with high efficacy. Although the results are impressive, the mechanism of siRNA encapsulation in neutral LNPs was not elucidated, and physicochemical properties of the LNPs were poorly characterized.

With the same aim in mind (to reduce cationic charge on the surface of liposome), Herscovici and colleagues designed a series of lipid molecules with thiourea headgroup on the basis of the observation that bis-thiourea moieties selectively bind to dihydrogen phosphate via multitopic hydrogen bonding [46]. Nevertheless, the transfection was unimpressive due to the hydrophobic nature of the lipopoly-thiourea liposome, which hinders the interaction with the cell membrane [46]. This feature, however, could be exploited to avoid nonspecific cell uptake and be overcome by postgrafting with targeting ligands to facilitate internalization [46]. The absence of cationic charges endowed the liposome with an improved PK profile *in vivo*, yet the benefit did not increase the tumor accumulation according to the proposed EPR effect [47]. It is also worth noticing that thiourea LNPs achieved a similar transfection level as cationic LNPs, although internalization of the thiourea LNPs was determined to be sixfold less than that of the cationic LNPs. This high expression level despite low uptake efficiency indeed revealed a favorable release profile for lipothiourea after internalization, which was measured to be three times faster than the cationic liposome. This fast unpackaging characteristic is attributable to the relatively low hydrogen bond-based interaction between phosphate group of nucleic acid and thiourea lipid [48] when compared with electrostatic interaction.

3.3.3 Anionic Lipid

Although anionic lipids alone cannot condense nucleic acid through electrostatic interaction, they are usually formulated into nanometric particle to reduce the nonspecific cellular uptake and adsorption by positively charged serum protein for *in vivo* nucleic acid delivery [49–52]. In this approach, nucleic acids are usually condensed with excessive polycations such as PEI or protamine to form a core complex with positive surface charge. Preformed anionic liposomes are subsequently added to the system, coating the core complexes with bilayer lipid membrane. Chen et al. formulated liposome/polycation/DNA (LPD)-II by coating liposome, composed of 1,2-dioleoyl-*sn*-glycero-3-phosphate (DOPA), 1,2-dioleoyl-*sn*-glycero-3-phosphoethanolamine (DOPE), and cholesterol, on prot-amine-condensed siRNA. The LPD-II formulation showed much lower immunoge-nicity compared with the cationic liposome coated LPD nanoparticles, which widened the therapeutic window for the therapeutic applications [52].

Another approach, reported by Srinivasan and Burgess, utilized anionic liposome as a gene carrier [53]. In their study, divalent cations (such as Ca^{2+}, Mg^{2+}, Mn^{2+}, and Ba^{2+}) were incorporated into the system to achieve the complexation between the nucleic acid and the anionic lipids. High percent of transfection was achieved *in vitro* in the presence and absence of serum and demonstrated minimal cytotoxicity. Although the hydrodynamic size of the complexed particles was at the submicron level, which was not suitable for *in vivo* applications, this formulation provided a novel way of using anionic liposome for transfection.

3.3.4 PEG–Lipid Conjugates

As proposed by Allen, "If you want to be invisible, look like water" [54]. For longer blood circulation, which is a prerequisite for efficient gene delivery after systemic administration, it is necessary to coat the surface of the LNPs with neutral and hydrophilic polymers such as PEG. Klibanov et al. have reported, for the first time, a prolonged liposome circulation after PEGylation back in 1990 [11], a landmark publication for long circulation liposomes. Later, Campbell observed an increased tumor accumulation of PEGylated cationic LNP due to EPR effect [55]. To date, PEG is the most appreciated hydrophilic polymer with a tunable size (normally from 300 to 25,000 Da), particularly high biocompatibility and versatility. Surface modification with PEG stabilizes liposome in the presence of serum and extends the blood circulation after administration, as a result of reduced level of protein adsorption to the hydrophobic area of the LNPs. This effect predominantly depends on the molecular weight, density, conformation, and flexibility of the PEG chain. Vonarbourg and Benoit did an excellent review on this subject [10]. It was demonstrated that the protein adsorption efficiency decreased with an increase of MW of PEG chain [56]. Most authors advocate an efficient MW in the range 1500–3500 Da [10]. In addition, the PEG density and conformation, which not only depend on the MW of PEG chain, but also the size of LNPs, are of great importance to resist opsonin penetration. Briefly, smaller LNPs with larger curvature, or smaller radius, need a higher density of PEG compared with larger ones. The conformation of PEG chain on the surface is dictated by the relationship between the distance (D) of the anchorage sites and the gyration radius (R_F). When $D > 2R_F$, the grafted PEG chain is in a mushroom-like conformation. However, when there is more PEG grafted on the surface of the lipid bilayer, which makes polymer chains closer (in such case, $D \ll R_F$), polymer chains tend to appear as a brush-like conformation. Brush conformation was believed to be most effective in repelling against the complement components, more resistant to phagocytosis (Figure 3.3).

Various lipid-PEG conjugates are inserted into the lipid bilayer membrane during or after the formation of the LNPs. The hydrophobic interaction between the anchor lipid of the conjugate and hydrophobic chain of the membrane lipid is the major force that keeps the PEG on the surface of the LNPs. Since the lipid-PEG conjugate is soluble in the aqueous bulk, the molecule also tends to shed from the liposome surface especially under a sink condition. Cullis et al. have taken the initiative to

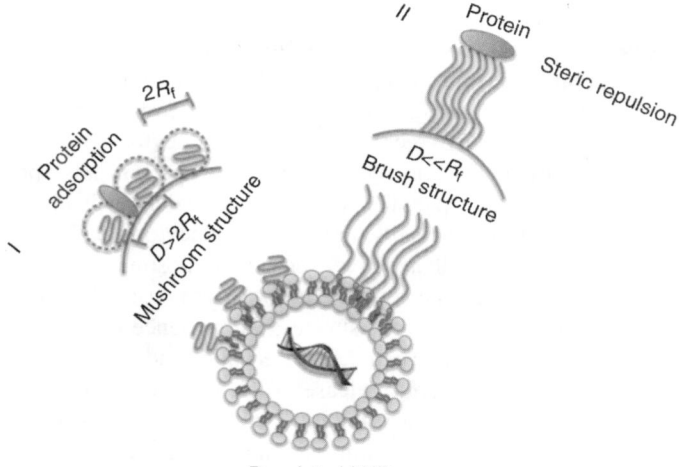

Pegylated LNP

FIGURE 3.3 Putative PEG conformation with respect to PEG concentration in the surface of LNPs. Conformations of the polymer chains are dictated by the relationship between the distance (D) between the anchorage sites of two chains and the gyration radius (R_f). When $D > 2R_f$, PEG chains form a mushroom-like structure. Protein tends to be adsorbed to the surface of the LNPs. When $D \ll R_f$, PEG chains assume brush-like conformation. The high-density, brush-like PEG configurations would sterically suppress the deposition of large proteins.

investigate and have tried to figure out the correlation between different lipid-PEG conjugates and transfection efficiency of the LNPs prepared by the dialysis method [57–60]. It was demonstrated that longer acyl chain in the lipid-PEG conjugates stabilized the particle, whereas the transfection efficiency was compromised due to less interaction between the cationic lipid and cell membrane. On the other hand, shorter acyl chain length resulted in a readily dissociable surface coating that could transform the LNPs from a stable form into a transfection-competent form. It was also demonstrated that the composition of acyl chain determined the rate and extent of PEG extraction, which was reflected on the elimination rate in the bloodstream. Adlakha-Hutcheon et al. reported the relationship between the anchor chain length and liposome circulation time [59]. In their study, a PEGylated programmable fusogenic vesicles (PFV) stabilized with PEG was employed to exhibit time-dependent destabilization in body. Three types of PEG-lipid with different acyl chain length were used to coat the surface of liposomes: 1,2-dimyristoyl-*sn*-glycero-3-phosphoethanolamine (DMPE, C14), 1,2-dipalmitoyl-*sn*-glycero-3-phosphoetha-nolamine (DPPE, C16), and 1,2-distearoyl-*sn*-glycero-3-phosphoethanolamine (DSPE, C18). The results revealed that formulation with the shortest acyl chain (C14) were rapidly eliminated from the circulation. On the other hand, formulations with DPPE–PEG (C16) and DSPE–PEG (C18) showed a prolonged half-life for 4 and 24 h, respectively [59]. A complete protection is favorable for long-circulating liposome in the blood stream.

Shedding of the PEG coating is preferred once the LNPs are in the proximity of targeted cells, because the hydrophilic PEG coating tends to prevent the interaction between cell membrane and liposome, impeding the internalization of the LNPs. Grafting of a ligand on the distal end of PEG chain can circumvent this dilemma, facilitating internalization through receptor-mediated endocytosis. However, caution needs to be taken, as the distal end ligand can lead to protein adsorption, which speeds up the blood clearance especially when the ligand density is high [61]. After the LNPs are internalized into the cells through an endocytotic pathway, either the cationic lipid needs to interact with negatively charged endosome membrane or some fusogenic lipid, such as DOPE, needs to interact with and destabilize the endosome membrane structure. Under such circumstances, the presence of PEG chain might sterically hinder the interaction between the carrier lipids and endosomal membrane lipid, substantially reducing cytosolic release efficiency. Therefore, it is desirable that PEG coating is shed away to allow the interaction between fusogenic lipid and endosome membrane. Although the continuous shedding effect of PEG takes place after systemic delivery due to the sink condition in the blood, it is difficult to have a complete control over the shedding rate to make sure that shedding is complete when LNPs are entrapped in the endosome of the targeted cells. This kind of "passive" shedding demands extensive characterization of lipid-PEG conjugates and their interaction with the *in vivo* environment. Therefore, a series of stimulus-responsive-lipid-PEG have been developed to realize a controlled shedding of PEG. The idea is to introduce a linker between the lipid and PEG chain, which is cleavable under certain stimulus exclusively in endosome such as low pH [62–65], reducing agents [66], or enzymes [67,68]. Some linkers are designed to respond to extracellular environment of the tumor cells, such as low pH and the presence of matrix metalloproteinases [69]. Such rational designs of shedding process significantly promote the drug release and target cell interaction process.

3.4   FORMULATIONS OF THE LNP FOR NUCLEIC ACIDS ENCAPSULATION

The most conventional way of encapsulating nucleic acids with cationic liposome is to mix the two components to form a lipoplex, which however does not have much control over the structure and homogeneity of the complex. Various approaches have been investigated and developed to form the LNPs with desired structure so as to have more control over the performance of nanoparticles *in vivo*.

3.4.1   Self-Assembly

One of the most exploited strategies of self-assembly is to use a polycation to condense nucleic acid into anionic compact core complex, upon which cationic liposomes are coated on the surface to form a membrane-core structure. In the mid-1990s, the Huang laboratory used poly(L-lysine) and protamine [70] to precondense pDNA before incorporating the pDNA into the preformed cationic liposome

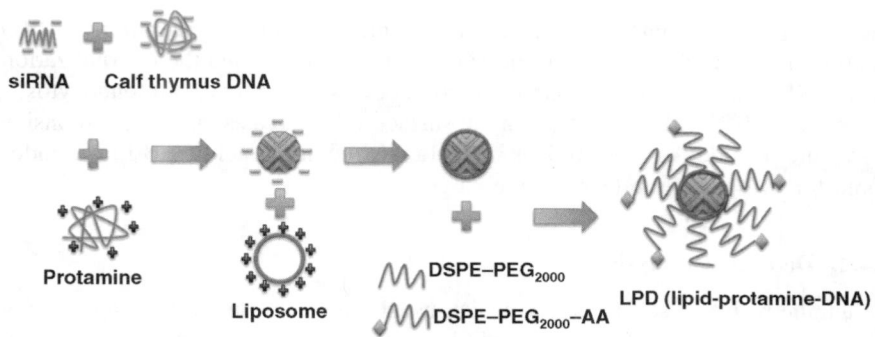

FIGURE 3.4 Schematic illustration of LPD preparation. Briefly, anionic mix of siRNA and calf thymus DNA is condensed by protamine. The negatively charged condensates are then coated by cationic DOTAP–cholesterol liposome to form the core-membrane structure complex. The nanoparticle is PEGylated by incubation with DSPE–PEG$_{2000}$–anisamide (AA) and DSPE–PEG$_{2000}$ micelles.

containing 3β-[N-(N',N'-dimethylaminoethane)-carbamoyl] cholesterol (DC-Chol) and DOPE. The membrane-core-structured LNPs were named as lipid-polycation-DNA (LPD) nanoparticles. The precondensation dramatically decreased the particle size of the pDNA–liposome complex and effectively protected DNA from nuclease degradation [71]. This formulation not only enhanced the transfection efficiency of the LNPs by up to 28-fold *in vitro* [71] but also demonstrated *in vivo* transfection capacity with minor modifications [70]. Later, Li et al. modified the formulation by coating the surface with DSPE–PEG and a sigma receptor targeting ligand DSPE–PEG–anisamide (Figure 3.4), which substantially increased *in vivo* siRNA delivery efficiency by fourfold and the gene silencing effect by two- to threefold [72]. Successful delivery of LPD to subcutaneous and lung metastatic tumors was believed to be due to reduced RES uptake by PEG coating, which resulted in higher tumor accumulation because of the EPR effect [73,74]. It is worth noticing that the negatively charged protamine/calf thymus DNA/siRNA core played a critical role in supporting the lipid bilayer, allowing more surfactant-like DSPE–PEG conjugates to be inserted in the membrane without ripping off the lipid membrane [75]. This high density of sheddable PEG significantly improved the PK of LPD nanoparticles in the blood stream [76].

With a similar approach, Harashima et al. have devised a series of lipid based core-shell structured nanocarriers, namely, multifunctional envelope-type nano-device (MEND) [68,77–93]. In these formulations, stearyl-octaarginine and cholesteryl-GALA(WEAALAEALAEALAEHLAEALAEALEALAA) were incorporated into the preformed lipid membrane mixed with DOPE and cholesteryl hemisuccinate (CHEMS) [81,83,84,90,91]. Slightly different from the LPD formulations, the pDNA/protamine or PEI condensates in water are added to the blown lipid film. The hydration of lipid membrane initiates the formation of liposome, which coats the condensates subsequently. The final nanoparticle is sonicated to ensure homogeneity. MEND has been employed to deliver genes to the liver [83] or the lung [82] with the surface modification of octaarginine or IRQ peptide (IRQRRRR). The octaarginine or IRQ peptide enabled the internalization of MEND nanoparticles through macropinocytosis and caveolar endocytosis, respectively. The GALA peptide on the surface was also used as a pH-responsive fusogenic peptide that destabilizes membrane structure at acidic pH in the endosome for cytoplasmic release of the cargo.

3.4.2 Detergent Dialysis

As an alternative to self-assembly, which usually involves preformed liposomes, a detergent dialysis method is used for lipid-based nanocarriers to encapsulate negatively charged nucleic acid [94]. The idea is to use nonionic detergent with a high critical micelle concentration to solubilize amphiphilic lipid molecule in aqueous solution in the presence of the nucleic acid so that nucleic acids were bound to free lipid instead of liposome. As the detergent is removed gradually by dialysis, cationic lipid and the associated nucleic acid are rearranged and form a vesicle structure that entraps the nucleic acid inside the liposomes. The name, stabilized plasmid-lipid particle (SPLP), is well recognized and represents a class of formulated LNPs prepared by the detergent dialysis method. This method forms 100 nm liposomes with higher trapping efficiency (60–70%) than those achieved by conventional methods such as reverse evaporation phase, extrusion, or dehydration-rehydration, which also usually result in large particle size due to aggregation [59,95].

With a similar idea, Jeffs et al. have developed a new procedure-spontaneous vesicle formation (SVF) to encapsulate pDNA [96]. A lipid solution containing cholesterol, distearoylphosphatidylcholine (DSPC), 1,2-dioleyloxy-N,N-dimethylaminopropane (DODMA), and 3-O-(2′(-methoxypolyethyleneglycol)$_{2000}$)-1,2-distearoyl-sn-glycerol (PEG–S–DSG) in 90% ethanol was used to mix with pDNA in the buffer. The mixing procedure resulted in the drop of ethanol concentration, which was insufficient to support lipid solubility. A lipid vesicle was subsequently formed to encapsulate pDNA. The pDNA-containing liposome was further diluted with buffer to further stabilize the metastable vesicles, resulting in a significant increase in DNA encapsulation from 60% to 90% [96]. The employment of a T-shaped mixing chamber ensured the reproducibility of SPLP, which showed the potential for scale-up manufacture for clinical studies.

Morrissey et al. adapted the detergent dialysis method to encapsulate chemically stabilized siRNA with a superb efficiency of 93% [97]. The specialized liposome, namely, stable nucleic-acid-lipid particle (SNALP), was composed of DSPC, cholesterol, PEG–C-DMA, and ionizable lipid DLinDMA. The LNP is formed with similar nucleic acid to lipid ratio and mean particle size (140 nm) regardless of the nucleic acid payload. The in vivo knockdown efficiency was evaluated in a mouse model of hepatitis B virus replication. The formulation drastically reduced the effective dose of siRNA delivery to 3 mg/kg/day and demonstrated a persistent knockdown effect up to 6 weeks with a weekly injection. Later, Zimmermann et al. has examined the siRNA delivery efficiency of SNALP by knocking down disease target apolipoprotein B (ApoB) in the liver of nonhuman primates. It was demonstrated that a single injection of 2.5 mg/kg resulted in a sustained 80% silencing of ApoB mRNA expression in the liver till 11 days after injection [36]. The degree and duration of the gene-silencing effect was way beyond the data acquired from rodent animal studies. The outstanding effects may be attributed to species differences with respect to turnover time of RNAi. The report supported the clinical potential to realize RNAi therapies with the aid of SNALP delivery system.

3.4.3 Nanometric Calcium Phosphate/siRNA Precipitation

Co-precipitates formed between calcium phosphate and DNA have been used as an efficient method to transfect cultured mammalian cells to express exogenous genes since it was first reported by Graham et al. [98]. However, the application of this technique has been limited to *in vitro* transfection as a result of uncontrollable size and batch-to-batch variability of transfection efficiency [99]. Li and Huang have harnessed the low toxicity and pH-induced degradability of the co-precipitates and formulated a nano-sized calcium phosphate particles coated with liposomes for siRNA systemic delivery, namely, lipid–calcium–phosphate (LCP) [100,101]. The calcium phosphate particle is precipitated in an aqueous phase of a water/oil (cyclohexane) reverse microemulsion system stabilized with surfactants (IGPAL-520 and DOPA) (Figure 3.5). Size of the formed calcium phosphate amorphous precipitate is well controlled, as the surfactants on the surface prevent aggregation of the precipitates. siRNA is entrapped within the nanoprecipitate by the interaction between phosphate groups of siRNA and calcium ions. The lipid monolayer of DOPA is stabilized on the surface of the nanoprecipitate as a result of interactions

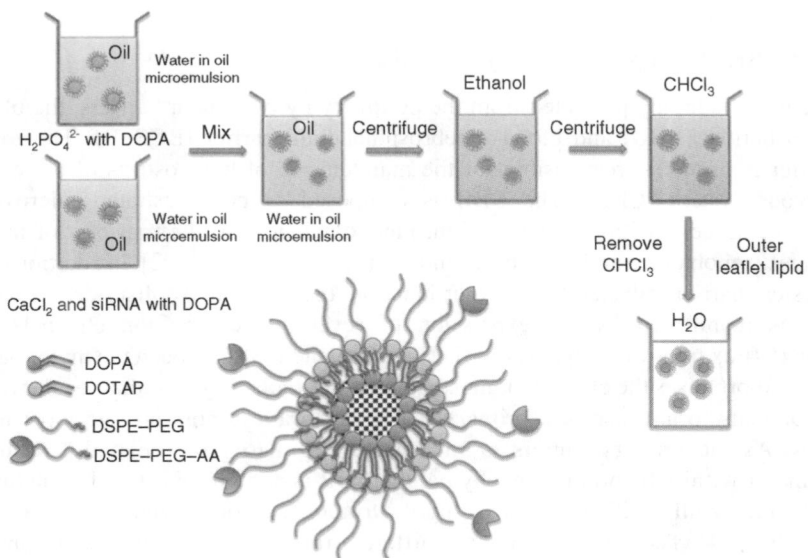

FIGURE 3.5 Preparation of liposome–calcium–phosphate (LCP) nanoparticles.

between the phosphate groups of DOPA and calcium ions. IGEPA-520 is subsequently removed by repeated centrifugation and washing with ethanol. The final LCP is formed by incubating DOPA-stabilized calcium phosphate nanoprecipitates with outer leaflet lipids in an aqueous solution [100,101].

The LCP demonstrates a pH-responsive siRNA release profile. As the pH in the endosome starts to drop, calcium phosphate precipitates dissolve and release cargo. Meanwhile, the released calcium ions increase the osmotic pressure of endosomal compartments, destabilize the endosomal membrane, and facilitate the release of siRNA into the cytosol. This property substantially enhances delivery efficiency of the LCP. The *in vitro* half maximal inhibitory concentration IC_{50} was determined to be 60 nM, which was three to four times lower than that of LPD (200 nM) [100,101]. The LCP nanoparticle has been demonstrated to knockdown the luciferase activity in a B16F10 lung metastasis model with an ED_{50} of 0.06 mg/ml. The co-delivery of VEGF, MDM2, and c-myc siRNA formulated in LCP nanoparticle significantly suppressed the lung metastasis ($\sim$70%) with a relatively low dose (0.36 mg/kg).

3.5   ORGAN-SPECIFIC DELIVERY OF NUCLEIC ACIDS WITH LNPs

The advancement of nanoparticles has triggered the site-specific delivery of nucleic acid based therapeutics. Moreover, owing to the different anatomical structures and physiological functions of major organs, particularly the endothelium status and the RES, various strategies harnessing the organ-specific features have been developed so as to specifically deliver nucleic acid loaded LNPs into targeted organs.

3.5.1   Brain-Targeted Delivery

The brain is highly protected from the periphery by two major barriers, the blood–brain barrier (BBB) and blood–cerebrospinal fluid barrier (BCSFB). The former barrier is primarily responsible for the maintenance of homeostasis of the central nervous system (CNS). The BBB is composed of nonfenestrated microvessel endothelial cells with continuous tight junctions that limit the transport of macrophages, xenobiotics, and some other endogenous metabolites [102]. In addition to the physical barrier reflected by the tight junction, there is a metabolism-driven barrier that is manifested by the expression of several receptors, ion channels, and influx/efflux transport proteins [103]. Therefore, the BBB has been the major barrier that compromises the efficacy of most therapeutics including macromolecular drugs.

Of many brain diseases, primary malignant brain tumors are the most lethal ones. As the primary tumors or metastases begin to grow beyond 1–2 mm in diameter within the brain parenchyma, the BBB is compromised both structurally and functionally with features of highly tortuous, disorganized, and permeable vessels [104]. These features are not different from tumors arising elsewhere and could be exploited for the delivery of therapeutics based on the EPR effect. Nevertheless, for other brain diseases such as neurodegenerative disorders or

ischemic stroke, the BBB still remains intact and is able to block the pharmacological reagents from crossing the endothelium to access the target site. The
advancement of nanotechnology has tackled the problem with great success for
CNS drug delivery since the nanocarriers could selectively interact with endothelial microvessels after modification with targeting moieties and result in a high
local concentration of therapeutics in the brain parenchyma [103]. The LNPs are
probably the most investigated and clinically tested due to the long circulation
time and low toxicity [105]. The major strategies for the development of LNPs to
cross the BBB is through transcytosis, initially observed as a mechanism by which
iron-loaded transferrin is transported to the parenchyma of the brain [106].
Transferrin or antibodies against transferrin receptor have been extensively
used for the surface modification of these LNPs for brain delivery [107–113].

The most outstanding LNP that has been well studied and applied to various
brain disease models is the Trojan horse liposome (THL), developed by Pardridge
et al. [108]. The surface modifications with targeting ligands for insulin receptor or
transferrin receptor expressed on the surface of BBB enable selective transport of
the LNP from the luminal side of the endothelium to the brain parenchyma.
Additional grafting of neuron-specific ligands facilitates the nanoparticle uptake
via receptor-mediated endocytosis. The delivery efficiency of the LNPs has been
demonstrated in gene therapy for the treatment of Parkinson's disease model in
rats [109,114,115], VII mucopolysaccharidosis model in mice [111], and RNAi
therapy targeting epidermal growth factor receptor with shRNA in mouse glioblastoma [116].

An alternative strategy to enhance brain delivery is to coat the nanoparticle with
cell-penetrating peptides such as TAT and octaarginine (R8) since these peptides
increase the selectivity of BBB and promote drug trafficking [117]. Although the
mechanism of action was not well elucidated, it was observed that the surface charge
of the particle played an important role in the brain delivery [117,118]. This implies
that electrostatic interaction between the positively charged particle and the lectins
on the microvessel endothelial cells might trigger the endocytosis.

3.5.2 Liver-Targeted Delivery

Among various organs, the liver has drawn most attention due to its critical role in
the metabolism. Moreover, the liver has a unique sinusoidal endothelium, which
processes open pores or fenestrae lacking a diaphragm and a basal lamina underneath
the endothelial lining with the pore size around 100 nm in diameter [119]. This
property of the well-perfused organ makes it easy for the LNPs to extravasate to
reach the interstitial space of the liver tissue. However, Kupffer cells, the macrophages of the liver, lay in ambush in the endothelium to capture and destroy foreign
particles. Any successful nanoparticle drug delivery vectors will have to evade the
Kupffer cells before accessing other cells of the liver. In this regard, a dense PEG
coating usually disguises the LNPs from being recognized by the RES. The presence
of a hepatocyte-specific targeting ligand diverts the LNPs from highly phagocytic
cells and enhances the uptake by hepatocytes. Since hepatocytes uniquely express

asialoglycoprotein receptor on the cell surface, galactose is widely exploited as a ligand for liver-specific gene transfer [120–122].

Another strategy to direct LNPs to hepatocytes is to take advantage of apolipoprotein (Apo), a component of high-density lipoprotein (HDL) that guides the transport of lipids to the liver for processing as a natural ligand for the scavenger receptor on hepatocytes. The hydrophobic domain in the Apo prompts the interaction between the Apo and the LNPs, essentially forming an artificial HDL. The incorporation of the Apo as a targeting ligand makes the LNPs predominantly taken up by the liver cells with well-characterized mechanism [123,124]. LNPs have proven to be highly efficient carriers for the liver delivery. This is probably because the circulating LNPs could interact with serum proteins and acquire protein components that guide the transport of the LNPs to specific cells [125,126]. Apolipoprotein E (ApoE), shown to be capable of enhancing the uptake into hepatocytes through low-density lipoprotein receptor (LDLR), was exploited as an "endogenous ligand" and an alternative to an exogenous targeting ligand [127]. The functionality of the proposed endogenous ligand was examined by Akinc et al. in LDLR−/− mice and ApoE−/− mice, respectively. It was demonstrated that the hepatocyte-specific delivery of the neutral liposome is ApoE mediated and LDLR dependent [127].

In addition to the aforementioned rational-based design of LNPs for the liver delivery, some molecules, which have demonstrated liver-specific properties, are also explored for the modification of LNPs for liver specificity. It was observed that R8-peptide-modified LNPs largely accumulated in the liver after intravenous administration and transfected hepatocytes with high efficiency. The uptake was proven to be R8 density dependent [128,129]. The liver-specific biodistribution is likely attributable to a unique property of R8 given that this type of peptide represents a promising device for specific liver delivery [130].

3.5.3 Lung-Targeted Delivery

Owing to the anatomical structure of the lung, rich in microcapillaries, gene transfer to the lung could be performed through systemic administration similar to other organs. Genetic therapeutics could also be administered noninvasively via airways. Because of the limited scope of this review, we focus the discussion mainly on the former route. The LNPs with cationic surface charge will preferably accumulate in the lung tissue without any modification since there is an excellent chance for them to interact with microcapillary endothelium and trigger the endocytosis [131]. Although the non-PEGylated LNPs are more likely to be opsonized and taken up by the macrophages in the liver and the spleen, it still results in a high level of transgene expression in the lung [131]. Recent studies have revealed that various tissue disorders are related to endothelium, particularly tumor angiogenesis. Therefore, the endothelium could serve as a therapeutic target for primary or metastatic lung tumors. Surface modifications with PEG and a tumor-specific targeting ligand can enhance the accumulation of liposome in the interstitial space of solid tumor and significantly increase the uptake of liposome by cancer cells [74,132,133].

3.5.4 Kidney-Targeted Delivery

Although the prevention and treatment of renal diseases is highly demanding, nonviral gene therapists have not paid as much attention to the kidney as to the liver and the brain. For one of the reasons, the kidney is not an organ that has the large accumulation of nanoparticles after systemic administration. Moreover, the signal from the kidney in the biodistribution study was often proven to be the fragments of dissociated nanoparticle, which was disposed through the glomerular filtration. Lai et al. tried to use Lipofectin formulation (DOTMA–DOPE) to deliver β-galactosidase gene to the kidney via intra-renal-arterial injection. The expression of the reporter gene was detected in the cortex and the outer medulla area and exclusively observed in the tubular epithelial cells [134]. Ito et al. used the same formulation to introduce inducible nitric oxide synthase (iNOS) gene into the kidney via intraureteral injection. The iNOS expression was detected in the collecting ducts, distal tubules, and glomerulus in the injected kidney and improved the renal function in unilateral ureteral obstruction [135]. Nevertheless, most of the kidney gene therapy has been conducted by viral vectors for high specificity and transgene expression.

3.6 CURRENT CHALLENGES TO THE DEVELOPMENT OF LNPs

The term *targeted gene delivery* indicated a predominant accumulation of genetic therapeutics in the intended tissues or even in the specific cell populations. This idea was originally conceived by Paul Ehrlich as the "magic bullet" that goes straight to the target cells to deliver cargos after systemic administration. The LNPs at the current stage, regardless of materials and formulations, are far from the ideal "magic bullet" because the biodistribution of most nanoparticles relies on both the properties of the nanoparticles and the vascular endothelium status of the organs or tumors. Although PEGylation could significantly improve the PK profiles of nanoparticles, drastically reducing the liver and spleen accumulations by the RES uptake and increasing the blood circulation time, the tumor accumulation of LNPs is reported to be around 5–10% of the injected dose. It has been noticed that the accumulation would not increase even if there were excessive LNPs circulating in the bloodstream. This suggested that a saturation state existed for the EPR-mediated tumor accumulation. Maeda et al. have investigated factors that affect tumor vasculature permeability and described a bradykinin-mediated pathway to account for the EPR effect [136]. Some intermediate molecules in the pathway such as bradykinin or NOS were targeted for the modulation of vascular permeability to enhance the EPR effect [136]. Nevertheless, there was only marginal increase in the tumor accumulation. Strictly speaking, there is no passive "targeting" if more than 90% of the administered dose goes elsewhere other than the targeted site. Furthermore, the degree of the leakiness of tumor vasculature is highly variable and heterogeneous even for the same type of xenograft implanted at the same site [137]. So far, the EPR effect has been the major driving force for nanoparticle delivery to solid tumors. It is imperative that some alternative mechanisms be exploited in order to enhance tumor accumulation of nanoparticles to meet the therapeutic challenge.

3.7 FUTURE PERSPECTIVES

The lack of satisfactory gene delivery systems impedes the application of the therapy in the medical practice. Liposome-mediated gene delivery appears to be most feasible and reliable compared with other nonviral vectors. This is by and large due to their relatively high rate of gene transfer efficiency, biodegradability, and cost-effectiveness. The emergence of ionizable catonic lipids, such as conditional catonic lipids, has revealed its translational potential. Apparently, the development of this class of lipids has been the focus of the research due to the promising therapeutic activity in animal disease models. Meanwhile, we begin to understand the factors determining the delivery efficiency such as the status of nanoparticles in the blood circulation and the impact of the major organs on the nanoparticles due to their anatomical structure and physiological function, tumor spatial and temporal heterogeneity, and conditions of tumor microenvironment. With this knowledge in hand, hundreds of rationally designed molecules including lipids and other macromolecules have been developed materialwise to improve the PK profile of the LNPs and transfection efficiency. All components are expected to contribute to the delivery efficiency according to their designs. Despite that these sophisticated components work wonderfully on their own, there was hardly any synergistic effect observed when combined together. Thus, in assembling all the components, it is important to ascertain that an optimization of formulation is conducted to achieve better delivery efficiency. It is envisaged that with the advent of novel lipids and formulation procedures, the seemingly insurmountable hurdle of gene therapy will be overcome in due course.

Acknowledgments

The original work in authors' laboratory has been supported by NIH grants CA129835, CA149363, CA151652, and CA151455.

REFERENCES

1. Mulligan, R. C. (1993). The basic science of gene therapy. *Science 260*, 926–932.
2. Fire, A., Xu, S., Montgomery, M. K., Kostas, S. A., Driver, S. E., Mello, C. C. (1998). Potent and specific genetic interference by double-stranded RNA in *Caenorhabditis elegans*. *Nature 391*, 806–811.
3. Heller, L. C., Ugen, K., Heller, R. (2005). Electroporation for targeted gene transfer. *Expert Opinion on Drug Delivery 2*, 255–268.
4. Knapp, J. E., Liu, D. (2004). Hydrodynamic delivery of DNA. *Methods in Molecular Biology 245*, 245–250.
5. Thomas, C. E., Ehrhardt, A., Kay, M. A. (2003). Progress and problems with the use of viral vectors for gene therapy. *Nature Reviews Genetics 4*, 346–358.

6. Cavazzana-Calvo, M., Hacein-Bey, S., de Saint Basile, G., Gross, F., Yvon, E., Nusbaum, P., Selz, F., Hue, C., Certain, S., Casanova, J. L., Bousso, P., Deist, F. L., Fischer, A. (2000). Gene therapy of human severe combined immunodeficiency (SCID)-X1 disease. *Science 288*, 669–672.

7. Kaiser, J. (2003). Gene therapy. Seeking the cause of induced leukemias in X-SCID trial. *Science 299*, 495.

8. Felgner, P. L., Gadek, T. R., Holm, M., Roman, R., Chan, H. W., Wenz, M., Northrop, J. P., Ringold, G. M., Danielsen, M. (1987). Lipofection: a highly efficient, lipid-mediated DNA-transfection procedure. *Proceedings of the National Academy of Sciences of the United States of America 84*, 7413–7417.

9. Radler, J. O., Koltover, I., Salditt, T., Safinya, C. R. (1997). Structure of DNA-cationic liposome complexes: DNA intercalation in multilamellar membranes in distinct inter-helical packing regimes. *Science 275*, 810–814.

10. Vonarbourg, A., Passirani, C., Saulnier, P., Benoit, J. P. (2006). Parameters influencing the stealthiness of colloidal drug delivery systems. *Biomaterials 27*, 4356–4373.

11. Klibanov, A. L., Maruyama, K., Torchilin, V. P., Huang, L. (1990). Amphipathic polyethyleneglycols effectively prolong the circulation time of liposomes. *FEBS Letters 268*, 235–237.

12. Hobbs, S. K., Monsky, W. L., Yuan, F., Roberts, W. G., Griffith, L., Torchilin, V. P., Jain, R. K. (1998). Regulation of transport pathways in tumor vessels: role of tumor type and microenvironment. *Proceedings of the National Academy of Sciences of the United States of America 95*, 4607–4612.

13. Whitehead, K. A., Langer, R., Anderson, D. G. (2009). Knocking down barriers: advances in siRNA delivery. *Nature reviews Drug Discovery 8*, 129–138.

14. Maeda, H. (2010). Tumor-selective delivery of macromolecular drugs via the EPR effect: background and future prospects. *Bioconjugate Chemistry 21*, 797–802.

15. Matsumura, Y., Maeda, H. (1986). A new concept for macromolecular therapeutics in cancer chemotherapy: mechanism of tumoritropic accumulation of proteins and the antitumor agent smancs. *Cancer Research 46*, 6387–6392.

16. Baxter, L. T., Jain, R. K. (1989). Transport of fluid and macromolecules in tumors. I. Role of interstitial pressure and convection. *Microvascular Research 37*, 77–104.

17. Rejman, J., Oberle, V., Zuhorn, I. S., Hoekstra, D. (2004). Size-dependent internalization of particles via the pathways of clathrin- and caveolae-mediated endocytosis. *The Biochemical Journal 377*, 159–169.

18. Maxfield, F. R., McGraw, T. E. (2004). Endocytic recycling. *Nature Reviews Molecular Cell Biology 5*, 121–132.

19. Tseng, Y. C., Mozumdar, S., Huang, L. (2009). Lipid-based systemic delivery of siRNA. *Advanced Drug Delivery Reviews 61*, 721–731.

20. Goodsell, D. S. (1991). Inside a living cell. *Trends in Biochemical Sciences 16*, 203–206.

21. Luby-Phelps, K. (2000). Cytoarchitecture and physical properties of cytoplasm: volume, viscosity, diffusion, intracellular surface area. *International Review of Cytology 192*, 189–221.

22. Leopold, P. L., Kreitzer, G., Miyazawa, N., Rempel, S., Pfister, K. K., Rodriguez-Boulan, E., Crystal, R. G. (2000). Dynein- and microtubule-mediated translocation of adenovirus serotype 5 occurs after endosomal lysis. *Human Gene Therapy 11*, 151–165.

23. Suh, J., Wirtz, D., Hanes, J. (2004). Real-time intracellular transport of gene nanocarriers studied by multiple particle tracking. *Biotechnology Progress 20*, 598–602.

24. Pichon, C., Billiet, L., Midoux, P. (2010). Chemical vectors for gene delivery: uptake and intracellular trafficking. *Current Opinion in Biotechnology 21*, 640–645.

25. Midoux, P., Pichon, C., Yaouanc, J. J., Jaffres, P. A. (2009). Chemical vectors for gene delivery: a current review on polymers, peptides and lipids containing histidine or imidazole as nucleic acids carriers. *British Journal of Pharmacology 157*, 166–178.

26. Aronsohn, A. I., Hughes, J. A. (1998). Nuclear localization signal peptides enhance cationic liposome-mediated gene therapy. *Journal of Drug Targeting 5*, 163–169.

27. Subramanian, A., Ranganathan, P., Diamond, S. L. (1999). Nuclear targeting peptide scaffolds for lipofection of nondividing mammalian cells. *Nature Biotechnology 17*, 873–877.

28. Matulis, D., Rouzina, I., Bloomfield, V. A. (2002). Thermodynamics of cationic lipid binding to DNA and DNA condensation: roles of electrostatics and hydrophobicity. *Journal of the American Chemical Society 124*, 7331–7342.

29. Morille, M., Passirani, C., Vonarbourg, A., Clavreul, A., Benoit, J. P. (2008). Progress in developing cationic vectors for non-viral systemic gene therapy against cancer. *Biomaterials 29*, 3477–3496.

30. Alexander, M. Y., Akhurst, R. J. (1995). Liposome-medicated gene transfer and expression via the skin. *Human Molecular Genetics 4*, 2279–2285.

31. Remy, J. S., Sirlin, C., Vierling, P., Behr, J. P. (1994). Gene transfer with a series of lipophilic DNA-binding molecules. *Bioconjugate Chemistry 5*, 647–654.

32. Zhang, S., Xu, Y., Wang, B., Qiao, W., Liu, D., Li, Z. (2004). Cationic compounds used in lipoplexes and polyplexes for gene delivery. *Journal of Controlled Release 100*, 165–180.

33. Gao, X., Huang, L. (1991). A novel cationic liposome reagent for efficient transfection of mammalian cells. *Biochemical and Biophysical Research Communications 179*, 280–285.

34. Vigneron, J. P., Oudrhiri, N., Fauquet, M., Vergely, L., Bradley, J. C., Basseville, M., Lehn, P., Lehn, J. M. (1996). Guanidinium-cholesterol cationic lipids: efficient vectors for the transfection of eukaryotic cells. *Proceedings of the National Academy of Sciences of the United States of America 93*, 9682–9686.

35. Semple, S. C., Klimuk, S. K., Harasym, T. O., Dos Santos, N., Ansell, S. M., Wong, K. F., Maurer, N., Stark, H., Cullis, P. R., Hope, M. J., Scherrer, P. (2001). Efficient encapsulation of antisense oligonucleotides in lipid vesicles using ionizable aminolipids: formation of novel small multilamellar vesicle structures. *Biochimica et Biophysica Acta 1510*, 152–166.

36. Zimmermann, T. S., Lee, A. C., Akinc, A., Bramlage, B., Bumcrot, D., Fedoruk, M. N., Harborth, J., Heyes, J. A., Jeffs, L. B., John, M., Judge, A. D., Lam, K., McClintock, K., Nechev, L. V., Palmer, L. R., Racie, T., Rohl, I., Seiffert, S., Shanmugam, S., Sood, V., Soutschek, J., Toudjarska, I., Wheat, A. J., Yaworski, E., Zedalis, W., Koteliansky, V., Manoharan, M., Vornlocher, H. P., MacLachlan, I. (2006). RNAi-mediated gene silencing in non-human primates. *Nature 441*, 111–114.

37. Semple, S. C., Akinc, A., Chen, J., Sandhu, A. P., Mui, B. L., Cho, C. K., Sah, D. W., Stebbing, D., Crosley, E. J., Yaworski, E., Hafez, I. M., Dorkin, J. R., Qin, J., Lam, K., Rajeev, K. G., Wong, K. F., Jeffs, L. B., Nechev, L., Eisenhardt, M. L., Jayaraman, M.,

Kazem, M., Maier, M. A., Srinivasulu, M., Weinstein, M. J., Chen, Q., Alvarez, R., Barros, S. A., De, S., Klimuk, S. K., Borland, T., Kosovrasti, V., Cantley, W. L., Tam, Y. K., Manoharan, M., Ciufolini, M. A., Tracy, M. A., de Fougerolles, A., MacLachlan, I., Cullis, P. R., Madden, T. D., Hope, M. J. (2010). Rational design of cationic lipids for siRNA delivery. *Nature Biotechnology 28*, 172–176.

38. Akinc, A., Zumbuehl, A., Goldberg, M., Leshchiner, E. S., Busini, V., Hossain, N., Bacallado, S. A., Nguyen, D. N., Fuller, J., Alvarez, R., Borodovsky, A., Borland, T., Constien, R., de Fougerolles, A., Dorkin, J. R., Narayanannair Jayaprakash, K., Jayaraman, M., John, M., Koteliansky, V., Manoharan, M., Nechev, L., Qin, J., Racie, T., Raitcheva, D., Rajeev, K. G., Sah, D. W., Soutschek, J., Toudjarska, I., Vornlocher, H. P., Zimmermann, T. S., Langer, R., Anderson, D. G. (2008). A combinatorial library of lipid-like materials for delivery of RNAi therapeutics. *Nature Biotechnology 26*, 561–569.

39. Heyes, J., Palmer, L., Bremner, K., MacLachlan, I. (2005). Cationic lipid saturation influences intracellular delivery of encapsulated nucleic acids. *Journal of Controlled Release 107*, 276–287.

40. Landen, C. N., Jr., Chavez-Reyes, A., Bucana, C., Schmandt, R., Deavers, M. T., Lopez-Berestein, G., Sood, A. K. (2005). Therapeutic EphA2 gene targeting *in vivo* using neutral liposomal small interfering RNA delivery. *Cancer Research 65*, 6910–6918.

41. Gray, M. J., Dallas, N. A., Van Buren, G., Xia, L., Yang, A. D., Somcio, R. J., Gaur, P., Mangala, L. S., Vivas-Mejia, P. E., Fan, F., Sanguino, A. M., Gallick, G. E., Lopez-Berestein, G., Sood, A. K., Ellis, L. M. (2008). Therapeutic targeting of Id2 reduces growth of human colorectal carcinoma in the murine liver. *Oncogene 27*, 7192–7200.

42. Villares, G. J., Zigler, M., Wang, H., Melnikova, V. O., Wu, H., Friedman, R., Leslie, M. C., Vivas-Mejia, P. E., Lopez-Berestein, G., Sood, A. K., Bar-Eli, M. (2008). Targeting melanoma growth and metastasis with systemic delivery of liposome-incorporated protease-activated receptor-1 small interfering RNA. *Cancer Research 68*, 9078–9086.

43. Landen, C. N., Merritt, W. M., Mangala, L. S., Sanguino, A. M., Bucana, C., Lu, C., Lin, Y. G., Han, L. Y., Kamat, A. A., Schmandt, R., Coleman, R. L., Gershenson, D. M., Lopez-Berestein, G., Sood, A. K. (2006). Intraperitoneal delivery of liposomal siRNA for therapy of advanced ovarian cancer. *Cancer Biology & Therapy 5*, 1708–1713.

44. Mangala, L. S., Han, H. D., Lopez-Berestein, G., Sood, A. K. (2009). Liposomal siRNA for ovarian cancer. *Methods in Molecular Biology 555*, 29–42.

45. Merritt, W. M., Lin, Y. G., Spannuth, W. A., Fletcher, M. S., Kamat, A. A., Han, L. Y., Landen, C. N., Jennings, N., De Geest, K., Langley, R. R., Villares, G., Sanguino, A., Lutgendorf, S. K., Lopez-Berestein, G., Bar-Eli, M. M., Sood, A. K. (2008). Effect of interleukin-8 gene silencing with liposome-encapsulated small interfering RNA on ovarian cancer cell growth. *Journal of the National Cancer Institute 100*, 359–372.

46. Leblond, J., Mignet, N., Leseurre, L., Largeau, C., Bessodes, M., Scherman, D., Herscovici, J. (2006). Design, synthesis, and evaluation of enhanced DNA binding new lipopolythioureas. *Bioconjugate Chemistry 17*, 1200–1208.

47. Leblond, J., Mignet, N., Largeau, C., Seguin, J., Scherman, D., Herscovici, J. (2008). Lipopolythiourea transfecting agents: lysine thiourea derivatives. *Bioconjugate Chemistry 19*, 306–314.

48. Breton, M., Leblond, J., Seguin, J., Midoux, P., Scherman, D., Herscovici, J., Pichon, C., Mignet, N. (2010). Comparative gene transfer between cationic and thiourea lipoplexes. *The Journal of Gene Medicine 12*, 45–54.

49. Lee, R. J., Huang, L. (1996). Folate-targeted, anionic liposome-entrapped polylysine-condensed DNA for tumor cell-specific gene transfer. *The Journal of Biological Chemistry 271*, 8481–8487.

50. Mastrobattista, E., Kapel, R. H., Eggenhuisen, M. H., Roholl, P. J., Crommelin, D. J., Hennink, W. E., Storm, G. (2001). Lipid-coated polyplexes for targeted gene delivery to ovarian carcinoma cells. *Cancer Gene Therapy 8*, 405–413.

51. Sun, X., Provoda, C., Lee, K. D. (2010). Enhanced *in vivo* gene expression mediated by listeriolysin O incorporated anionic LPDII: Its utility in cytotoxic T lymphocyte-inducing DNA vaccine. *Journal of Controlled Release 148*, 219–225.

52. Chen, Y., Bathula, S. R., Li, J., Huang, L. (2010). Multifunctional nanoparticles delivering small interfering RNA and doxorubicin overcome drug resistance in cancer. *The Journal of Biological Chemistry 285*, 22639–22650.

53. Srinivasan, C., Burgess, D. J. (2009). Optimization and characterization of anionic lipoplexes for gene delivery. *Journal of Controlled Release 136*, 62–70.

54. Allen, T. M. (1994). The use of glycolipids and hydrophilic polymers in avoiding rapid uptake of liposomes by the mononuclear phagocyte system. *Advanced Drug Delivery Reviews 13*, 285–309.

55. Campbell, R. B., Fukumura, D., Brown, E. B., Mazzola, L. M., Izumi, Y., Jain, R. K., Torchilin, V. P., Munn, L. L. (2002). Cationic charge determines the distribution of liposomes between the vascular and extravascular compartments of tumors. *Cancer Research 62*, 6831–6836.

56. Moghimi, S. M., Hunter, A. C. (2001). Capture of stealth nanoparticles by the body's defences. *Critical Reviews in Therapeutic Drug Carrier Systems 18*, 527–550.

57. Mok, K. W., Lam, A. M., Cullis, P. R. (1999). Stabilized plasmid-lipid particles: factors influencing plasmid entrapment and transfection properties. *Biochimica et Biophysica Acta 1419*, 137–150.

58. Tam, P., Monck, M., Lee, D., Ludkovski, O., Leng, E. C., Clow, K., Stark, H., Scherrer, P., Graham, R. W., Cullis, P. R. (2000). Stabilized plasmid-lipid particles for systemic gene therapy. *Gene Therapy 7*, 1867–1874.

59. Adlakha-Hutcheon, G., Bally, M. B., Shew, C. R., and Madden, T. D. (1999). Controlled destabilization of a liposomal drug delivery system enhances mitoxantrone antitumor activity. *Nat Biotechnol 17*, 775–779.

60. Zhang, Y. P., Sekirov, L., Saravolac, E. G., Wheeler, J. J., Tardi, P., Clow, K., Leng, E., Sun, R., Cullis, P. R., Scherrer, P. (1999). Stabilized plasmid-lipid particles for regional gene therapy: formulation and transfection properties. *Gene Therapy 6*, 1438–1447.

61. Allen, T. M., Brandeis, E., Hansen, C. B., Kao, G. Y., Zalipsky, S. (1995). A new strategy for attachment of antibodies to sterically stabilized liposomes resulting in efficient targeting to cancer cells. *Biochimica et Biophysica Acta 1237*, 99–108.

62. Guo, X., Szoka, F. C. Jr. (2001). Steric stabilization of fusogenic liposomes by a low-pH sensitive PEG–diortho ester–lipid conjugate. *Bioconjugate Chemistry 12*, 291–300.

63. Li, W., Huang, Z., MacKay, J. A., Grube, S., Szoka, F. C. Jr. (2005). Low-pH-sensitive poly(ethylene glycol) (PEG)-stabilized plasmid nanolipoparticles: effects of PEG chain

length, lipid composition and assembly conditions on gene delivery. *The Journal of Gene Medicine 7*, 67–79.

64. Choi, J. S., MacKay, J. A., Szoka, F. C. Jr. (2003). Low-pH-sensitive PEG-stabilized plasmid-lipid nanoparticles: preparation and characterization. *Bioconjugate Chemistry 14*, 420–429.

65. Guo, X., MacKay, J. A., Szoka, F. C. Jr. (2003). Mechanism of pH-triggered collapse of phosphatidylethanolamine liposomes stabilized by an ortho ester polyethyleneglycol lipid. *Biophysical Journal 84*, 1784–1795.

66. Maeda, T., Fujimoto, K. (2006). A reduction-triggered delivery by a liposomal carrier possessing membrane-permeable ligands and a detachable coating. *Colloids and Surfaces B: Biointerfaces 49*, 15–21.

67. Zhang, J. X., Zalipsky, S., Mullah, N., Pechar, M., Allen, T. M. (2004). Pharmaco attributes of dioleoylphosphatidylethanolamine/cholesterylhemisuccinate liposomes containing different types of cleavable lipopolymers. *Pharmacological Research 49*, 185–198.

68. Hatakeyama, H., Akita, H., Kogure, K., Oishi, M., Nagasaki, Y., Kihira, Y., Ueno, M., Kobayashi, H., Kikuchi, H., Harashima, H. (2007). Development of a novel systemic gene delivery system for cancer therapy with a tumor-specific cleavable PEG-lipid. *Gene Therapy 14*, 68–77.

69. Romberg, B., Hennink, W. E., Storm, G. (2008). Sheddable coatings for long-circulating nanoparticles. *Pharmaceutical Research 25*, 55–71.

70. Li, S., Huang, L. (1997). *In vivo* gene transfer via intravenous administration of cationic lipid-protamine-DNA (LPD) complexes. *Gene Therapy 4*, 891–900.

71. Gao, X., Huang, L. (1996). Potentiation of cationic liposome-mediated gene delivery by polycations. *Biochemistry 35*, 1027–1036.

72. Li, S. D., Huang, L. (2006). Surface-modified LPD nanoparticles for tumor targeting. *Annals of the New York Academy of Sciences 1082*, 1–8.

73. Li, S. D., Chen, Y. C., Hackett, M. J., Huang, L. (2008). Tumor-targeted delivery of siRNA by self-assembled nanoparticles. *Molecular Therapy 16*, 163–169.

74. Li, S. D., Chono, S., Huang, L. (2008). Efficient gene silencing in metastatic tumor by siRNA formulated in surface-modified nanoparticles. *Journal of Controlled Release 126*, 77–84.

75. Li, S. D., Huang, L. (2009). Nanoparticles evading the reticuloendothelial system: role of the supported bilayer. *Biochimica et Biophysica Acta 1788*, 2259–2266.

76. Li, S. D., Huang, L. (2010). Stealth nanoparticles: high density but sheddable PEG is a key for tumor targeting. *Journal of Controlled Release 145*, 178–181.

77. Akita, H., Kudo, A., Minoura, A., Yamaguti, M., Khalil, I. A., Moriguchi, R., Masuda, T., Danev, R., Nagayama, K., Kogure, K., Harashima, H. (2009). Multi-layered nanoparticles for penetrating the endosome and nuclear membrane via a step-wise membrane fusion process. *Biomaterials 30*, 2940–2949.

78. El-Sayed, A., Masuda, T., Khalil, I., Akita, H., Harashima, H. (2009). Enhanced gene expression by a novel stearylated INF7 peptide derivative through fusion independent endosomal escape. *Journal of Controlled Release 138*, 160–167.

79. Hatakeyama, H., Akita, H., Harashima, H. (2011). A multifunctional envelope type nano device (MEND) for gene delivery to tumours based on the EPR effect: a strategy for overcoming the PEG dilemma. *Advanced Drug Delivery Reviews 63*, 152–160.

80. Hatakeyama, H., Akita, H., Ito, E., Hayashi, Y., Oishi, M., Nagasaki, Y., Danev, R., Nagayama, K., Kaji, N., Kikuchi, H., Baba, Y., Harashima, H. (2011). Systemic delivery of siRNA to tumors using a lipid nanoparticle containing a tumor-specific cleavable PEG-lipid. *Biomaterials 32*, 4306–4316.

81. Hatakeyama, H., Ito, E., Akita, H., Oishi, M., Nagasaki, Y., Futaki, S., Harashima, H. (2009). A pH-sensitive fusogenic peptide facilitates endosomal escape and greatly enhances the gene silencing of siRNA-containing nanoparticles *in vitro* and *in vivo*. *Journal of Controlled Release 139*, 127–132.

82. Ishitsuka, T., Akita, H., Harashima, H. (2011). Functional improvement of an IRQ-PEG-MEND for delivering genes to the lung. *Journal of Controlled Release 154*, 77–83.

83. Khalil, I. A., Hayashi, Y., Mizuno, R., Harashima, H. (2011). Octaarginine- and pH sensitive fusogenic peptide-modified nanoparticles for liver gene delivery. *Journal of Controlled Release 156*, 374–380.

84. Khalil, I. A., Kogure, K., Futaki, S., Hama, S., Akita, H., Ueno, M., Kishida, H., Kudoh, M., Mishina, Y., Kataoka, K., Yamada, M., Harashima, H. (2007). Octaarginine-modified multifunctional envelope-type nanoparticles for gene delivery. *Gene Therapy 14*, 682–689.

85. Kogure, K., Moriguchi, R., Sasaki, K., Ueno, M., Futaki, S., Harashima, H. (2004). Development of a non-viral multifunctional envelope-type nano device by a novel lipid film hydration method. *Journal of Controlled Release 98*, 317–323.

86. Kuramoto, H., Park, Y. S., Kaji, N., Tokeshi, M., Kogure, K., Shinohara, Y., Harashima, H., Baba, Y. (2008). On-chip fabrication of mutifunctional envelope-type nanodevices for gene delivery. *Analytical and Bioanalytical Chemistry 391*, 2729–2733.

87. Masuda, T., Akita, H., Nishio, T., Niikura, K., Kogure, K., Ijiro, K., Harashima, H. (2008). Development of lipid particles targeted via sugar-lipid conjugates as novel nuclear gene delivery system. *Biomaterials 29*, 709–723.

88. Moriguchi, R., Kogure, K., Akita, H., Futaki, S., Miyagishi, M., Taira, K., Harashima, H. (2005). A multifunctional envelope-type nano device for novel gene delivery of siRNA plasmids. *International Journal of Pharmaceutics 301*, 277–285.

89. Mudhakir, D., Akita, H., Tan, E., Harashima, H. (2008). A novel IRQ ligand-modified nano-carrier targeted to a unique pathway of caveolar endocytic pathway. *Journal of Controlled Release 125*, 164–173.

90. Nakamura, Y., Kogure, K., Futaki, S., Harashima, H. (2007). Octaarginine-modified multifunctional envelope-type nano device for siRNA. *Journal of Controlled Release 119*, 360–367.

91. Nakamura, Y., Kogure, K., Yamada, Y., Futaki, S., Harashima, H. (2006). Significant and prolonged antisense effect of a multifunctional envelope-type nano device encapsulating antisense oligodeoxynucleotide. *The Journal of Pharmacy and Pharmacology 58*, 431–437.

92. Sasaki, K., Kogure, K., Chaki, S., Kihira, Y., Ueno, M., Harashima, H. (2005). Construction of a multifunctional envelope-type nano device by a SUV*-fusion method. *International Journal of Pharmaceutics 296*, 142–150.

93. Shaheen, S. M., Akita, H., Nakamura, T., Takayama, S., Futaki, S., Yamashita, A., Katoono, R., Yui, N., Harashima, H. (2011). KALA-modified multi-layered nanoparticles as gene carriers for MHC class-I mediated antigen presentation for a DNA vaccine. *Biomaterials 32*, 6342–6350.

94. Wang, C. Y., Huang, L. (1987). pH-sensitive immunoliposomes mediate target-cell-specific delivery and controlled expression of a foreign gene in mouse. *Proceedings of the National Academy of Sciences of the United States of America 84*, 7851–7855.

95. Hofland, H. E., Shephard, L., Sullivan, S. M. (1996). Formation of stable cationic lipid/DNA complexes for gene transfer. *Proceedings of the National Academy of Sciences of the United States of America 93*, 7305–7309.

96. Jeffs, L. B., Palmer, L. R., Ambegia, E. G., Giesbrecht, C., Ewanick, S., MacLachlan, I. (2005). A scalable, extrusion-free method for efficient liposomal encapsulation of plasmid DNA. *Pharmaceutical Research 22*, 362–372.

97. Morrissey, D. V., Lockridge, J. A., Shaw, L., Blanchard, K., Jensen, K., Breen, W., Hartsough, K., Machemer, L., Radka, S., Jadhav, V., Vaish, N., Zinnen, S., Vargeese, C., Bowman, K., Shaffer, C. S., Jeffs, L. B., Judge, A., MacLachlan, I., Polisky, B. (2005). Potent and persistent *in vivo* anti-HBV activity of chemically modified siRNAs. *Nature Biotechnology 23*, 1002–1007.

98. Graham, F. L., van der Eb, A. J. (1973). A new technique for the assay of infectivity of human adenovirus 5 DNA. *Virology 52*, 456–467.

99. Jordan, M., Schallhorn, A., Wurm, F. M. (1996). Transfecting mammalian cells: optimization of critical parameters affecting calcium-phosphate precipitate formation. *Nucleic Acids Research 24*, 596–601.

100. Li, J., Chen, Y. C., Tseng, Y. C., Mozumdar, S., Huang, L. (2010). Biodegradable calcium phosphate nanoparticle with lipid coating for systemic siRNA delivery. *Journal of Controlled Release 142*, 416–421.

101. Li, J., Yang, Y., Huang, L. (2012). Calcium phosphate nanoparticles with an asymmetric lipid bilayer coating for siRNA delivery to the tumor. *Journal of Controlled Release 158*, 108–114.

102. Weksler, B. B., Subileau, E. A., Perriere, N., Charneau, P., Holloway, K., Leveque, M., Tricoire-Leignel, H., Nicotra, A., Bourdoulous, S., Turowski, P., Male, D. K., Roux, F., Greenwood, J., Romero, I. A., Couraud, P. O. (2005). Blood-brain barrier-specific properties of a human adult brain endothelial cell line. *The FASEB Journal 19*, 1872–1874.

103. Wong, H. L., Wu, X. Y., Bendayan, R. (2011). Nanotechnological advances for the delivery of CNS therapeutics. *Advanced Drug Delivery Reviews 64*, 686–700.

104. Jain, R. K., di Tomaso, E., Duda, D. G., Loeffler, J. S., Sorensen, A. G., Batchelor, T. T. (2007). Angiogenesis in brain tumours. *Nature Reviews Neuroscience 8*, 610–622.

105. Costantino, L., Tosi, G., Ruozi, B., Bondioli, L., Vandelli, M. A., Forni, F. (2009). Chapter 3—Colloidal systems for CNS drug delivery. *Progress in Brain Research 180*, 35–69.

106. Fishman, J. B., Rubin, J. B., Handrahan, J. V., Connor, J. R., Fine, R. E. (1987). Receptor-mediated transcytosis of transferrin across the blood-brain barrier. *Journal of Neuroscience Research 18*, 299–2304.

107. Ko, Y. T., Bhattacharya, R., Bickel, U. (2009). Liposome encapsulated polyethylenimine/ODN polyplexes for brain targeting. *Journal of Controlled Release 133*, 230–237.

108. Zhang, Y., Schlachetzki, F., Zhang, Y. F., Boado, R. J., Pardridge, W. M. (2004). Normalization of striatal tyrosine hydroxylase and reversal of motor impairment in experimental parkinsonism with intravenous nonviral gene therapy and a brain-specific promoter. *Human Gene Therapy 15*, 339–350.

109. Zhang, Y., Calon, F., Zhu, C., Boado, R. J., Pardridge, W. M. (2003). Intravenous nonviral gene therapy causes normalization of striatal tyrosine hydroxylase and reversal of motor impairment in experimental parkinsonism. *Human Gene Therapy 14*, 1–12.

110. Boado, R. J. (2007). Blood-brain barrier transport of non-viral gene and RNAi therapeutics. *Pharmaceutical Research 24*, 1772–1787.

111. Zhang, Y., Wang, Y., Boado, R. J., Pardridge, W. M. (2008). Lysosomal enzyme replacement of the brain with intravenous non-viral gene transfer. *Pharmaceutical Research 25*, 400–406.

112. Boado, R. J., Pardridge, W. M. (2011). The Trojan horse liposome technology for nonviral gene transfer across the blood-brain barrier. *Journal of Drug Delivery 2011*, 296151.

113. Pardridge, W. M. (2010). Preparation of Trojan horse liposomes (THLs) for gene transfer across the blood-brain barrier. *Cold Spring Harbor Protocols 2010*, pdb prot5407.

114. Zhang, Y., Pardridge, W. M. (2009). Near complete rescue of experimental Parkinson's disease with intravenous, non-viral GDNF gene therapy. *Pharmaceutical Research 26*, 1059–1063.

115. Xia, C. F., Boado, R. J., Zhang, Y., Chu, C., Pardridge, W. M. (2008). Intravenous glial-derived neurotrophic factor gene therapy of experimental Parkinson's disease with Trojan horse liposomes and a tyrosine hydroxylase promoter. *The Journal of Gene Medicine 10*, 306–315.

116. Zhang, Y., Zhang, Y. F., Bryant, J., Charles, A., Boado, R. J., Pardridge, W. M. (2004). Intravenous RNA interference gene therapy targeting the human epidermal growth factor receptor prolongs survival in intracranial brain cancer. *Clinical Cancer Research 10*, 3667–3677.

117. Qin, Y., Chen, H., Yuan, W., Kuai, R., Zhang, Q., Xie, F., Zhang, L., Zhang, Z., Liu, J., He, Q. (2011). Liposome formulated with TAT-modified cholesterol for enhancing the brain delivery. *International Journal of Pharmaceutics 419*, 85–95.

118. Qin, Y., Zhang, Q., Chen, H., Yuan, W., Kuai, R., Xie, F., Zhang, L., Wang, X., Zhang, Z., Liu, J., He, Q. (2012). Comparison of four different peptides to enhance accumulation of liposomes into the brain. *Journal of Drug Targeting 20*, 235–245.

119. Braet, F., Wisse, E. (2002). Structural and functional aspects of liver sinusoidal endothelial cell fenestrae: a review. *Comparative Hepatology 1*, 1.

120. Hashida, M., Nishikawa, M., Yamashita, F., Takakura, Y. (2001). Cell-specific delivery of genes with glycosylated carriers. *Advanced Drug Delivery Reviews 52*, 187–196.

121. Kawakami, S., Fumoto, S., Nishikawa, M., Yamashita, F., Hashida, M. (2000). *In vivo* gene delivery to the liver using novel galactosylated cationic liposomes. *Pharmaceutical Research 17*, 306–313.

122. Nishikawa, M., Kawakami, S., Yamashita, F., Hashida, M. (2003). Glycosylated cationic liposomes for carbohydrate receptor-mediated gene transfer. *Methods in Enzymology 373*, 384–399.

123. Kim, S. I., Shin, D., Choi, T. H., Lee, J. C., Cheon, G. J., Kim, K. Y., Park, M., Kim, M. (2007). Systemic and specific delivery of small interfering RNAs to the liver mediated by apolipoprotein A-I. *Molecular Therapy 15*, 1145–1152.

124. Kim, S. I., Shin, D., Lee, H., Ahn, B. Y., Yoon, Y., Kim, M. (2009). Targeted delivery of siRNA against hepatitis C virus by apolipoprotein A-I-bound cationic liposomes. *Journal of Hepatology 50*, 479–488.

125. Chonn, A., Semple, S. C., Cullis, P. R. (1992). Association of blood proteins with large unilamellar liposomes *in vivo*. Relation to circulation lifetimes. *The Journal of Biological Chemistry 267*, 18759–18765.

126. Cullis, P. R., Chonn, A., Semple, S. C. (1998). Interactions of liposomes and lipid-based carrier systems with blood proteins: relation to clearance behaviour *in vivo*. *Advanced Drug Delivery Reviews 32*, 3–17.

127. Akinc, A., Querbes, W., De, S., Qin, J., Frank-Kamenetsky, M., Jayaprakash, K. N., Jayaraman, M., Rajeev, K. G., Cantley, W. L., Dorkin, J. R., Butler, J. S., Qin, L., Racie, T., Sprague, A., Fava, E., Zeigerer, A., Hope, M. J., Zerial, M., Sah, D. W., Fitzgerald, K., Tracy, M. A., Manoharan, M., Koteliansky, V., Fougerolles, A., Maier, M. A. (2010). Targeted delivery of RNAi therapeutics with endogenous and exogenous ligand-based mechanisms. *Molecular Therapy 18*, 1357–1364.

128. Khalil, I. A., Hayashi, Y., Mizuno, R., Harashima, H. (2011). Octaarginine- and pH sensitive fusogenic peptide-modified nanoparticles for liver gene delivery. *Journal of Controlled Release 156*, 374–380.

129. Mudhakir, D., Akita, H., Khalil, I. A., Futaki, S., Harashima, H. (2005). Pharmacokinetic analysis of the tissue distribution of octaarginine modified liposomes in mice. *Drug Metabolism and Pharmacokinetics 20*, 275–281.

130. Sarko, D., Beijer, B., Garcia Boy, R., Nothelfer, E. M., Leotta, K., Eisenhut, M., Altmann, A., Haberkorn, U., Mier, W. (2010). The pharmacokinetics of cell-penetrating peptides. *Molecular Pharmacology 7*, 2224–2231.

131. Li, S., Rizzo, M. A., Bhattacharya, S., Huang, L. (1998). Characterization of cationic lipid-protamine-DNA (LPD) complexes for intravenous gene delivery. *Gene Therapy 5*, 930–937.

132. Chen, Y., Bathula, S. R., Yang, Q., Huang, L. (2010). Targeted nanoparticles deliver siRNA to melanoma. *The Journal of Investigative Dermatology 130*, 2790–2798.

133. Chen, Y., Zhu, X., Zhang, X., Liu, B., Huang, L. (2010). Nanoparticles modified with tumor-targeting scFv deliver siRNA and miRNA for cancer therapy. *Molecular Therapy 18*, 1650–1656.

134. Lai, L. W., Moeckel, G. W., Lien, Y. H. (1997). Kidney-targeted liposome-mediated gene transfer in mice. *Gene Therapy 4*, 426–431.

135. Ito, K., Chen, J., Khodadadian, J. J., Seshan, S. V., Eaton, C., Zhao, X., Vaughan, E. D., Jr., Lipkowitz, M., Poppas, D. P., Felsen, D. (2004). Liposome-mediated transfer of nitric oxide synthase gene improves renal function in ureteral obstruction in rats. *Kidney International 66*, 1365–1375.

136. Maeda, H., Wu, J., Sawa, T., Matsumura, Y., Hori, K. (2000). Tumor vascular permeability and the EPR effect in macromolecular therapeutics: a review. *Journal of Controlled Release 65*, 271–284.

137. Yuan, F., Salehi, H. A., Boucher, Y., Vasthare, U. S., Tuma, R. F., Jain, R. K. (1994). Vascular permeability and microcirculation of gliomas and mammary carcinomas transplanted in rat and mouse cranial windows. *Cancer Research 54*, 4564–4568.

4

PHOTOSENSITIVE LIPOSOMES AS POTENTIAL TARGETED THERAPEUTIC AGENTS

DAVID H. THOMPSON, POCHI SHUM, OLEG V. GERASIMOV, AND MARQUITA QUALLS

Department of Chemistry, Purdue University, West Lafayette, IN, USA

4.1 INTRODUCTION

Many biological processes, including those involved in endosomal uptake, tumor growth and inflammation, occur in locally low pH or hypoxic environments [1,2]. In the case of rapidly growing tumors or inflammatory sites, the tissues also are often characterized by porous or poorly formed vasculature, providing limited nanoparticle access to these target tissues via the enhanced permeability and retention effect [3,4]. Taken together, these features have led to the development of a vast array of materials designs and delivery schemes to enhance the efficacy of therapeutic agents in acidic tumors and inflamed tissue. Herein we report recent efforts to utilize photochemical processes to initiate drug release from liposomal carrier systems in a controlled and potentially localized manner.

Liposomes are clinically proven, commercially successful nanoscale carrier systems for delivering high concentrations of small molecule agents such as doxorubicin [5], amphotericin B [6], verteporfin [7], cytarabin [8], and morphine [9]. Widespread clinical application of liposomes to other classes of drugs, however, has been hindered by both their limited tissue penetration [10,11] and their premature release of contents before they reach the target site. Acid-sensitive liposomes bearing targeting ligands can promote site-specific internalization via receptor-mediated

Nanoparticulate Drug Delivery Systems: Strategies, Technologies, and Applications, First Edition.
Edited by Yoon Yeo.
© 2013 John Wiley & Sons, Inc. Published 2013 by John Wiley & Sons, Inc.

endocytosis [12] and exploit the time-dependent decrease in endosomal pH to promote escape of the liposomal contents into the cytosol [13–15]. They are particularly effective if they are designed with fusogenic lipids; however, if the formulation lacks an efficient triggering and endosomal escape mechanism, the drug cargo is destined for lysosomal degradation [16,17]. Bioactivation that promotes liposome–endosome membrane fusion is an effective strategy to promote cargo escape from both the liposomal and endosomal compartments. This is commonly achieved using phospholipid 1,2-dioleoyl-*sn*-glycero-3-phosphoethanolamine (DOPE), which adopts a bilayer state only at pH above the pK_a of the phosphoethanolamine headgroup (i.e., >pH 8.0). At physiological pH and temperature, pure DOPE liposomes form an H_{II} phase (i.e., an inverted micelle structure) that is "nonbilayer" and unable to retain the contents [18]. It is possible to increase the stability of DOPE-rich liposomes at physiological pH and temperature by formulating the liposome with stabilizing components, such as phosphoserine (PS) lipid [19,20], cholesterol hemisuccinate (CHEMS) [21,22], cleavable polyethylene glycol (PEG) lipids [23–26], or other cleavable polymers [27]. *In vitro* studies at neutral pH have shown that DOPE dispersions with CHEMS, PS, or PEG–vinyl ether–lipids [25,28,29] retain encapsulated fluorophores, suggesting that these liposomes should be structurally stable during circulation but remain responsive to low pH media. A family of PEG–vinyl ether–lipids recently reported by our group provide tunable pH-sensitivity over 10 orders of magnitude, making it now possible to engineer the liposomal release properties to display the desired release properties for a given target pH and timescale [30,31].

A greater control of acid-induced drug release may be achieved by photochemical acidification of target tissues. Light-activated dendrimeric photosensitizers [32], sensitizer-grafted polymer nanoparticles [33], gold nanorods [34], gold nanoshells [35], and liposomal-borne photosensitizers [16,36–38] have been described as agents to exert photodynamic effects on tumor cells. Liposomes offer many advantages over these systems as delivery vehicles because of the modularity of their design that offers a large aqueous compartment for cargo entrapment and a membrane shell that provides an adjustable surface-to-volume ratio as well as a site for hosting targeting ligands, fusogenic species, or other functional components. Thompson and coworkers have studied plasmenylcholines [39] and PEG–vinyl ether–lipids [25,28–31] for their potential as gene and drug delivery vehicles that are activated under low pH conditions and hydrolyzed via vinyl ether protonation reaction leading to cargo release from the carrier system [40]. It is well known that various onium salts are capable of producing Brønsted acids upon photo-induced intramolecular electron transfer bond cleavage [41–43]. One class of onium salts that exhibits this property is an anthracene-based photoacid generator (An-PAG), which liberates protons on exposure to UV illumination. Incorporation of lipid-modified An-PAG into liposomal bilayers, therefore, was anticipated to provide a means to stimulate contents release from these vehicles via acid-catalyzed vinyl ether hydrolysis in a light intensity-dependent manner. Incorporation of photosensitizers that can efficiently generate singlet oxygen within the vicinity of vinyl ether lipid membranes is another method of stimulating liposome release on aerobic irradiation.

This work describes the synthesis of a hexadecyl-modified An-PAG for use as a light-activated trigger, which will induce contents release from DOPE:lysoplasmenylcholine liposomes. We also describe the liposomes containing photosensitizers such as bacteriochlorophyll a (Bchl a) or aluminum phthalocyanine tetrasulfonate ($AlPcS_4^{4-}$). Finally, biodistribution of $AlPcS_4^{4-}$-loaded liposomes in a murine tumor xenograft model is described.

4.2　EXPERIMENTAL PROCEDURES

4.2.1　Materials

DOPE and lysoplasmenylcholine (LysoPlsC) were obtained from Avanti Polar Lipids; all chemicals were obtained from Aldrich Chemical Company. Synthetic intermediates and final products were characterized by 300 MHz ^{1}H NMR and MALDI MS using the Purdue University Center for Cancer Research's Interdepartmental NMR Facility and Campus-Wide Mass Spectrometry Center, respectively.

4.2.2　Synthesis of 1-Bromo-3-hexadecylthiobenzene (1)

To a solution of 1-bromo-3-thiophenol (2 ml, 19.4 mmol) in 20 ml of dry THF at 20°C was added NaH (20.0 mmol) with stirring under an Ar atmosphere. After gas evolution ceased ($\sim$10 min), 1-bromohexadecane (6.1 ml, 20 mmol) was added, along with a trace of Bu_4NI as phase transfer catalyst, to the reaction mixture and heated at reflux for 20 h. The reaction mixture was then concentrated by rotary evaporation, the unreacted NaH destroyed by addition of H_2O (5 ml), and the product extracted with CH_2Cl_2 (3 × 20 ml). The combined CH_2Cl_2 extracts were dried over $MgSO_4$ before filtration, evaporation, and purification by silica gel column chromatography with 95:5 hexane:EtOAc as eluent.

4.2.3　Synthesis of 9-[3-Hexadecylthiophenyl] anthracene (2)

Compound (1) (1 g, 2.42 mmol) was dissolved in 20 ml dry THF and Mg turnings (57.6 mg, 2.5 mmol) added along with a small I_2 crystal. The mixture was stirred under an Ar atmosphere for 10 min and then heated at reflux until all of the Mg had dissolved (about 1 h). Anthrone (580 mg, 3 mmol) dissolved in 10 ml dry benzene was then added all at once and the mixture stirred at 20°C for 18 h. The reaction was quenched by addition of 30% H_2SO_4 in water (10 ml) and the mixture extracted with $CHCl_3$ (3 × 20 ml). The extracts were combined, dried over $MgSO_4$, filtered, and the filtrate concentrated by rotary evaporation.

4.2.4　Synthesis of 9-[3-Hexadecyl-3'-cyanobenzylsulfonium] phenylanthracene Trifluoromethanesulfonate (C16-An-PAG) (3)

Compound (2) was dissolved in CH_2Cl_2 with 3-cyanobenzyl bromide. $Ag(OTf)_2$-dioxane complex was added to the reaction mixture as a solid and the reaction stirred for 18 h at 20°C before filtration and evaporation of the solvent to give a crude residue that was purified by silica gel column chromatography using hexane:Et_2O as eluent to give C_{16}-An-PAG.

4.2.5   Liposome Preparation

Stock solutions of DOPE (20 mg/ml), LysoPlsC (20 mg/ml), and C_{16}-An-PAG (4 mg/ml) were prepared in 95:5 benzene:MeOH and combined to give 10 mmol total of a 35:35:30 molar ratio of components. The lipid stock solutions were combined at the desired ratios and vortexed for 2 min before flash freezing in LN_2 for 3 min and lyophilization (15 h) to produce a powder. The lyophilized lipid mixtures were then hydrated in 1 ml of 100 mM $CaCl_2$ solution via 10 freeze–thaw–vortex cycles. The lipid suspension then was extruded 10 times through 100 nm poly-carbonate membranes filters at 45°C. The liposome solution was passed through a Sephadex G-50 column equilibrated with a 150 mM NaCl solution to remove unentrapped Ca^{2+} and the liposomes used immediately. Remote loading of doxoru-bicin was performed using the ammonium sulfate gradient method as described previously [37]. $AlPcS_4^{4-}$ was loaded using the passive entrapment method using DSPE–PEG2000–folate as targeting ligand as previously described [16].

4.2.6   Dynamic Light Scattering

Liposome sizes were measured by dynamic light scattering using a Coulter N4-Plus instrument and the distributions calculated using the manufacturer's supplied software. The experiments were run in triplicate and the average diameter reported.

4.2.7   Photoexcitation Method

Samples of the liposome solution (3 ml) were placed in a 1 cm path length quartz cuvette and placed in a temperature-controlled cuvette holder with stirring. A frequency-doubled continuous wave, mode-locked Nd:YAG laser provided 10 mW/s at 368 nm with a maximum energy of 1 mJ/pulse was used to activate the C_{16}-An-PAG-containing liposomes. Lenses were used to focus the beam to a typical spot size of 6 mm diameter. Bchl a- and $AlPcS_4^{4-}$ -containing liposomes were irradiated with an Al/Ga/As diode array laser emitting at 800 nm and coupled into a 1 mm optical fiber that was focused onto a temperature-controlled cuvette containing the stirred solution of liposomes.

4.2.8   Calcium Ion Release Assay

Absorption spectra were measured with a Hewlett-Packard 8453 UV–Vis spectro-photometer equipped with Chem Station software. Calcium-loaded liposomes were mixed with 2.85 ml Arsenazo III in 20 mM HEPES, pH 7.4. The absorption at 656 nm was measured and then Triton X-100 added to disrupt the liposomes and the absorption at 656 nm re-measured. The percent release was quantified using the ratio method: % release $= 100 \times [(T - T_0)/(1 - T_0)]$, where $T =$ (fluorescence before Triton addition/fluorescence after Triton addition) at time T; $T_0 =$ fluorescence ratio at $T = 0$ min [44].

4.2.9 Negative Stain Transmission Electron Microscopy

A small drop of liposome solution was placed on a carbon-coated G-300 mesh copper grid and the excess liquid blotted away using filter paper after 1 min. A 1% solution of UO_2OAc_2 was then added for 1 min and the excess again removed using the filter paper. The sample was dried in a desiccator overnight before imaging on a Philips EM-400 transmission electron microscope operating at 80 kV accelerating voltage.

4.2.10 Animal Studies

Balb/c mice on folate-deficient diets were implanted with 5×10^5 M109 lung tumor cells via subcutaneous injection in a 100 µl volume with folate-deficient DMEM. The implants were allowed to grow for 3 week before imaging using a custom-built long-focal length microscope and camera system.

4.3 RESULTS AND DISCUSSION

4.3.1 DOPE:LysoPlsC Liposomes Using C_{16}-An-PAG as a Photosensitizer

4.3.1.1 Synthesis of C_{16}-An-PAG. Indirect photoinitiation of liposomal content release was sought by using a membrane-anchored photoacid generator. We elected to prepare a hexadecyl-modified version of the anthracene-based photoacid generator [41–43] that transfers an excited π^* electron from the anthracene moiety to the σ^* lowest unoccupied molecular orbital localized on the phenylsulfonium moiety on irradiation at 368 nm, leading to bond rearrangement and production of a Brønsted acid. Previous work also suggests that the cyanobenzyl-substituted sulfonium salts are cleaved by a concerted electron transfer-bond fragmentation process (Scheme 4.1). The rate of C—S bond cleavage is exceedingly fast and dependent on several factors, including the thermodynamic stability of the cation radical–radical pair. Compound (**3**) was prepared as shown in Scheme 4.2 using a modified procedure reported for the methyl analog [41].

4.3.1.2 Solution Acidification on Photolysis of C_{16}-An-PAG. C_{16}-An-PAG was dissolved in dry acetonitrile at a concentration of 1.5 mM and the absorption spectrum recorded, revealing peaks at 334, 350, 369, and 388 nm (Figure 4.1). Irradiation of this solution at 368 nm (35 mW/cm^2 illumination), with periodic sampling and analysis on 1:10 dilution into water, produced a rapid acidification over 2.5 pH units within 2 min at 37°C (Figure 4.2). The rapid acidification produced by this photoacid generator suggested that it could be useful for initiating acid-triggered release

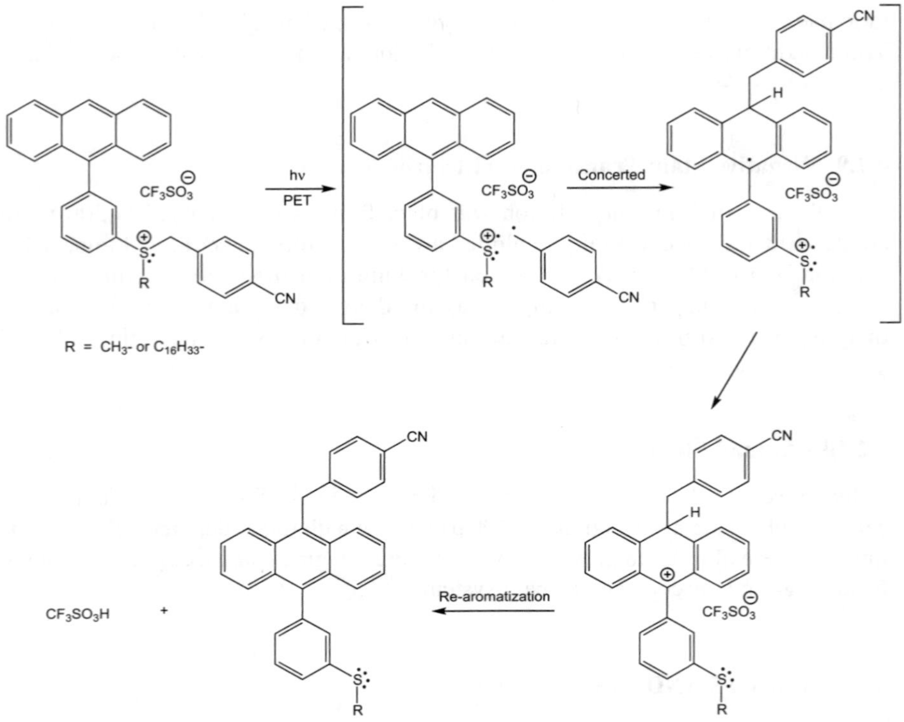

SCHEME 4.1 Photoinduced electron transfer process leading to the formation of triflic acid.

from appropriately designed pH-sensitive liposomes. Prior work by Thompson and coworkers [36,37] suggested that plasmenyl-type lipids could be used to effect rapid contents release from liposomes on acidification. This potential was tested using blends of LysoPlsC and DOPE.

4.3.1.3 Light-Induced Liposome Leakage and Phase Change Using C16-An-PAG.

The performance of 100 nm 35:35:30 DOPE:LysoPlsC:C_{16}-An-PAG liposomes was evaluated using a Ca^{2+}-Arsenazo III assay [45]. This formulation was selected because the H_{II} phase of DOPE can be transformed into a stable lamellar phase by blending it with LysoPlsC, a pH-sensitive vinyl ether lipid of opposite curvature propensity that ultimately produces fatty acid degradation products upon acid-catalyzed hydrolysis in aerobic solution (Scheme 4.3). Cleavage of the LysoPlsC fraction within the liposome membrane causes a change in the intrinsic curvature of the fatty acid fragments, leading to a loss of L_α phase stability and a collapse of the liposome into the preferred H_{II} phase of DOPE with concomitant release of the liposomal contents. Irradiation of the extruded, unbuffered liposome solution at 368 nm lead to 75% Ca^{2+} release within 4 h, with accompanying change in the lipid morphology from spherical L_α phase liposomes to collapsed H_{II} phase aggregates (Figure 4.3). Nonirradiated and irradiated buffered controls produced

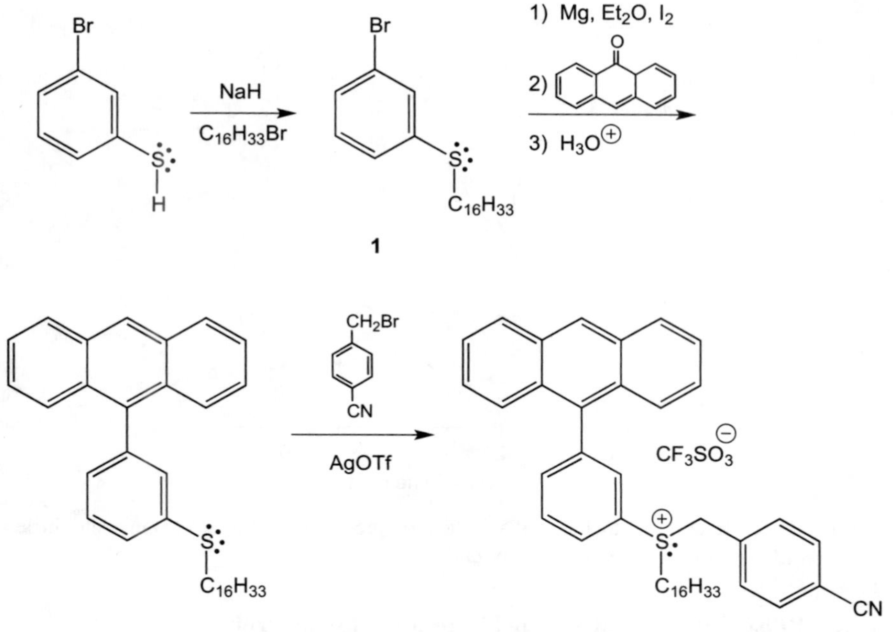

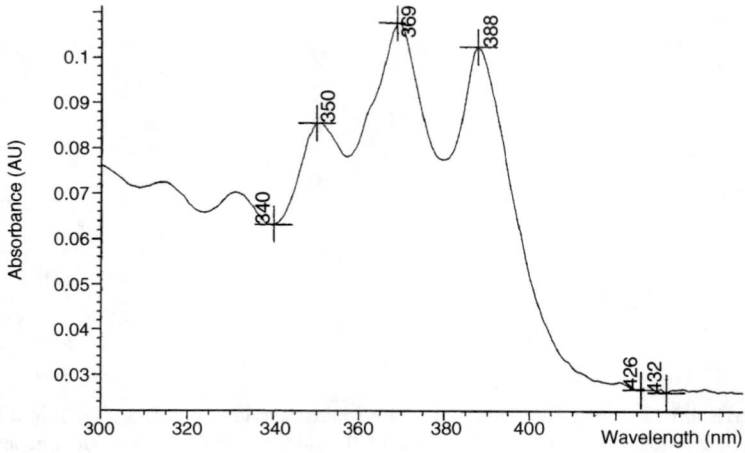

SCHEME 4.2 Synthesis pathway for C_{16}-An-PAG.

13% and 20% release, respectively, under the same illumination conditions. While the system containing C_{16}-An-PAG shows light-activated content release in an unbuffered condition, a faster release kinetics was desired. Therefore, we investigated another light-sensitive liposome systems.

FIGURE 4.1 Absorption spectrum of C_{16}-An-PAG (**3**) in acetonitrile at 20°C.

FIGURE 4.2   Rate of photochemical acidification produced by 1.5 mM (**3**) in acetonitrile at 37°C on illumination at 368 nm, 35 mW/cm^2.

4.3.2   PPlsC Liposomes Using Bchl a as a Photosensitizer

4.3.2.1   Light-Induced Liposome Leakage and Phase Change Using Photosensitized Production of Singlet Oxygen (${}^1O_2$).   Prior work has shown that rapid calcein release occurs on photosensitized oxidation of palmitoylplasmenylcholine (PPlsC) [36,37] and dipalmitoylplasmenylcholine (DPPlsC) [46,47] liposomes at neutral pH using a variety of different chromophores. We evaluated the potential of PPlsC liposomes for drug delivery using Bchl a (Figure 4.4a) as a membrane-soluble

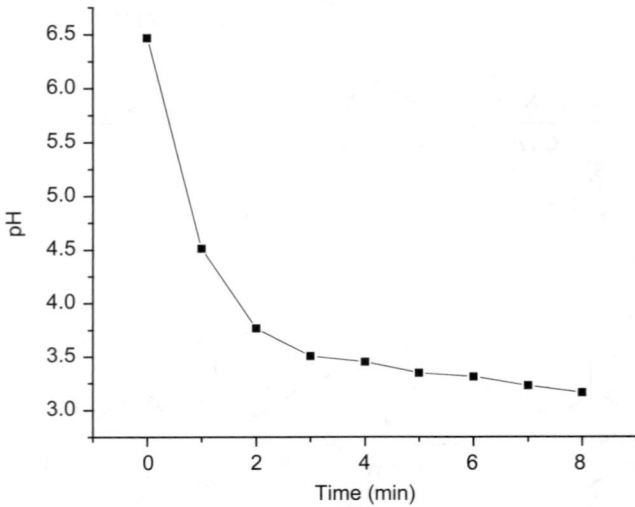

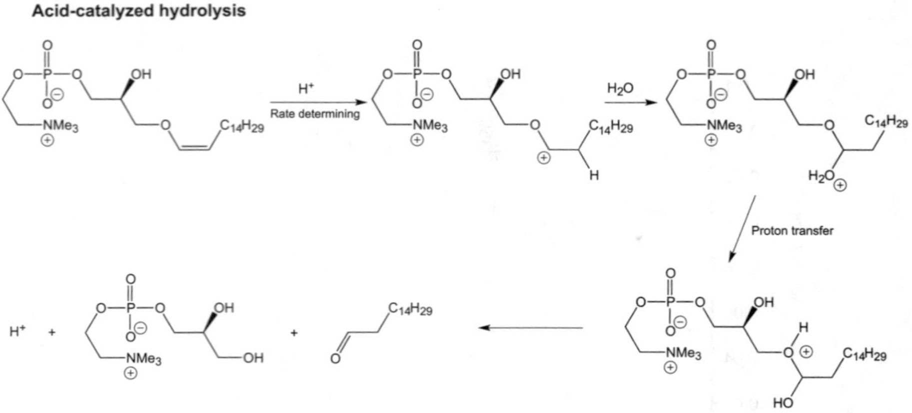

SCHEME 4.3   Pathway for acid-catalyzed lysoplasmenylcholine (LysoPlsC) lipid hydrolysis. Aerobic oxidation of the initially formed fatty aldehyde hydrolysis product generates the corresponding fatty acid.

(a)

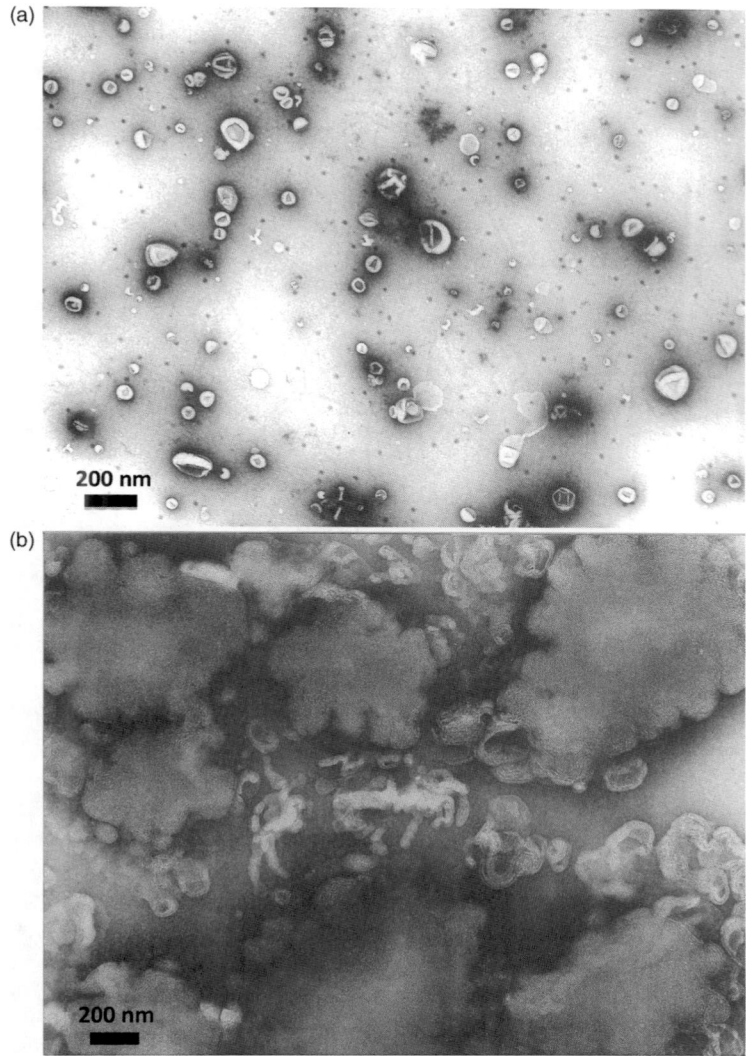

FIGURE 4.3 Negative-stain TEM of 35:35:30 DOPE:LysoPlsC:C_{16}-An-PAG liposomes before (a) and 4 h after (b) irradiation at 368 nm.

photosensitizer. Since Bchl a is a near-infrared sensitizer that efficiently generates $^1O_2$ on aerobic irradiation, it was envisioned that rapid drug release should be possible on irradiation of drug-loaded PPlsC liposomes doped with Bchl a.

Liposome samples were prepared by evaporating a $CHCl_3$ solution of PPlsC and Bchl a into a thin film before hydrating the sample in ammonium citrate buffer and extruding it through 100 nm track-etch membranes. After buffer exchange to HEPES buffered saline via gel chromatography, doxorubicin was added and the solution incubated at 50°C for 10 min. Analysis of both the lipid and doxorubicin content indicated that a drug:lipid ratio of 0.085 was achieved, or approximately

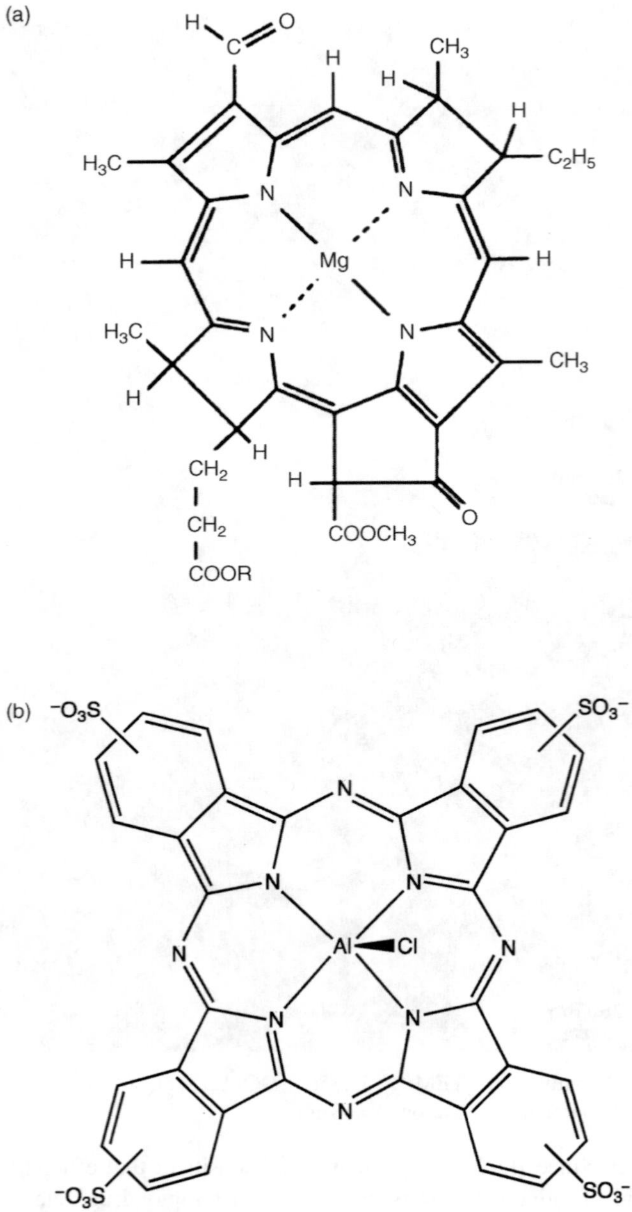

(a)

(b)

FIGURE 4.4 Photosensitizers used in this study. Bacteriochlorophyll a; (Bchl a, a) and chloroaluminum phthalocyaninetetrasulfonate ($AlPcS_4^{4-}$, b).

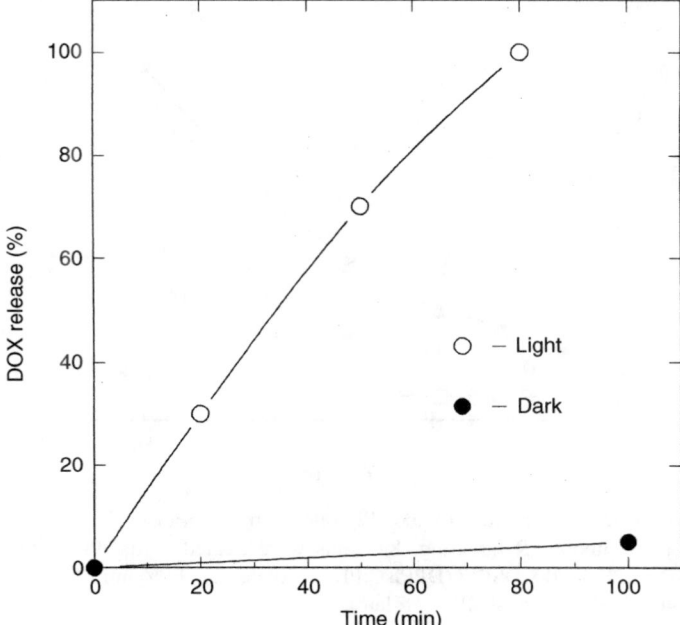

FIGURE 4.5 Rate of doxorubicin release from remote-loaded 2-palmitoylplasmenylcholine (PPlsC) liposomes sensitized with Bchl a. [PPlsC] = 1.12 mM; [dihydrocholesterol] = 0.28 mM; [DOX] = 120 μM; [Bchl a] = 34 μM; 500 mW illumination at 800 nm, 38°C, 20 mM HEPES, 150 mM NaCl, pH 7.4.

2000 doxorubicin molecules per liposome. Illumination of this sample at 38°C with an Al/Ga/As diode laser emitting 800 nm excitation at 500 mW intensity led to complete doxorubicin release within 80 min (Figure 4.5). Since Bchl a is known to photobleach rapidly within the first 10 min of irradiation [37], the observed rate of drug release is actually a minimum drug efflux rate under these conditions since the extent of PPlsC photooxidation (estimated to be less than ~25%) is lower than desired to effect rapid release. The photoreleased doxorubicin remained cytotoxic within experimental error of the measurement [37].

4.3.3 Folate Receptor (FR)-Targeted DPPlsC Liposomes Using AlPcS$_4^{4-}$ as a Photosensitizer

4.3.3.1 Rate of Liposome Internalization and Cargo Leakage Using the Folic Acid-Based Receptor-Mediated Endocytosis Pathway. Low and coworkers have demonstrated the utility of folate conjugates for targeting folate receptor (FR) positive tumor cells *in vitro* and *in vivo* [48]. Folate-modified liposomes have also been used to deliver cytosine arabinose to the cytoplasm of target cells [13]. In order to assess the kinetic timescale of this process, we loaded FR-targeted DPPlsC liposomes with propidium iodide (PI) as a probe that produces little fluorescence

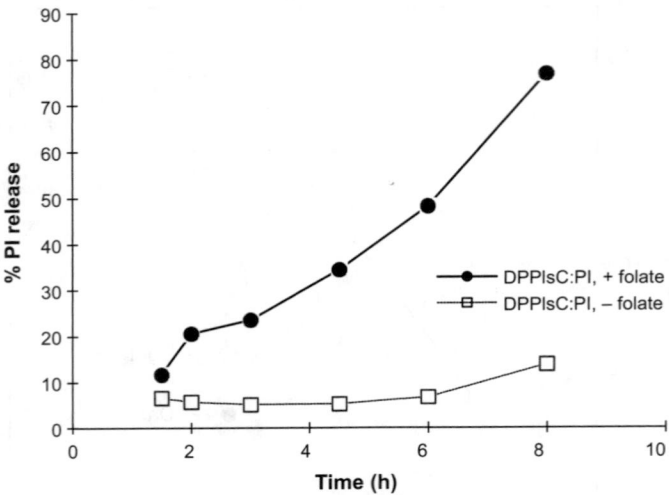

FIGURE 4.6 Rate of propidium iodide (PI) release from DPPlsC liposomes in KB cell culture with and without FR targeting. KB cells were treated with DPPlsC:PI liposomes containing 5% mPEG2000–DSPE (DPPlsC:PI, − folate) or 4.9% mPEG2000 and 0.1% folate–PEG3400–DSPE (DPPlsC:PI, + folate).

until it intercalates into DNA. Once inside the acidic endosomal pathway, we anticipated that the vinyl ether lipid would be hydrolyzed in a manner similar to Scheme 4.3 and the corresponding fatty aldehyde/fatty acid fragments would facilitate PI escape from the endosome and intercalate into cellular DNA. As the data in Figure 4.6 show, nearly 80% of the encapsulated PI is gradually released from FR-targeted liposomes over a 2–8 h timescale, whereas ≤14% PI staining of KB cells over the same time period when nontargeted DPPlsC:PI liposomes were used. These results show that the FR-targeted liposomes can deliver a cargo to FR-positive cells using receptor-mediated internalization.

4.3.3.2 Pharmacokineticss and Biodistribution of FR-Targeted DPPlsC Liposomes Bearing AlPcS$_4^{4-}$ as a Photosensitizer Cargo.

We sought to combine other water-soluble cargo, such as chloroaluminum phthalocyaninetetrasulfonate (AlPcS$_4^{4-}$) (Figure 4.4b), in the FR-targeted liposomes for labeling the cytoplasm of target cells for photodynamic therapy applications. We investigated the potential of this approach by loading FR-targeted DPPlsC liposomes with AlPcS$_4^{4-}$ at self-quenching concentrations so that the internalized sensitizer would be retained in a nonphotoactive state until reaching the target cell and escaping from the endosomes into the surrounding cytoplasm [16]. This photosensitizer was chosen because it is (i) much more photostable than Bchl a; (ii) rapidly cleared from blood in its free form (i.e., will be rapidly excreted if lost from the liposomal carrier during systemic circulation, thereby limiting the dermatologic sensitivity that is commonly observed for hematoporphyrin derivative formulations); (iii) capable of triggering efflux from DPPlsC liposomes as shown in Scheme 4.4; and (iv) highly water soluble, thus

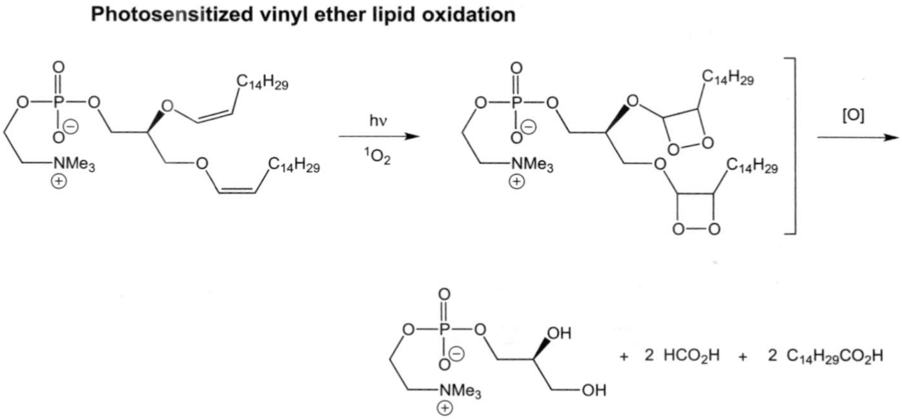

Photosensitized vinyl ether lipid oxidation

SCHEME 4.4 Pathway for photooxidative cleavage of the divinyl ether lipid, diplasmenylcholine (DPPlsC). Stepwise addition of $^1O_2$ to the electron rich vinyl ether bond leads to the thermally unstable, putative bis-dioxetane intermediate shown in brackets. Initial fragmentation of this species produces 2 equiv. of glycerophosphocholine bisformate ester that are rapidly hydrolyzed to formate and glycerophosphocholine. Aerobic oxidation of the initially formed fatty aldehyde photolysis product generates the corresponding fatty acid.

potentially limiting its volume of distribution *in vivo* to only those cells that have internalized the $AlPcS_4^{4-}$-loaded liposomes, such as FR-positive tumor cell targets and nontarget cells of the reticuloendothelial system that commonly scavenge nanoparticulate carriers (i.e., macrophages within the liver and spleen).

As the data in Figure 4.7 show, substantial $AlPcS_4^{4-}$ fluorescence appears rapidly in the 1–2 mm subcutaneous M109 FR-positive xenograft tumors within 2 h compared to nontarget surrounding skin tissue. Approximately 35% of this tumor-localized $AlPcS_4^{4-}$ fluorescence is lost within 5 h, with slow depletion of an additional 12% occurring over the next 67 h, to give a total retained dose of ~53% 72 h after injection (<0.3% of skin-localized dose was detectable in healthy skin at this time point). Since the $AlPcS_4^{4-}$ cargo was encapsulated at self-quenching concentrations, we cannot entirely exclude the possibility that normal skin was capable of harboring liposomes but not stimulating $AlPcS_4^{4-}$ release as in FR-targeted KB cells [16]. However, in the absence of specific adhesion mechanisms that would promote liposome accumulation in vital skin, we infer from these findings that the low observed $AlPcS_4^{4-}$ fluorescence is due to low levels of skin-accumulated $AlPcS_4^{4-}$. This is an important advantage since one of the most common side effects of the hydrophobic photosensitizers typically used in photodynamic therapy is cutaneous photosensitivity. The low levels of skin-accumulated $AlPcS_4^{4-}$ fluorescence suggest that this issue may not be as problematic as other hydrophobic photosensitizers that display a long persistence in skin.

The biodistribution of $AlPcS_4^{4-}$ in the mouse model was evaluated 2 h after infusion of FR-targeted DPPlsC/$AlPcS_4^{4-}$ liposomes via tail vein injection (Figure 4.8). Liver, tumor, spleen, and kidney were found to have the highest $AlPcS_4^{4-}$

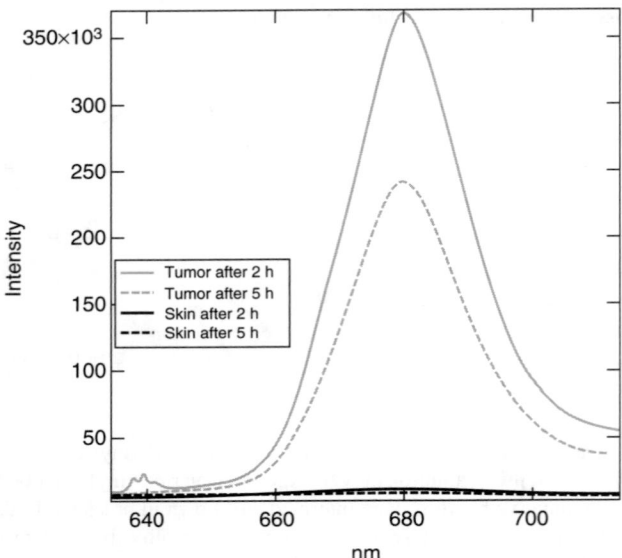

FIGURE 4.7 Fluorescence kinetics of FR-targeted DPPlsC liposomes loaded with $AlPcS_4^{4-}$ in a 1–2 mm subcutaneous folate-receptor positive M109 FR-positive tumor murine xenograft relative to nontarget skin tissue. DPPlsC:Chol:mPEG2000–DSPE:folate–PEG3400–DSPE liposomes (59.9:35:5:0.1) containing 9.8 μM $AlPcS_4^{4-}$ were injected via tail vein and imaged using a home-built fluorescence imaging microscope at the times indicated.

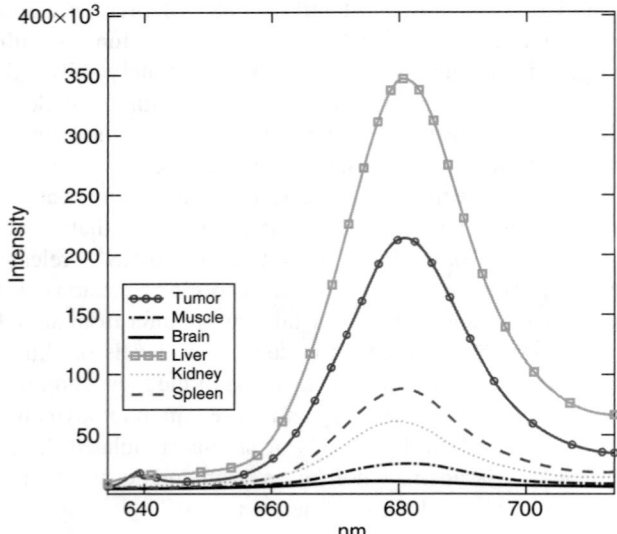

FIGURE 4.8 Biodistribution of $AlPcS_4^{4-}$ fluorescence 48 h after infusion of FR-targeted DPPlsC/$AlPcS_4^{4-}$ liposomes as described in Figure 4.7.

fluorescence levels, with liver and tumor displaying 2–3 times higher accumulation than spleen and kidney over this time period. Extensive uptake by liver and spleen is commonly observed for nanoparticles [8], as is kidney uptake due to scavenger receptors for folic acid in these tissues [49]. It is noteworthy that the levels present in tumor were comparatively high, however, additional experiments are needed to determine their specific localization (i.e., deeply penetrated into tumor tissue or near the extravasation site) and whether they have been internalized by FR-positive tumor cells or are simply adherent to the surface of the cells.

4.4   CONCLUSIONS

Several different photochemical schemes were developed and employed using a variety of vinyl ether-based liposome systems. Photoacid generation using (3), a membrane-immobilized, anthracene-based sulfonium salt, induced cargo release on the hours timescale even though the rate of H^+ release occurs on the minutes timescale. This suggests that, in spite of membrane localization of the C_{16}-An-PAG derivative, the protons are released into bulk solution rather than remain associated with the membrane where vinyl ether hydrolysis must occur. Photooxidative drug release was also demonstrated on the minutes timescale using the near infrared photosensitizer, Bchl a. FR targeting of liposomal PI and $AlPcS_4^{4-}$ to FR-positive KB cells was also demonstrated in cell culture and *in vivo*, with the latter displaying promising characteristics with regard to tumor:nontarget tissue distributions. Taken together, these results suggest that it is possible to design light-responsive liposomes to provide a level of functional control and reaction kinetics that are difficult to achieve with other triggering modalities.

4.5   PROSPECTS FOR LIGHT-ACTIVATED LIPOSOMAL DRUG DELIVERY

Our findings suggest that opportunities remain for the development of effective phototherapeutic strategies using specific targeting ligands that can selectively deliver cytotoxic agents to tumor cells *in vivo*. Major challenges remain, however, in the development of (i) chromophores that can be activated in the near infrared to elicit their action in the absence of oxygen, since most solid tumors have regions that are substantially hypoxic; (ii) strategies that display large bystander effects due to tumor heterogeneity; and (iii) techniques that enable the real-time assessment of therapeutic progress so that the treatment can be modulated as required for individual patient outcome.

Acknowledgments

We gratefully acknowledge financial support by National Institutes of Health (GM087016) and the 21st Century Research and Technology Fund of Indiana. We would like to thank the Purdue University Center for Cancer Research

Interdepartmental NMR Facilities and Campus-Wide Mass Spectrometry Center (supported by NCI CCSG CA23168 to Purdue University Center for Cancer Research) for their materials characterization support. We also thank Mike Kennedy, Karim Jallad, Dor Ben-Amotz, and Phil Low for fruitful discussions.

REFERENCES

1. Mellman, I., Fuchs, R., Helenius, A. (1986). Acidification of the endocytic and exocytic pathways. *Annual Review of Biochemistry 55*, 663–700.
2. Gerweck, L. E. (1998). Tumor pH: implications for treatment and novel drug design. *Seminars in Radiation Oncology 8*, 176–182.
3. Maeda, H. (2010). Tumor-selective delivery of macromolecular drugs via the EPR effect: background and future prospects. *Bioconjugate Chemistry 21*, 797–802.
4. Jain, R. K. (2001). Delivery of molecular and cellular medicine to solid tumors. *Advanced Drug Delivery Reviews 46*, 149–168.
5. Drummond, D. C., Noble, C. O., Hayes, M. E., Park, J. W., Kirpotin, D. B. (2008). Pharmacokinetics and in vivo drug release rates in liposomal nanocarrier development. *Journal of Pharmaceutical Sciences 97*, 4696–4740.
6. Guo, L. S. S. (2001). Amphotericin B colloidal dispersion: an improved antifungal therapy. *Advanced Drug Delivery Reviews 47*, 149–163.
7. Cruess, A. F., Zlateva, G., Pleil, A. M., Wirostko, B. (2009). Photodynamic therapy with verteporfin in age-related macular degeneration: a systematic review of efficacy, safety, treatment modifications and pharmacoeconomic properties. *Acta Opthalmologica 87*, 118–132.
8. Duncan, R., Gaspar, R. (2011). Nanomedicine(s) under the microscope. *Molecular Pharmaceutics 8*, 2101–2141.
9. Allen, T. M., Cullis, P. R. (2004). Drug delivery systems: entering the mainstream. *Science 303*, 1818–1822.
10. Kirpotin, D. B., Drummond, D. C., Shao, Y., Shalaby, M. R., Hong, K. L., Nielsen, U. B., Marks, J. D., Benz, C. C., Park, J. W. (2006). Antibody targeting of long-circulating lipidic nanoparticles does not increase tumor localization but does increase internalization in animal models. *Cancer Research 66*, 6732–6740.
11. Wong, C., Stylianopoulos, T., Cui, J., Martin, J., Chauhan, V. P., Jiang, W., Popovic, Z., Jain, R. K., Bawendi, M. G., Fukumura, D. (2011). Multistage nanoparticle delivery system for deep penetration into tumor tissue. *Proceedings of the National Academy of Sciences of the United States of America 108*, 2426–2431.
12. Mayor, S., Pagano, R. E. (2007). Pathways of clathrin-independent endocytosis. *Nature Reviews Molecular Cell Biology 8*, 603–612.
13. Rui, Y., Wang, S., Low, P. S., Thompson, D. H. (1998). Diplasmenylcholine-folate liposomes: an efficient vehicle for intracellular drug delivery. *Journal of the American Chemical Society 120*, 11213–11218.
14. Drummond, D. C., Noble, C. O., Guo, Z., Hayes, M. E., Connolly-Ingram, C., Gabriel, B. S., Hann, B., Liu, B., Park, J. W., Hong, K. L., Benz, C. C., Marks, J. D., Kirpotin, D. B. (2010). Development of a highly stable and targetable nanoliposomal formulation of topotecan. *Journal of Controlled Release 141*, 13–21.

15. Biswas, S., Dodwadkar, N. S., Sawant, R. R., Torchilin, V. P. (2011). Development of the novel PEG–PE-based polymer for reversible attachment of specific ligands to liposomes: synthesis and in vitro characterization. *Bioconjugate Chemistry 22*, 2005–2013.

16. Qualls, M. M., Thompson, D. H. (2001). Chloroaluminum phthalocyanine tetrasulfonate delivered via acid-labile diplasmenylcholine-folate liposomes: intracellular localization and synergistic phototoxicity. *International Journal of Cancer 93*, 384–392.

17. Coon, B. G., Crist, S., Gonzalez-Bonet, A., Kim, H.-K., Sowa, J., Thompson, D. H., Ratliff, T. L., Aguilar, R. C. (2011). Fibronectin attachment protein (FAP) from Bacillus Calmette-Guerin as a targeting agent for bladder tumor cells. *International Journal of Cancer 131*, 591–600.

18. Ellens, H., Bentz, J., Szoka, F. C. (1986). Fusion of phosphatidylethanolamine-containing liposomes and mechanism of the La-HII phase transition. *Biochemistry 25*, 4141–4147.

19. Bally, M. B., Tilcock, C. P. S., Hope, M. J., Cullis, P. R. (1983). Polymorphism of phosphatidylethanolamine phosphatidylserine model systems: influence of cholesterol and Mg^{2+} on Ca^{2+}-triggered bilayer to hexagonal (HII) transitions. *Canadian Journal of Biochemistry & Cell Biology 61*, 346–352.

20. Ellens, H., Bentz, J., Szoka, F. C. (1984). pH-Induced destabilization of phosphatidyle-thanolamine-containing liposomes: role of bilayer contact. *Biochemistry 23*, 1532–1538.

21. Hafez, I. M., Ansell, S., Cullis, P. R. (2000). Tunable pH-sensitive liposomes composed of mixtures of cationic and anionic lipids. *Biophysical Journal 79*, 1438–1446.

22. Slepushkin, V. A., Simoes, S., deLima, M. C. P., Duzgunes, N. (2004). Sterically stabilized pH-sensitive liposomes. *Methods in Enzymology 387*, 134–147.

23. Kirpotin, D., Hong, K. L., Mullah, N., Papahadjopoulos, D., Zalipsky, S. (1996). Liposomes with detachable polymer coating: destabilization and fusion of dioleoylphos-phatidylethanolamine vesicles triggered by cleavage of surface-grafted poly(ethylene glycol). *FEBS Letters 388*, 115–118.

24. Zalipsky, S., Qazen, M., Walker, J. A., Mullah, N., Quinn, Y. P., Huang, S. K. (1999). New detachable poly(ethylene glycol) conjugates: cysteine-cleavable lipopolymers regenerat-ing natural phospholipid, diacylphosphatidylethanolamine. *Bioconjugate Chemistry 10*, 703–707.

25. Boomer, J. A., Inerowicz, H. D., Zhang, Z.-Y., Bergstrand, N., Edwards, K., Kim, J.-M., Thompson, D. H. (2003). Acid-triggered release from sterically-stabilized fusogenic vesicles via a hydrolytic dePEGylation strategy. *Langmuir 19*, 6408–6415.

26. Shin, J., Shum, P., Thompson, D. H. (2003). Acid-triggered release via dePEGylation of DOPE liposomes containing acid-labile vinyl ether PEG-lipids. *Journal of Controlled Release 91*, 187–200.

27. Sakaguchi, N., Kojima, C., Harada, A., Kono, K. (2008). Preparation of pH-sensitive poly (glycidol) derivatives with varying hydrophobicities: their ability to sensitize stable liposomes to pH. *Bioconjugate Chemistry 19*, 1040–1048.

28. Bergstrand, N., Arfvidsson, M. C., Kim, J.-M., Thompson, D. H., Edwards, K. (2003). Interactions between pH-sensitive liposomes and model membranes. *Biophysical Chem-istry 104*, 361–379.

29. Boomer, J. A., Qualls, M. M., Inerowicz, H. D., Haynes, R. H., Patri, G. V. S., Kim, J.-M., Thompson, D. H. (2009). Cytoplasmic delivery of liposomal contents mediated by an acid-labile cholesterol–vinyl ether–PEG conjugate. *Bioconjugate Chemistry 20*, 47–59.

30. Shin, J., Shum, P., Grey, J., Fujiwara, S., Malhotra, G. S., González-Bonet, A. M., Moase, E., Allen, T. M., Thompson, D. H. (2012). Acid-labile mPEG-vinyl ether-1,2-dioleylglycerol lipids with tunable pH sensitivity: synthesis and structural effects on hydrolysis rates, DOPE liposome release performance and pharmacokinetics. *Molecular Pharmaceutics 9*, 3266–3276.

31. Kim, H.-K., Van den Bossche, J., Hyun, S.-H., Thompson, D. H. (2012). Acid-triggered release via dePEGylation of fusogenic liposomes mediated by heterobifunctional phenyl substituted vinyl ethers with tunable pH-sensitivity. *Bioconjugate Chemistry 23*, 2071–2077.

32. Nishiyama, N., Morimoto, Y., Wang, W. D., Kataoka, K. (2009). Design and development of dendrimer photosensitizer-incorporated polymeric micelles for enhanced photodynamic therapy. *Advanced Drug Delivery Reviews 61*, 327–338.

33. Kuruppuarachchi, M., Savoie, H., Lowry, A., Alonso, C., Boyle, R. W. (2011). Polyacrylamide nanoparticles as a delivery system in photodynamic therapy. *Molecular Pharmaceutics 8*, 920–931.

34. Huff, T. B., Tong, L., Zhao, Y., Hansen, M. N., Cheng, J.-X., Wei, A. (2007). Hyperthermic effects of gold nanorods on tumor cells. *Nanomedicine 2*, 125–132.

35. Loo, C., Lowery, A., Halas, N. J. J. W., Drezek, R. (2005). Immunotargeted nanoshells for integrated cancer imaging and therapy. *Nano Letters 5*, 709–711.

36. Thompson, D. H., Gerasimov, O. V., Wheeler, J. J., Anderson, V. C. (1996). Triggerable plasmalogen liposomes: improvement of system efficiency. *Biochimica et Biophysica Acta 1279*, 25–34.

37. Thompson, D. H., Rui, Y., Gerasimov, O. V, Triggered release from liposomes mediated by physically- and chemically-induced phase transitions, In *Vesicles: Surfactant Science Series*, ed. Rosoff, M., Marcel-Dekker, New York, NY, Vol. *62*, 1996, pp. 679–746.

38. Shum, P., Kim, J.-M., Thompson, D. H. (2001). Phototriggering of liposomal drug delivery systems. *Advanced Drug Delivery Reviews 53*, 273–284.

39. Gerasimov, O., Schwan, A., Thompson, D. H. (1997). Acid-catalyzed plasmenylcholine hydrolysis and its effect on bilayer permeability: a quantitative study. *Biochimica et Biophysica Acta 1324*, 200–214.

40. Keeffe, J. R., Kresge, A. J., In *The Chemistry of Enols*, ed., Rappoport, Z., Wiley, New York, NY, 1990, p. 437.

41. Saeva, F. D., Breslin, D. T., Luss, H. R. (1991). Intramolecular photoinduced rearrangements via electron transfer-induced, concerted bond cleavage and cation radical coupling. *Journal of the American Chemical Society 113*, 5333–5337.

42. Saeva, F. D., Martic, P. A., Garcia, E. (1993). Intramolecular photoinduced electron transfer (PET) bond cleavage in some sulfonium salt derivatives: effect of distance and thermodynamics on PET rate. *Journal of Physical Organic Chemistry 6*, 333–340.

43. Saeva, F. D., Garcia, E., Martic, P. A. (1995). Comparative photochemical behavior of some anthracenyl and naphthacenyl sulfonium salt derivatives. *Journal of Photochemistry and Photobiology A: Chemistry 86*, 149–154.

44. Allen, T. M., Hong, K., Papahadjopoulos, D. (1990). Membrane contact, fusion and hexagonal (HII) transitions in phosphatidylethanolamine liposomes. *Biochemistry 29*, 2976–2985.

45. Wymer, N., Gerasimov, O. V., Thompson, D. H. (1998) Cascade liposomal triggering: light-induced Ca^{2+} release from plasmenylcholine liposomes triggers PLA2-catalyzed

hydrolysis and contents leakage from DPPC liposomes. *Bioconjugate Chemistry 9*, 305–308.

46. Collier, J. H., Hu, B.-H., Ruberti, J. W., Zhang, Z.-Y., Shum, P., Thompson, D. H., Messersmith, P. B. (2001). Thermally and photochemically triggered self-assembly of peptide hydrogels. *Journal of the American Chemical Society 123*, 9463–9464.

47. Zhang, Z.-Y., Shum, P., Yates, M., Messersmith, P. B., Thompson, D. H. (2002). Formation of fibrinogen-based hydrogels using phototriggerable diplasmalogen liposomes. *Bioconjugate Chemistry 13*, 640–646.

48. Sega, E. I., Low, P. S. (2008). Tumor detection using folate receptor-targeted imaging agents. *Cancer and Metastasis Reviews 27*, 655–664.

49. Kennedy, M. D., Jallad, K., Lu, J., Low, P. S., Ben-Amotz, D. (2003). Evaluation of folate conjugate uptake and transport by the choroid plexus of mice. *Pharmaceutical Research 20*, 714–719.

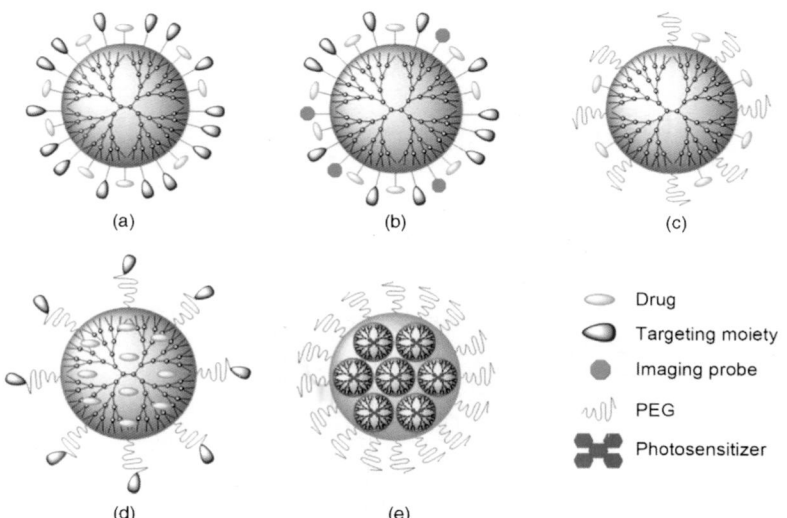

FIGURE 5.4 Recently studied dendrimer-based anticancer drug delivery systems. (*See text for full caption.*)

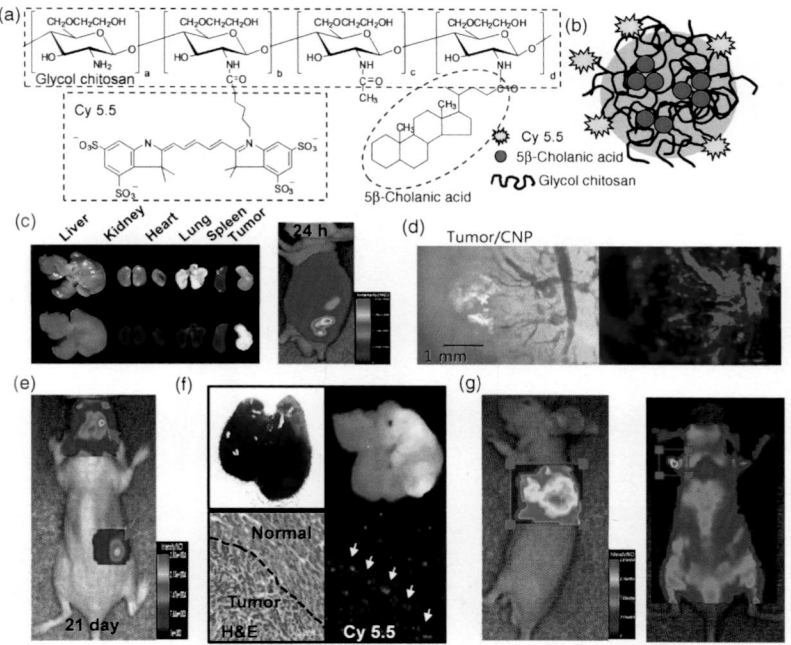

FIGURE 6.1 Chitosan nanoparticle for optical imaging. (*See text for full caption.*)

Nanoparticulate Drug Delivery Systems: Strategies, Technologies, and Applications, First Edition.
Edited by Yoon Yeo.

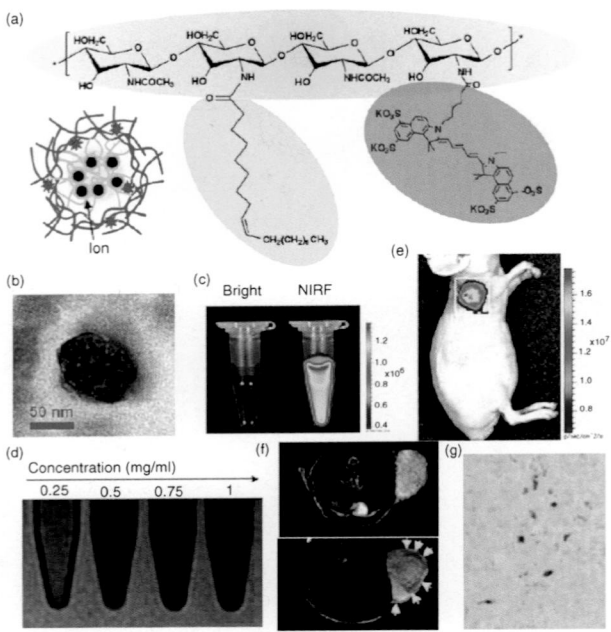

FIGURE 6.2 Chitosan nanoparticle for optical/MR imaging. (*See text for full caption.*)

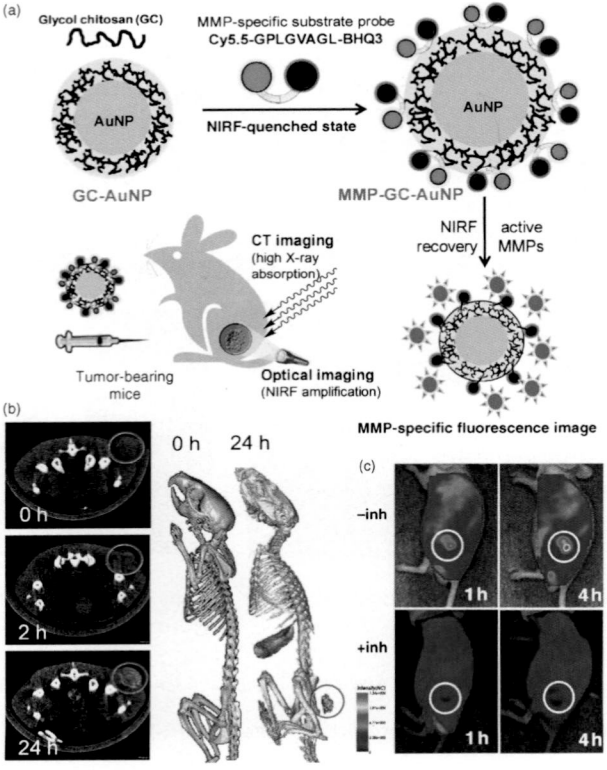

FIGURE 6.3 Chitosan nanoparticle for optical/CT imaging. (*See text for full caption.*)

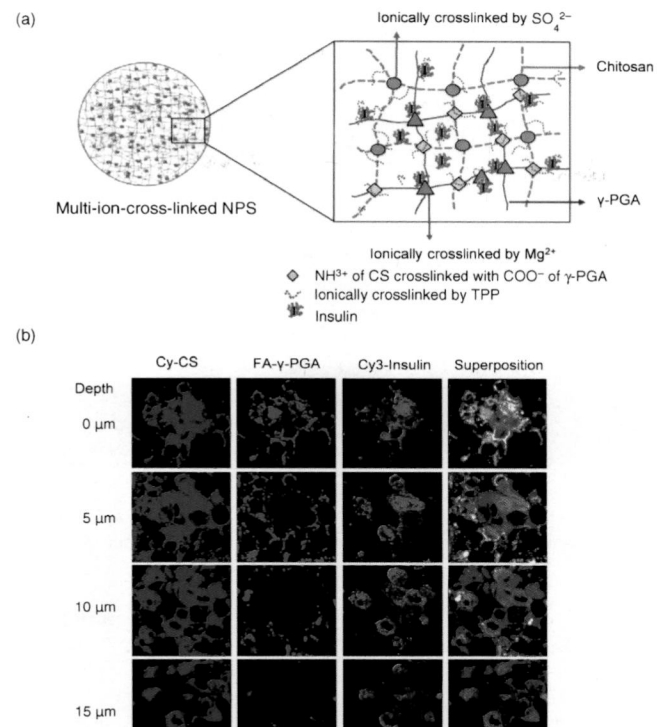

FIGURE 6.4 Chitosan nanoparticle for drug delivery. (*See text for full caption.*)

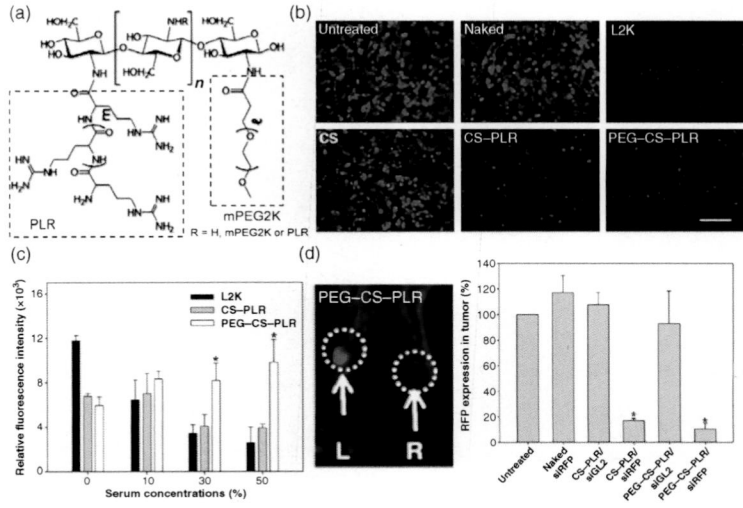

FIGURE 6.5 Chitosan nanoparticle for gene delivery. (*See text for full caption.*)

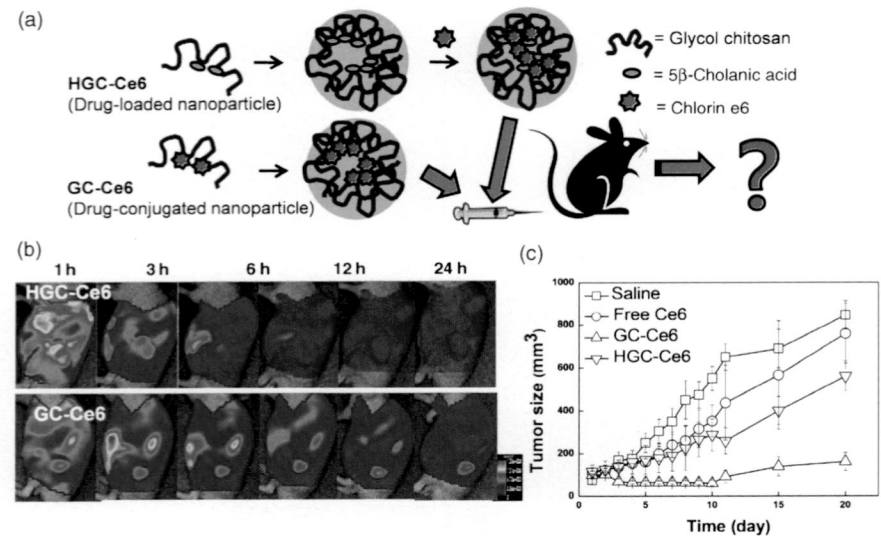

FIGURE 6.6 Chitosan nanoparticle for photodynamic therapy. (*See text for full caption.*)

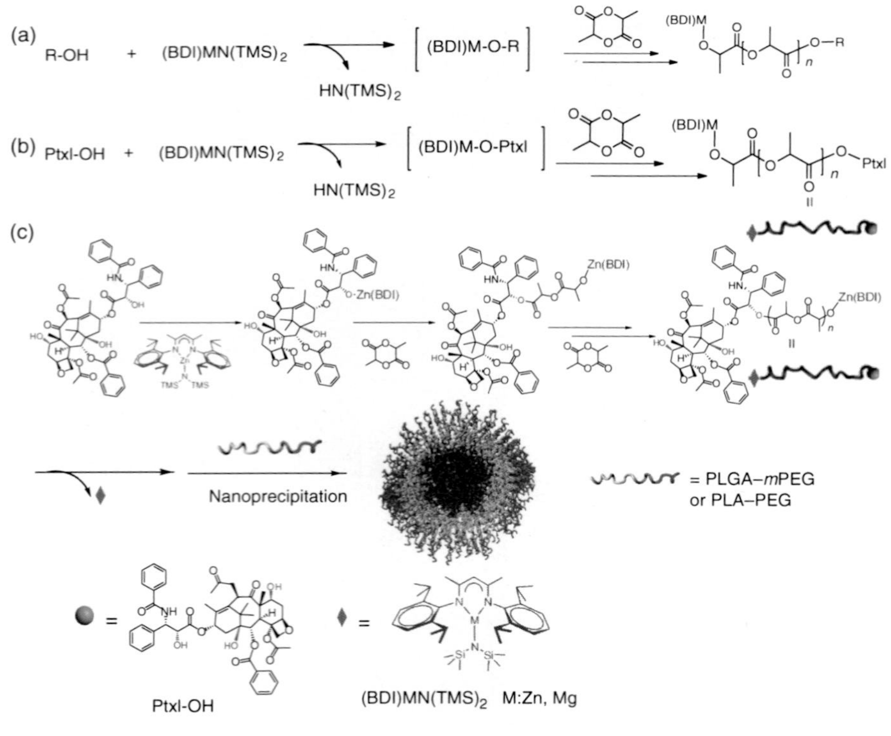

FIGURE 7.3 *See text for full caption.*

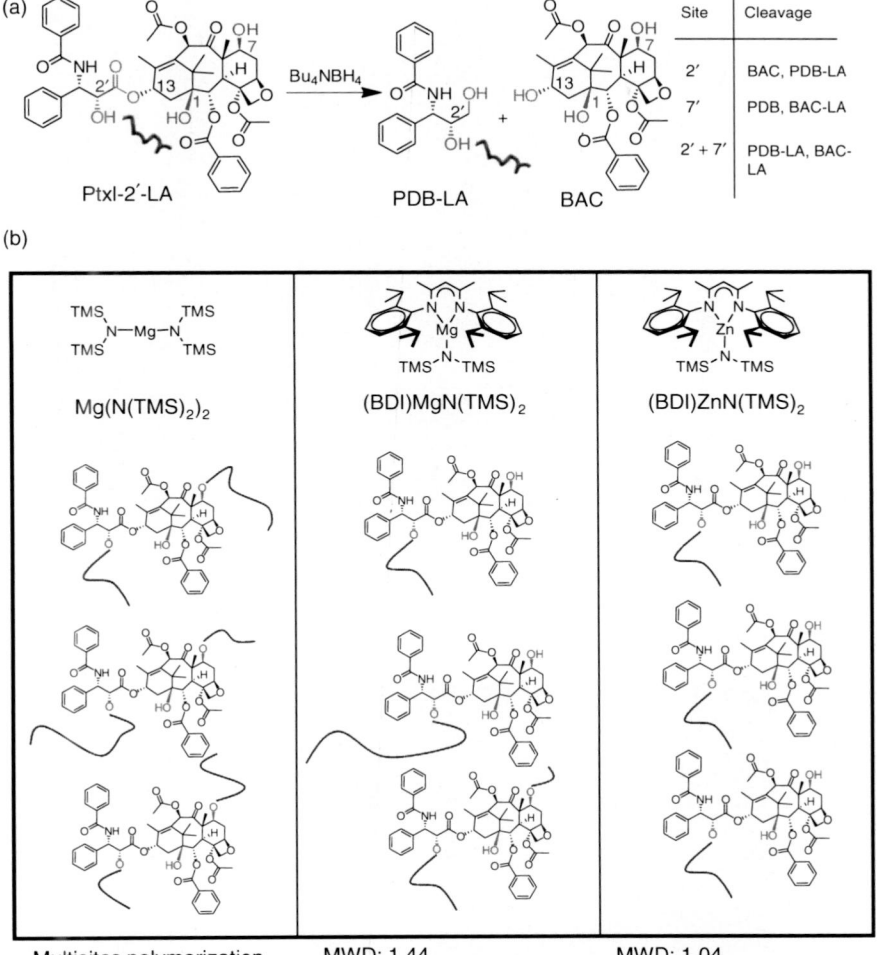

(a)

Ptxl-2′-LA

Bu₄NBH₄

PDB-LA   BAC

Site	Cleavage
2′	BAC, PDB-LA
7′	PDB, BAC-LA
2′ + 7′	PDB-LA, BAC-LA

(b)

Mg(N(TMS)₂)₂

(BDI)MgN(TMS)₂

(BDI)ZnN(TMS)₂

Multisites polymerization   MWD: 1.44   MWD: 1.04

FIGURE 7.5 (a) Bu₄NBH₄-induced site-specific degradation of Ptxl for the formation of PDB and baccatin (BAC). (b) Scheme of Ptxl–PLA conjugates mediated by different catalyst, with the indication of regioselectivity and molecular weight distribution (MWD).

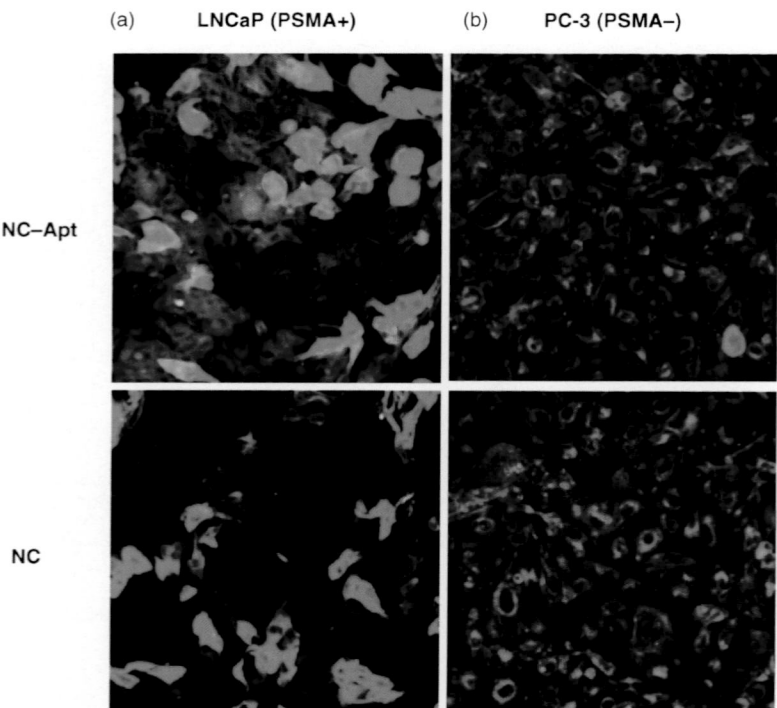

FIGURE 7.8 Confocal images of LNCaP (a) and PC-3 (b) cells treated with aptamer-functionalized nanoconjugates (NC–Apt, top) and nanoparticles without aptamer (NC, bottom). (*See text for full caption.*)

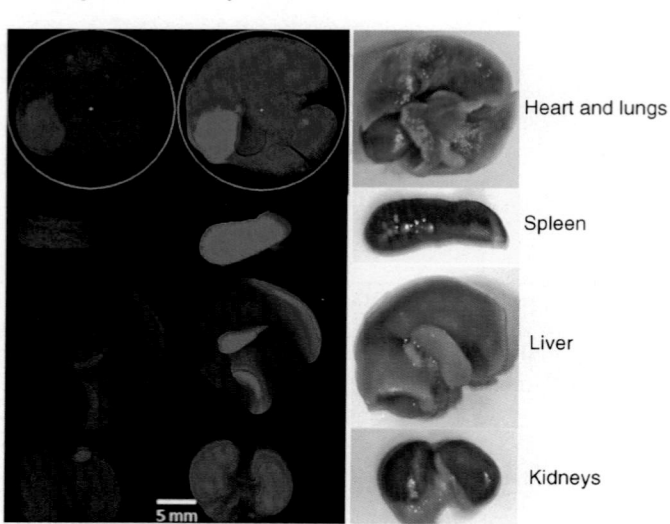

FIGURE 7.9 Distribution of Cy5 dye-labeled PLA NCs in various visceral mice organs. Mice were sacrificed 24 h after i.v. injection of Cy5–PLA NCs.

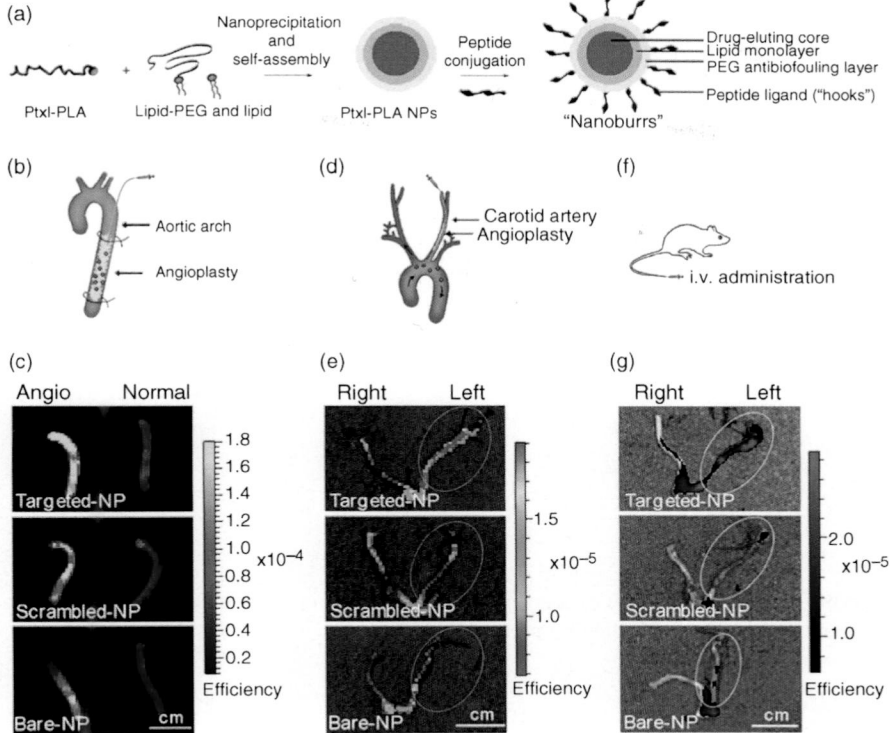

FIGURE 7.10 *See text for full caption.*

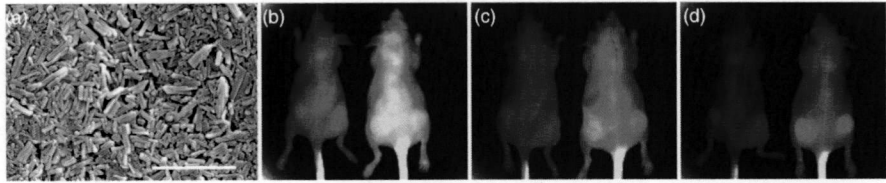

FIGURE 8.4 SEM image of paclitaxel/fluorophore (FPR-749) nanocrystals (a, scale bar: 2 μm), and fluorescence images obtained by IVIS of two mice bearing MCF-7 xenograft tumors in rear flanks. FPR-749 solution (left) and the hybrid nanocrystals (right) were intravenously injected via tail vein and images were captured at 25 min (b), 4 h (c) and 48 h (d), respectively. Exposure time was 0.25 s, F-stop 2, and binning medium.

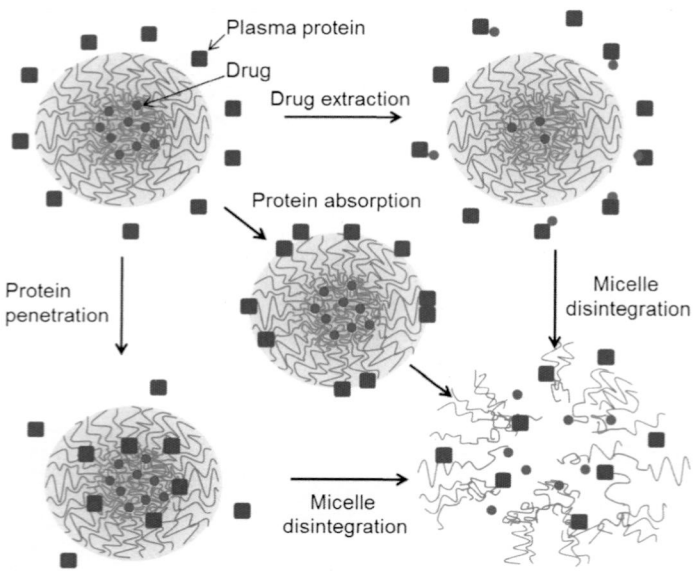

FIGURE 9.2 Disintegration of nanoparticles by plasma protein; protein adsorption, protein penetration, and drug extraction by proteins.

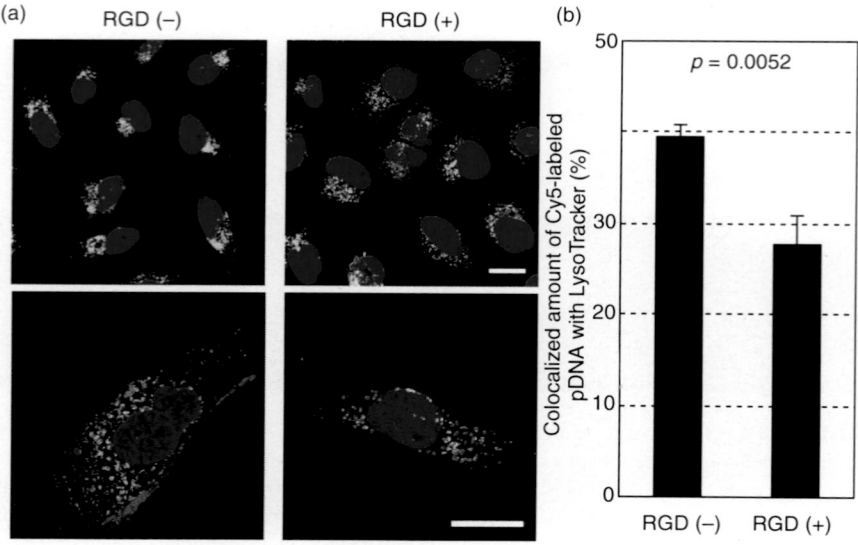

FIGURE 11.3 Distribution of RGD (+) and RGD (−) polyplex micelles in late endosomes and lysosomes. Polyplex micelles loading Cy5-labeled pDNA (red) were incubated with HeLa cells for 1 h. (*See text for full caption.*)

5

MULTIFUNCTIONAL DENDRITIC NANOCARRIERS: THE ARCHITECTURE AND APPLICATIONS IN TARGETED DRUG DELIVERY

RYAN M. PEARSON, JIN WOO BAE, AND SEUNGPYO HONG

Department of Biopharmaceutical Sciences, College of Pharmacy, University of Illinois, Chicago, IL, USA

5.1 RECENT ADVANCES IN DENDRITIC NANOMATERIALS

Dendrimers are hyperbranched, monodispersed macromolecules with chemically well-defined structures. Synthesized by either the convergent or divergent method, a dendrimer consists of three functional domains (core, interior, and periphery) as shown in Figure 5.1. Each domain of a dendrimer can be tailored to precisely and easily control its molecular weight, size (hydrodynamic radius), and surface charge and functionality [1]. The highly controllable physical properties of dendrimers enable the fine-tuning of their biological properties, such as cytotoxicity, biodistribution, and intracellular uptake/trafficking/fate [2–4]. Furthermore, the large number of the surface groups also allows facile multifunctionalization of dendrimers through incorporation of a variety of functional molecules such as therapeutic and diagnostic moieties. The high deformability and the capability to precisely control their surface functional groups make dendrimers ideally suited for mediating strong multivalent interactions between ligands and receptors [5–7]. These unique properties of dendrimers have led these nanomaterials to become one of the most promising platforms for targeted drug delivery. Over the past two

Nanoparticulate Drug Delivery Systems: Strategies, Technologies, and Applications, First Edition.
Edited by Yoon Yeo.
© 2013 John Wiley & Sons, Inc. Published 2013 by John Wiley & Sons, Inc.

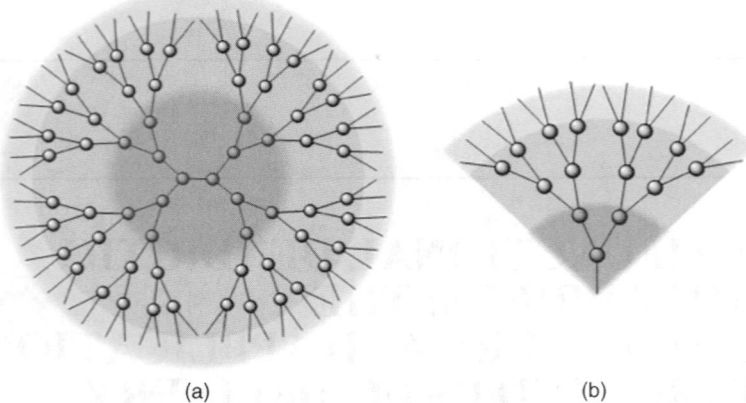

(a)   (b)

FIGURE 5.1   General structures of a dendrimer (a) and dendron (b). Dark to light shaded regions represent the characteristic regions of the dendrimer (core, interior, and periphery) and characteristic regions of the dendron (focal point, interior, and periphery).

decades, significant progress has been made toward the development of dendrimer-based nanotherapeutics for cancer treatments, as highlighted in several recent reviews [8–12]. Although dendrimers have shown great promise in improving cancer treatments and diagnoses, these materials have some limitations that may negatively impact the outcomes of clinical trials and hinder their advancement to the clinic. The drawbacks include complex, multistep synthetic procedures that present scalability issues and limited options for carrying drug molecules, that is, covalent conjugation. Conjugation of drug molecules to the dendrimer often results in low drug payload, in addition to introducing unnecessary chemical modifications to drug molecules [1,13]. Furthermore, extensive surface conjugation with many functional molecules often induces the deterioration of the material properties and structural regularity of the dendrimer [5].

In parallel with the progress of dendrimers, dendron-based nanomaterials have also attracted a great deal of scientific interests due to their capability to hybridize the multifunctional and multivalent properties of dendrimers with various organic and inorganic materials [14]. Dendrons are monodisperse wedge-shaped, hyperbranched sections that comprise dendrimers. Their unique molecular architecture enables chemical modifications of the focal point and/or the periphery, allowing a multitude of complex structures to be realized [14,15]. The use of dendrons for cancer treatments has not yet reached the same level as dendrimers; however, because of their versatility and modular characteristics, they have rapidly emerged as a potential targeted drug delivery platform. In particular, amphiphilic dendrons capable of self-assembly can be engineered to form dendron-based copolymer micelles in aqueous solutions at polymer concentrations higher than the critical micelle concentration (CMC). These micelles can carry significantly higher numbers of drug molecules within their core than dendrimers, while maintaining multiple functional groups on their surface available for chemical modifications [16–18].

However, another level of structural complexity arises from their nonunimolecular structure. Optimization of each structural component is thus necessary because the morphology of self-assembled structures is difficult to predict and often has a marked effect on its biological properties [19].

In this chapter, we highlight promising dendritic systems and discuss their limitations and recent approaches to overcome those drawbacks. The discussion progresses through sections describing important architectural parameters used to rationally design various dendritic materials. Dendritic structure-related toxicity, biodistribution, cellular uptake, and multivalency are also described in detail. Additionally, we discuss through examples how dendrimers and dendrons have been designed and used as targeted drug delivery vehicles for cancer diagnosis and treatments in particular. We anticipate that dendritic materials possessing multi-functional architectures will have a significant impact on the way by which cancer is prevented, diagnosed, treated, and ultimately cured in the future.

5.2 EFFECTS OF DENDRITIC ARCHITECTURES ON BIOLOGICAL SYSTEMS

Many types of dendritic macromolecules have been developed spanning various fields since 1978 [20,21]. However, the clinical translation of these exciting materials has been disappointingly slow due to the issues related to nonspecific toxicity and biodistribution. To date, VivaGel^{TM}, a topically applied vaginal microbicide [22], is the only dendrimer-based product that has been approved by the FDA for the clinical trials. Other promising dendrimer-based therapeutics that are in preparation for clinical trials include Gadomer-17, a magnetic resonance imaging (MRI) contrast agent [23], and a colony-stimulating factor 1 receptor (CSF1R)-targeted dendrimer for the treatment of rheumatoid arthritis [24]. To design highly effective dendrimers, a number of parameters must be considered and optimized. In this section, we discuss how architectural parameters, such as generation (size), surface charge, and chemical composition, can be balanced to control dendrimer-related toxicity and biodistri-bution, and to achieve highly specific biological interactions.

5.2.1 Structure-Related Toxicity and Biodistribution

A review by Duncan and Izzo described the biocompatibility and structure-related toxicity of dendrimers with various backbone structures [25]. The most important factors found to affect the biocompatibility of dendrimers were surface charges and generation. Malik et al. studied the effects that dendrimer surface functionality and generation had on hemolysis and cell morphology [26]. Amine-terminated poly(amidoamine) (PAMAM) and poly(propylenimine) (PPI) dendrimers exhib-ited generation-dependent toxicity. The increased toxicity of higher generation dendrimers has been attributed to (i) the large number of surface groups at the same molar concentration, that is, greater charge density; and (ii) the increased radius of curvature, which increases the contact area at the dendrimer–cell membrane interface [2]. The toxicity by the amine-terminated (positively charged) dendrimers can be substantially reduced by acetylation, carboxylation, or PEGylation (covering the surface with a hydrophilic nonfouling layer of poly (ethylene glycol) (PEG)) of the dendrimer surfaces [27]. Chen et al. modified melamine-based dendrimers to amine, guanidine, carboxylate, sulfonate, phosphate, and PEG-terminated ones [28]. The dendrimers with amine groups exhibited significantly higher cytotoxicity than all other types of dendrimers, whereas PEGylated dendrimers had significantly reduced cytotoxicity after 24 h incubation at a concentration of 10 mg/ml.

Amine-terminated dendrimers are generally known to cause cytotoxicity through destabilization of cell membranes because of the electrostatic interactions, but the underlying mechanism has not been fully understood. To study these nanoscale interactions between PAMAM dendrimers and lipid bilayers, Hong et al. used atomic force microscopy (AFM), a series of enzymatic assays, a fluorescence-activated cell sorter (FACS), and fluorescence microscopy [2,3,29]. For AFM observations *in situ*, 1,2-dimyristoyl-*sn*-glycero-3-phosphocholine (DMPC) was prepared as supported lipid bilayers and subsequently treated with various PAMAM dendrimers. It was observed that cationic dendrimers induced the formation of nanoscale holes in the supported DMPC lipid bilayers in a size-dependent manner. Generation 7 (G7) PAMAM dendrimers were able to initiate hole formation but G5 PAMAM dendrimers could only expand the size of pre-existing defects. Leakage of cytosolic proteins luciferase (Luc) and lactate-dehydrogenase (LDH) from KB and Rat2 cells also demonstrated that PAMAM dendrimers permeabilized cell membranes with size dependency, that is, G7 induced more enzyme leakage than G5 [2]. The membrane integrity was restored 2 h after the removal of G5 PAMAM dendrimers from the incubation medium, indicating the dendrimer-induced membrane permeabilization was reversible. On the other hand, acetamide-terminated (charge neutral) dendrimers neither caused defects nor induced enzyme leakage at the concentrations tested with either generation. This study provided direct evidence that amine-terminated PAMAM dendrimers could disrupt lipid bilayers in a concentration- and generation-dependent manner.

Given that the high toxicity and poor degradability of PAMAM and PPI dendrimers hinder their translation into clinical applications, several new dendrimers have been synthesized and tested as alternatives. For example, hydroxyl-terminated, biodegradable polyester dendrimers based on a 2,2-bis(hydroxymethyl) propionic acid (polyester) backbone were synthesized [30,31]. While IC_{50} (half maximal inhibitory concentration) of amine-terminated dendrimers have been reported as low as 50 μg/ml after 72 h incubation [26], polyester dendrimers have shown much less toxicity, where 60% of the growth rate of B16F10 cells was maintained at a concentration of 20 mg/ml after 48 h incubation [32]. The relatively low toxicity of polyester dendrimers is attributed to hydroxyl groups on the surface, which likely decreased nonspecific cellular interactions [33]. To address the nondegradability of dendrimers, stimuli-responsive functional groups have been incorporated into the dendrimer structure that respond to enzymes, light, reduction, or pH to ensure degradation [34].

Biodistribution reflects the *in vivo* fate of dendrimers. The biodistribution of dendrimers has been traditionally studied in the context of diagnostic applications, since their multiple surface functional groups offer conjugation sites to create multimodal imaging probes [35]. As observed in numerous *in vitro* studies, the generation and surface properties of dendrimers largely affects their biodistribution profiles. Margerum et al. first reported a generation-dependent biodistribution of PAMAM dendrimers, radiolabeled with gadolinium (Gd) using MRI [36]. G5 dendrimers accumulated more preferentially in the liver after 7 days as compared to lower generations G2–4. However, the biodistribution of these Gd-dendrimer chelates cannot be universally applied to other dendrimers because the Gd ions may influence the molecular weight and surface composition. ^{125}I-labeled cationic and anionic PAMAM dendrimers have shown structure and size dependent biodistribution profiles [26]. Anionic PAMAM dendrimers administered intravenously to Wistar rats at a concentration of 10 μg/ml showed a 10- to 20-fold longer circulation time than cationic PAMAM dendrimers. It was also noted that lower generation dendrimers exhibited longer circulation times. Both anionic and cationic dendrimers were found to primarily accumulate in the liver (30–90%) 1 h after administration, indicating that the dendrimer surface groups need to be modified to avoid hepatic uptake if tumor targeting is intended. In this regard, PEGylation has been exploited as a means to enhance the blood circulation half-life and decrease opsonization [37]. Kojima et al. have shown using lysine-modified PEGylated PAMAM dendrimers that the increased circulation time was most probably attributable to a large increase in molecular weight and full surface coverage of PEG on the dendrimer [38].

As for dendrons and dendron-based hybrid materials, it is difficult to make a general conclusion with respect to structure-related toxicity and biodistribution due to a large variety of the possible structures. For this reason, each type of dendron should be evaluated independently to ensure that it elicits appropriate biological responses. Because of the versatility and uniqueness of the structures that can be created, a vast number of different architectures have been synthesized [14]. To date, there have been only a few biodistribution studies of dendron-based materials. ^{111}In-labeled linear dendritic block-copolymer micelles comprised of a hydrophobic polypeptide (β-benzyl-L-aspartate) conjugated to a carboxyl-terminated PEGylated G4 polyester dendron were evaluated in tumor-bearing nude mice [39]. The micelles were shown to accumulate to all vessel-rich organs initially, and most of the micelles were cleared from the body by day 5. Gillies et al. synthesized polyester "bow-tie" hybrid dendrimers comprised of two asymmetric polyester dendrons and evaluated their biodistribution profiles in both tumored and nontumored mice [40]. Polymers with molecular weights of 40,000 Da or greater were found to have significantly enhanced circulation times compared to lower molecular weight polymers. The degree of branching was found to play a role in determining circulation times as well. Higher generation bow-tie dendrimers circulated for a longer time and showed relatively less renal clearance, likely due to decreased molecular flexibility and difficulty to pass glomerular filtration [40].

5.2.2   Mechanisms for Cell Entry of Dendrimers

The ability of dendrimers to interact with cells and cross cell membranes is critical to the effective delivery of therapeutic and diagnostic payloads. Multiple pathways have been proposed as potential mechanisms of cell entry of dendrimers, which are largely affected by molecular size, shape, charge, and surface chemistry [41]. It is particularly important to understand the cellular uptake, intracellular trafficking, and fate of dendrimers when designing novel nanocarriers with subcellular targeting capabilities [42]. Many groups have revealed various mechanisms of PAMAM dendrimer internalization as summarized in Table 5.1. Cationic PAMAM dendrimers have high affinities for plasma membranes that are negatively charged because of proteoglycans and phospholipids, resulting in a nonspecific adsorptive uptake mechanism. Therefore, modification of the dendrimer surface is necessary to achieve fine control over their cellular interactions.

The mechanism by which dendrimers are internalized into cells is a subject of considerable debate. For example, in Caco-2 cells, cationic PAMAM and carboxylated PAMAM dendrimers were shown to internalize by a clathrin-dependent endocytosis mechanism [43]. These results were further confirmed in studies performed using HeLa cells, where cationic PAMAM and partially acetylated PAMAM dendrimers were internalized via both clathrin-mediated endocytosis and macropinocytosis [44]. In other studies, carboxylated PAMAM dendrimers were internalized into A549 lung epithelial cells by a caveolae-mediated pathway, and neutral or cationic PAMAM dendrimers were taken up by noncla-thrin/noncaveolae-mediated mechanisms [45]. Recently, Hong et al. reported a systematic study regarding the cellular internalization mechanisms of G7 PAMAM dendrimers with various surface charges such as amine, acetyl, and carboxyl groups [3]. Fluorescently labeled cholera-toxin subunit B (CTB), transferrin (Tf), and ganglioside GM_1-pyrene were employed as endocytic markers to be co-incubated with PAMAM dendrimers. G7 PAMAM dendrimers were found to co-localize with CTB in KB and Rat2 cells whereas acetylated and carboxylated dendrimers did not interact with the cells. Co-localization with CTB is generally taken as indirect evidence for GM_1 interaction and a caveolae-mediated uptake mechanism since CTB does not internalize without GM_1. To further assess the role that GM_1 may have in facilitating the uptake of G7 PAMAM dendrimers, similar experiments were performed using C6 cells, which are GM_1 deficient. Interestingly, C6 cells that lack GM_1 still allowed G7 PAMAM dendrimers to be internalized, indicating that caveolae-mediated endocytosis is not the only mechanism for dendrimer internalization. Additionally, G7 PAMAM dendrimers were internalized into cells even at $4°C$, suggesting that the internalization mechanism may not be energy-dependent. This study, together with the series of publications [2,3,29], presented strong evidence that nanoscale hole formation is a mechanism of cellular inter-nalization for cationic dendrimers, in addition to other endocytotic mechanisms. Results of these studies strongly suggest that the mechanism of dendrimer internalization is multifaceted and depends on numerous factors including material properties (i.e., charge, generation), concentration, and cell type. It is most likely

TABLE 5.1 Proposed Mechanisms for Cellular Internalization of Dendrimers

Materials	Endocytic Pathway	Cells	References
PAMAM G4-(NH₂)₆₄	Cholesterol-dependent pathway	Caco-2 cells	[43]
PAMAM G2-(NH₂)₁₆	Clathrin-dependent pathway	Caco-2 cells	[116]
PAMAM G2-(NH₂)₁₆	Clathrin-dependent and macropinocytosis	HeLa cells	[44]
PAMAM G4-(NH₂)₆₄	pathway (G6 > G4 > G2)		
PAMAM G6-(NH₂)₂₅₆			
PAMAM G4-(NH₂)₆₄	Nonclathrin, noncaveolae-mediated mechanism	A549 lung epithelial cells	[45]
PAMAM G4-(OH)₆₄	involving electrostatic interactions or other		
	nonspecific fluid-phase endocytosis		
	(G4-NH₂ > G3.5-COOH > G4-OH)		
PAMAM G3.5-COOH	Caveolae-mediated endocytosis	A549 lung epithelial cells	[45]
PAMAM G3.5-NH₂	Clathrin-dependent and dynamin-dependent	Caco-2 cells	[117]
	pathway		
PAMAM G4-(NH₂)₆₄	Cholesterol-dependent pathway	B16F10 melanoma cells	[118]
PAMAM G3-(NH₂)₃₂	(G4 >> G3 > G2)		
PAMAM G2-(NH₂)₁₆			
PAMAM G3-(NH₂)₃₀-(Propanolol)₂	Caveolae-dependent and macropinocytosis	HT-29 cells	[119]
	pathways (G3 >> G3P2)		
PAMAM G3-(NH₂)₃₀-(Lauroyl)₂	Caveolae-dependent, clathrin-dependent, and	HT-29 cells	[119]
	macropinocytosis pathways (G3L2 >> G3)		
PAMAM G3-(NH₂)₂₈-(Lauroyl)₂-(Propanolol)₂	Caveolae-dependent, and possibly clathrin-	HT-29 cells	[119]
	dependent, endocytosis pathways		
	(G3 = G3L2P2)		
PAMAM G5-(NH₂)₁₂₅.₆-(PEG-lactoferrin)₂.₄/	Clathrin-dependent endocytosis, caveolae-	Brain capillary endothelial cells	[120]
pDNA complex	mediated endocytosis, and macropinocytosis		
PEPE G2-(acrylate)₁₆	Caveolae-dependent, clathrin-dependent	Brain vascular endothelial cells	[121]
	pathways		
PAMAM G7-(NH₂)₅₁₂	Nanoscale hole formation	KB, Rat2, and C6 cells	[3]

that multiple cellular uptake mechanisms are involved and govern the cellular uptake of dendrimers.

5.2.3 Multivalent Interactions

Multivalent interactions, which naturally occur in many biological systems, are characterized by simultaneous binding of multiple ligands to multiple receptors [46,47]. These interactions are critical to many pathological and physiological processes and can confer several advantages, such as improved and tight binding through amplification of an existing interaction, increased specificity of binding, efficient signaling communication with cells, and prevention of undesired interactions [46,48,49]. Therefore, understanding of multivalent ligand–receptor interactions is important for designing dendritic nanocarriers that are highly specific to target cells. Several possible mechanisms for multivalent interactions have been proposed. These include the chelate effect, subsite binding, steric stabilization, statistical rebinding, and receptor clustering [46,50]. In particular, the statistical effect (rebinding by a high local concentration of ligands) and the multivalent effect (multivalent binding between multiple ligands and receptors, which may lead to receptor clustering) have been mainly investigated [51,52]. Figure 5.2 depicts the two major mechanisms.

Early studies of multivalent targeting were primarily performed to understand carbohydrate–protein interactions and to gain insight into how proteins present on

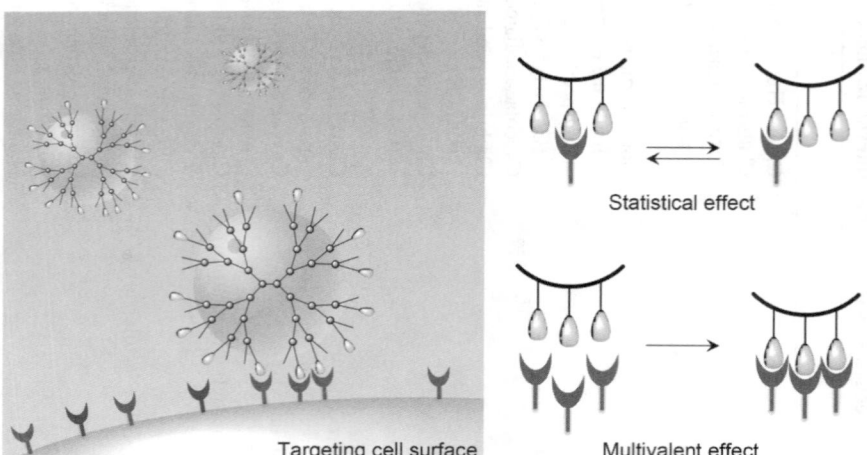

FIGURE 5.2 Schematic representation of multivalent binding of ligand-conjugated dendrimers to target cell surfaces. Dendrimers functionalized with multiple copies of targeting ligands can bind to specific receptors overexpressed on cell surfaces via multivalent interactions. Multiple ligand presentation on the dendrimer surface can lead to the statistical binding effect if their complementary receptors are not clustered on the cell surface. Simultaneous binding of multiple ligands to multiple receptors is responsible for the multivalent effect, which can affect signaling functions of receptors.

cell surfaces can recognize carbohydrates. These studies focused on interactions of mannoside-concanavalin A (Con A) with pea lectin [53], sialic acid with CTB with heat labile enterotoxin of *E. coli* [54], sialic acid with virus hemagglutinin [55], lactose with CTB [56], and hyaluronan with CD44 [57]. Carbohydrate–protein interactions are dependent upon multiple factors including hydrogen bonding and hydrophobic interactions. Additionally, these interactions are known to be involved in a number of important biological phenomena such as cell–cell interactions, antibody–antigen interactions, and viral infection by strengthening their interactions through receptor clustering [58–60]. Since then, many researchers have attempted to establish a rational dendritic design that captures the essence of the effective multivalent binding observed between carbohydrates and proteins.

Some of the studies to understand how dendritic structures may affect multivalent interactions were performed using glycodendrimers exploiting the well-known carbohydrate–protein interactions. Glycodendrimers with 16 mannoside residues possess much higher binding avidities toward Con A and pea lectins compared to single carbohydrate residues [53]. However, fully mannose-substituted larger glycodendrimers (36-mer with three mannose residues per arm) did not achieve a significant increase in binding. It appeared to be a result of steric hindrance of neighboring carbohydrates and/or saturation of binding sites available on the cell surface [61]. It implies that there should be an optimal number of ligands that can participate in multivalent interactions. Andre et al. further supported the notion of an optimal ligand number using lactose-functionalized G5 PAMAM dendrimers [62]. These dendrimers displayed an optimal level of carbohydrate density for effective multivalent interactions with different proteins. Similar results have been reported elsewhere [63–67], providing additional evidence that multivalent binding is highly dependent on the number of surface ligands.

Further modifications to the dendrimer structure were shown to enhance the multivalent binding events of carbohydrates. Simple modifications of the dendrimer surface with short and long chemical linkers increased the conformational flexibility of carbohydrates bound dendrimers, which might increase the mobility of ligands and probability of binding to its appropriate receptor. Flexible linkers can facilitate optimal geometric orientation of ligands toward binding sites in a thermodynamically favorable manner, compared to rigid linkers [68,69]. Although the effect of chemical nature and the length of linkers on multivalent interactions have not been fully understood, the use of a PEG linker has proven effective in inducing efficient multivalent binding [70,71].

In addition to the number and the flexibility of bound carbohydrates, a variety of architectural parameters such as size, shape, and density of binding elements can affect multivalent interactions. Gestwicki et al. explored the effect of ligand architecture on receptor binding and clustering by synthesizing a series of 28 mannose-containing ligands that varied in size, shape, and density [72]. These ligands were derived from five structurally different classes of materials (low-molecular weight compound, PAMAM dendrimer, bovine serum albumin, linear, and polydisperse polymers). The multivalent binding modes of the various mannosylated ligands with the lectin Con A were explored using high-throughput assays. As Con A can be clustered by multivalent ligands (e.g., mannose), Con A is an excellent model protein for examining different aspects of multivalent binding, such as the number of clustered receptors, receptor-clustering rate, and receptor proximity. It was found that ligand architectures resulted in different degrees of Con A inhibition and clustering. Although dendrimers did not generate the most effective ligand architecture for either Con A inhibition or clustering, it was shown that the dendrimer with higher mannose content was a more effective inhibitor of Con A than the dendrimer with a lower mannose content. In addition, the rapid receptor-clustering rate from the highly functionalized, large dendrimers was observed. However, further investigation is necessary to determine how architectural parameters influence multivalent binding modes in other systems.

On the basis of these initial efforts on multivalent targeting, many researchers have sought ways to treat and detect cancer effectively by ligand-conjugated dendritic materials. There are several studies supporting that appropriate presentation of tumor targeting ligands on dendritic surfaces enables the nanocarriers to achieve high binding avidity and specificity toward cancer cells [5,7,73–76]. Hong et al. first reported a quantitative and systematic study of the multivalent effect using G5 PAMAM dendrimers with different numbers of folate as a targeting ligand [5]. The measurements using surface plasmon resonance (SPR) revealed that increasing the number of folate per dendrimer can dramatically enhance the binding avidity toward folate receptors (FRs) up to 170,000-fold, compared to free folic acid (FA). However, high-folate conjugation (over ~7 folate molecules per dendrimer) compromised the material properties of the dendrimer (increased polydispersity index (PDI) and decreased solubility). Saturation behaviors were observed in cell-level experiments using FACS. When the number of folate per dendrimer was increased, dendrimer binding to FR-overexpressing KB cells was increased; however, the improvement became marginal when more than seven folates were conjugated to a dendrimer molecule. This study clearly demonstrated that an optimal number of targeting agents (five folate molecules per dendrimer in this particular case) needs to be identified, in order to achieve the maximal binding avidity without compromising materials properties.

Despite a number of studies on multivalent interactions, it is still unclear which architectural factors are most crucial to designing novel dendritic structures. Obviously, ligand number and flexibility of ligand presentation should be considered in optimizing multivalent interactions; however, evaluation of dendritic constructs with multivalent targeting should vary on a case-by-case basis depending on the types of ligands and target receptors.

5.3 FUNCTIONALIZATION OF DENDRITIC STRUCTURES VIA SURFACE MODIFICATION

Controlling biological properties of dendritic materials via surface modification has been well established as an effective method to create multifunctional architectures of dendritic materials. Depending on the purpose of each study, various molecules have been covalently conjugated to dendrimers at varying numbers and densities. To date, a variety of conjugation methods have been reported to introduce drugs,

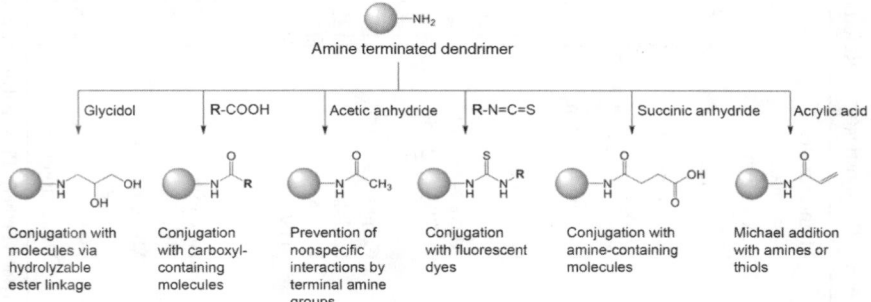

FIGURE 5.3 Various surface modification methods of amine-terminated dendrimers. Starting from amine-terminated dendrimers, many types of surface functional groups can be easily introduced to dendrimers for further applications.

hydrophilic spacers, targeting molecules, and imaging modalities onto dendritic surfaces. Although the conjugation chemistry depends on the molecular functionality to be linked, ideally the method for surface modification should be simple and well defined without causing any loss of inherent reactant properties. Various reagents are commercially available for chemical coupling of molecules and dendritic materials. Figure 5.3 depicts frequently used methods for surface modification of amine-terminated dendrimers such as PAMAM, PLL, and PPI dendrimers. These strategies allow conjugation of a variety of functional groups onto the dendrimer surface, enabling further modification of dendrimers. As previously mentioned, PEGylation is one of the effective surface modification methods that can increase the blood circulation time of dendritic materials through prevention of nonspecific interactions with biological substances. In general, two types of PEGylation strategies have been widely adopted. N-Hydroxysuccinimide (NHS)- and p-nitrophenyl chloroformate (p-NPC)-based chemistries have been commonly used for water- and organic solvent-based reactions, respectively. Zhu et al. compared the PEGylation efficiency of G3 PAMAM dendrimers using these two strategies [77]. Dendrimers PEGylated using p-NPC-PEG achieved significantly higher levels of PEGylation compared to those using PEG–NHS [78].

5.4 DENDRIMER-BASED THERAPEUTIC AND DIAGNOSTIC APPROACHES

Various dendrimer-based nanocarriers have been extensively exploited for effective cancer treatment. Table 5.2 lists the most recent achievements in cancer therapy and diagnosis. The surface properties of the dendrimers were rationally modified with bioactive entities (drugs, genes, targeting ligands, and imaging agents) and their *in vivo* efficacy was studied. In this section, we discuss how dendritic properties affect *in vivo* efficacy using examples in three major delivery applications: anticancer drugs, genes, and imaging agents.

TABLE 5.2 Recent Progress in Cancer Research Using Various Dendrimers

Purpose of Study	Dendritic Materials	Therapeutics or Diagnostics	References
Folate-mediated targeting of conjugated drugs	PAMAM dendrimer	Methotrexate, paclitaxel	[86,112]
Tumor accumulation by different PEGylation degrees and drug conjugation style	PAMAM dendrimer	Doxorubicin	[77]
Convection enhanced delivery of boronated dendrimer-epidermal growth factor bioconjugate for boron neutron capture therapy	PAMAM dendrimer	Boron	[122]
Folate-mediated targeting of drug encapsulated PEG-dendrimer	PAMAM dendrimer	5-Fluorouracil	[114]
Targeted siRNA delivery of cRGD conjugated dendrimers through spheroid model of malignant glioma	PAMAM dendrimer	siRNA	[123]
In vivo tumor imaging by activatable cell penetrating peptide-coated dendrimer labeled with Cy5 and gadolinium	PAMAM dendrimer	Cy5, gadolinium	[124]
Folate-mediated in vivo tumor imaging using gadolinium-conjugated dendrimer	PAMAM dendrimer	Gadolinium	[125]
In vivo imaging using multimodal and multicolor nanoprobes	PAMAM dendrimer	^{111}In, Cy5, Alexa 660, 680, 700, and 750	[126]
Systemic antiangiogenic activity of dendrimer itself	PLL dendrimer	—	[127]
Delivery of anticancer drug using PEG-dendrimer	PLL dendrimer	Camptothecin	[113]
In vivo tumor imaging using gadolinium-conjugated dendrimer	PLL dendrimer	Gadolinium	[128]
Targeted siRNA delivery using cross linked PEG-dendrimers with LHRH peptide	PPI dendrimer	siRNA	[129]
In vivo visualization of gene transfer mediated by systemic injection of dendrimer-pDNA complex	PPI dendrimer	pDNA	[130]
In vivo imaging and biodistribution of radiolabeled dendrimer	PE dendrimer	^{99m}Tc	[131]
Biodistribution of drug-conjugated dendrimer and comparison with drug-loaded liposome	PE dendrimer	Doxorubicin	[132]
Noninvasive imaging of angiogenesis using PEG-dendrimer coated with cRGD	PE dendrimer	^{76}Br	[133]
Targeting of glial tumor and enhanced permeability across blood brain barrier using glucosylated dendrimer	PEPE dendrimers	Methotrexate	[134]
Photodynamic therapy using dendrimer phthalocyanine-encapsulated polymeric micelles	PAE dendrimer	Phthalocyanine	[135]
In vivo efficacy of conjugated drugs using PEG-dendrimer	PEA dendrimer	Doxorubicin	[136]

PAMAM, polyamidoamine; PLL, poly(L-lysine); PPI, poly(propylenimine); PE, polyester; PEPE, polyether-copolyester; PAE, poly(acryl ether); PEA, poly(ester-amide); LHRH, luteinizing hormone-releasing hormone; cRGD, cyclic (Arg–Gly–Asp) peptide.

5.4.1 Delivery of Anticancer Drugs

Dendrimers have been investigated on numerous occasions for the delivery of small molecule anticancer drugs. Through implementation of rational design parameters, their efficacy has been reported to be superior to traditional chemotherapeutics. There are two major methods utilized to incorporate anticancer drugs with the dendrimer [18]. The first method is through encapsulation [79–81], where drugs noncovalently interact with the nonpolar interior core of the dendrimer forming a host-guest complex; and the second, drugs can be covalently attached to one of the many surface groups at the dendrimer periphery [82–84].

As drug encapsulation into dendrimers has several limitations such as low drug loading and uncontrollable drug release, conjugation of therapeutics to dendrimers has been preferred and well established as a means to increase drug solubility, circulating time, and accumulation in tumor tissue *in vivo* [9]. However, it should be noted that there is a limit to the number of molecules that can be conjugated to the surface of dendrimers to maintain structural regularity and solubility [5]. Ideal dendrimer-based carriers should mitigate any surface charges that cause cytotoxicity, possess an imaging capability for *in vivo* visualization, and carry multiple synergistic bioactive compounds to treat the target disease. Figure 5.4 illustrates a set of dendrimer-based

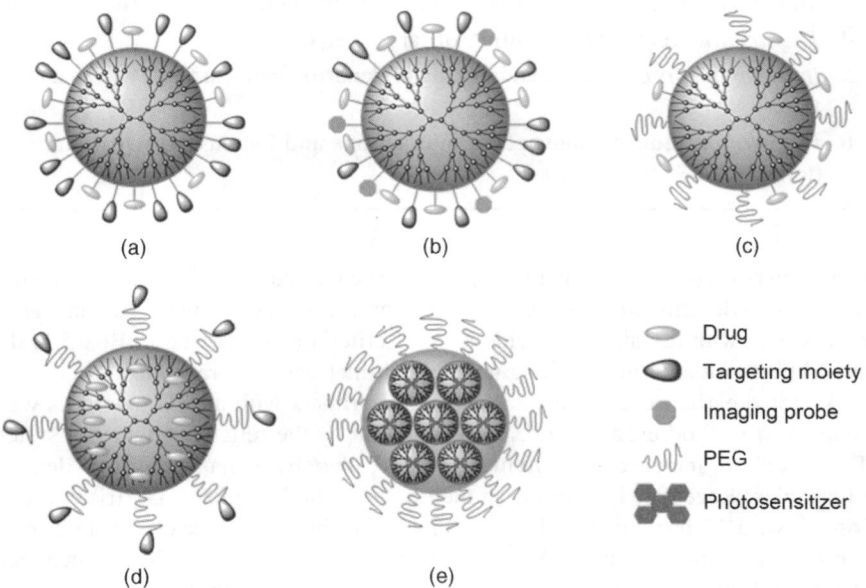

FIGURE 5.4 Recently studied dendrimer-based anticancer drug delivery systems. (a) PAMAM dendrimer–folic acid–methotrexate [111], (b) PAMAM dendrimer–folic acid–paclitaxel–FITC [112], (c) camptothecin conjugated PLL dendrimer–PEG [113], (d) 5-fluorouracil encapsulated PAMAM dendrimer–PEG–folic acid [114], (e) phthalocyanine dendrimer encapsulated PEG–PLL micelle [115]. (*See insert for color representation of the figure.*)

**BOX 5.1   IMPORTANT DESIGN CONSIDERATIONS FOR
THE DEVELOPMENT OF DENDRIMER-BASED
THERAPEUTICS**

Structure

1. *Generations* determine sizes and numbers of surface functional groups, which affects toxicity.
2. *Surface charges* determine cellular interactions and toxicity.
3. *Backbone compositions* determine degradation rates and drug encapsulation potential.
4. *Cores* determine branching degrees and impart potential to add additional functionalities.

Function

1. *Targeting ligands* enable selectivity and promote cellular uptake.
2. *Multivalent binding* substantially improves binding avidity by orders of magnitude.
3. *Stimuli-responsiveness* enhances drug delivery through site-specific degradation.
4. *Therapeutic agents* allow killing of target cells.
5. *Imaging agents* enable visual tracking of dendritic nanocarriers *in vitro* and *in vivo*.
6. *PEGylation* reduces nonspecific interactions and increases *in vivo* circulation time.

---

nanocarriers used for anticancer therapy that have had successful pre-clinical results. Here we will describe some of the most recent progress in dendrimer-based anticancer therapeutics with the aim to provide a set of criteria (summarized in Box 5.1) that should be considered in designing novel dendrimer-based therapeutics.

As noted earlier, coating the surface of dendrimers with a PEG layer is a well-established method used to decrease recognition by the reticuloendothelial system (RES) and to increase the circulation time, thereby enhancing drug delivery efficiencies. Recently, Lim et al. studied a series of PEGylated G2 triazine dendrimers with 12 paclitaxel molecules conjugated through either cleavable ester or ester/disulfide linkages using PC-3 prostate cancer xenografted SCID mice [85]. These dendrimers were well tolerated *in vivo* and could be administered at doses up to 200% of the maximum tolerated dose of free paclitaxel. However, low-level tumor localization and reduced circulation times were observed, suggesting that conjugation of multiple drug molecules may negatively influence the *in vivo* properties of the parent dendrimers. Thus, a key factor in designing dendrimers for drug delivery is to optimize the number of conjugated drug molecules so that important features such as tumor localization and circulating half-life times are minimally deteriorated. To further enhance the specific delivery of drug molecules from the dendrimer to the tumor site, environmentally sensitive dendrimer-drug linkages have been incorporated. Zhu et al. compared various degrees of PEGylation of PAMAM dendrimers conjugated to doxorubicin (DOX) through acid-sensitive *cis*-aconityl and acid-insensitive amide linkages [77]. Although increasing degrees of PEGylation resulted in a decrease in cytotoxicity, higher PEGylation led to lower cellular uptake of DOX-conjugated dendrimers. At the same PEGylation degree, the acid-labile DOX–dendrimer conjugates were more effective in killing cells than the amide-linked DOX-dendrimers because of the facilitated release of DOX in the lysosomes. *In vivo* evaluation showed that the acid-labile DOX–dendrimers had an improved antitumor activity with an increase in the PEGylation degree because of longer circulation time and more effective tumor accumulation. These results show that the optimally designed conjugation of the drug to a PEGylated dendrimer via a cleavable bond may be beneficial for the treatment of a solid tumor.

One major drawback of dendrimer-based drug delivery systems is their multistep synthesis, which is difficult to scale up. Multifunctional dendrimer-based therapeutics can still be produced with batch-to-batch consistency and specificity toward cancer cells [86]. To improve the processing speed of multifunctional dendrimers, a "one-pot" synthesis method was devised where G5 PAMAM dendrimers were conjugated to predetermined ratios of methotrexate (MTX) and FA. Utilizing this simple approach, similar therapeutic results were obtained to those based on the multistep approach [87]. Although the creation of multifunctional dendrimers is complex, these innovative designs, along with the novel synthetic methodologies, make the dendrimer-based nanocarriers highly promising to reach clinical implementation.

With the rapid advances in genetic analysis using DNA-microarray technologies, the conventional approach to cancer therapy has shifted toward a personalized approach [88]. Gene-expression profiling can now be used to identify prognostic and predictive profiles for each patient, which can give clinicians vital information about which course of treatment would be most advantageous. Furthermore, investigation of cancer-specific surface markers and development of genome-wide association studies provide the tools necessary to bring cancer therapy to a next level by identifying genetic predispositions and their influences on the progression of a disease state [89]. To ultimately succeed in clinical translation, dendrimer-based cancer therapeutics must adapt to utilize the information provided from these biochemical studies.

5.4.2 Gene Delivery

The transfer of generic material into a cell to induce, inhibit, or replace gene expression is an additional therapeutic strategy that can be utilized as an effective option for the treatment of multiple diseases. Traditionally, viruses have been used due to their high transfection efficiencies; however, owing to the many safety concerns (potential for mutagenicity, oncogenicity, immunogenicity, and cytotoxicity), a lack of target-specificity, and high-production costs, alternatives that are safe and yet effective are required [90,91]. Cationic nonviral gene delivery vectors that form polyplexes with negatively charged genes are viable options due to their low cost of preparation, low immunogenicity, and pathogenicity [92]. Unfortunately, the transfection efficiency of nonviral vectors is much lower than that of viruses, and the presence of a large positive charge density often induces nonspecific cytotoxicity. Extensive optimization in nonviral vectors is thus necessary.

PAMAM dendrimers were first reported to mediate the transfection of mammalian cells in culture by Haensler and Szoka [93]. Since then, researchers have designed numerous dendrimers to deliver a wide range of nucleic acids such as antisense oligonucleotides [94], microRNA [95], small interfering RNA [96], and plasmid DNA [97]. In a follow-up study by Szoka, "activated" dendrimers, or dendrimers partially degraded by solvolysis, had a greater than 50-fold enhancement in transfection efficiencies compared to intact dendrimers, which was attributed to an increase in flexibility [98]. Kuo et al. measured the expression of 44,000 genes found in HeLa cells treated with "activated" or intact dendrimers alone or in complex with pDNA [99]. Interestingly, it was found that the differences in molecular architecture induced almost completely different gene-expression profiles. Not surprisingly, surface modification and/or spatial modulation of the positive charges can enhance transfection efficiencies. Kim et al. synthesized G2 PPI dendrimers terminated with arginine. At concentrations as high as 150 μg/ml, a significantly improved transfection efficiency (by fourfold) was obtained over control polyethylenimine (PEI25K), while maintaining 80–90% cell viability [100]. Another group attempted to remove the positive surface charge by synthesizing a PAMAM-like dendrimer with internally quaternized amines and a fully acetylated surface [101]. The modification of the surface and addition of positively charged internal groups drastically reduced the cytotoxicity and enhanced cellular uptake. Surface modification of dendrimers with targeting ligands allows highly effective delivery of genes to their target site. Huang et al. developed a Tf-conjugated PAMAM-PEG dendrimer for brain targeting that showed 2.25-fold higher brain uptake than PAMAM and PAMAM-PEG *in vivo* [102]. Additionally, some groups have designed stimuli responsive dendrimers to increase cellular uptake and transfection efficiency. When short, highly charged oligocations were attached to a PEI-based dendrimeric vector via acid-cleavable linkers, a nearly fivefold increase in transfection efficiency was achieved without increasing cytotoxicity [103]. Rapid progression toward the development of better, more biocompatible, highly effective dendrimer-based gene delivery vectors illuminates the promising impact that dendrimers will have in the future for gene therapy applications.

5.4.3 Imaging

The necessity to obtain accurate imaging for *in vitro* and *in vivo* applications has led to the development of many novel nanoparticle-based imaging agents that can be utilized with current micro- and macroscale imaging technologies [104]. The inclusion of imaging agents for nanoparticle tracking and visualization of tumor sites is of great interest to both preclinical researchers and clinicians. In particular, dendrimer-based imaging agents have potential to deliver bioactive compounds and imaging agents simultaneously. Fluorescent imaging agents such as rhodamines, fluoresceins, and AlexaFluors have been conjugated to dendrimers to allow visualization in cell culture and within tumor sections using fluorescence/confocal microscopy. For *in vivo* imaging, there is a need to develop highly effective and safe contrast agents for imaging that can be used for early detection using MRI. Dendrimers are well suited for the uses as novel MRI contrast agents because of their multivalency, monodisperse characteristics, and relatively large size, compared to low molecular weight gadolinium (Gd) chelates [105]. Several groups have modified the surface of dendrimers with Gd(III) chelates (dendrimer–Gd(III)) and found large proton relaxation enhancements and high relaxivities. For example, PEGylated dendrimers were modified with a tumor-targeting peptide HAIYPRH and modified to chelate Gd(III) [106]. It was found that 98 Gd(III) ions could be loaded per PAMAM dendrimer molecule and that this system could increase the diagnostic efficiency to identify liver cancer. Luo et al. synthesized a set of multifunctional peptide dendritic probes of differing generations that utilized a Gd(III) chelate for imaging and clustered galactose moieties for targeting to liver hepatocytes [107]. The inclusion of targeting ligands with dendrimer-based imaging agents represents a method that can be used to increase image contrast and specificity as opposed to nontargeted molecules. The multifunctionality is a clear advantage of the dendrimers, which allows them to be developed as multifunctional imaging agents that both enhance and expedite the detection and diagnosis of cancers.

5.5 DENDRON-BASED HYBRID NANOMATERIALS

It is clear that dendrimers possess many advantageous properties for developing novel therapeutics. Along those lines, each individual dendritic wedge-shaped section, or dendron, that comprises the dendrimer can be used individually or hybridized with other materials to allow creation of a multitude of architectures. Dendrons possess the unique ability to merge the highly desirable properties of dendrimers with the advantages of linear-block copolymers. Unlike the unimolecular structured dendrimers, when engineered with an appropriate hydrophile-lipophile balance (HLB) [108], dendron-based copolymers form self-assembled supramolecular structures. For drug delivery, a variety of dendron-based architectures have been developed and are depicted in Figure 5.5.

Some of the most promising types of dendron-based materials are "bow-tie" hybrid dendrimers [40], linear dendritic block copolymers [16], and dendrimersomes [109]. These materials possess many highly favorable characteristics necessary for drug delivery over comparable dendrimers and linear-block copolymers. First, dendron-based copolymer micelles can overcome the low drug payload disadvantage of dendrimer systems because physical encapsulation into the core is an additional option for the micellar systems. Note that the stability and controllability are the issues in dendrimer encapsulation and the *in vivo* feasibility of the so-called host–guest system is questionable [110]. Encapsulation itself has a significant advantage that there is no need to chemically modify the structure of the drug molecule. Second, the presence of the dendron and its multiple surface functional groups allow

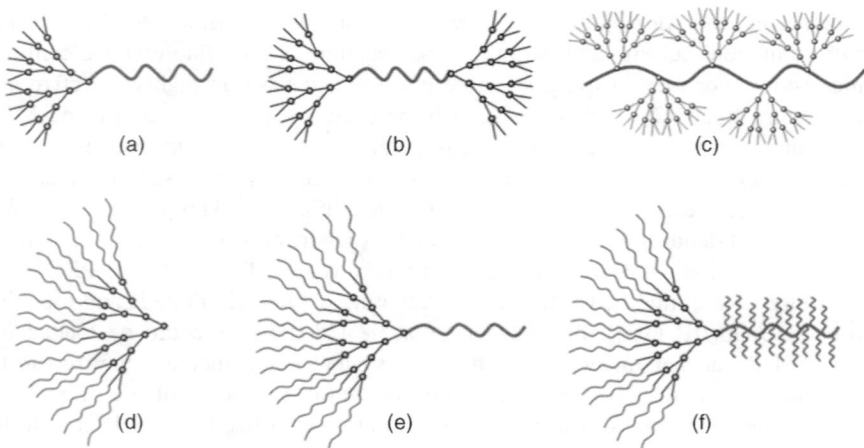

FIGURE 5.5 A variety of dendron-based architectures for drug delivery. (a) A linear polymer linked to the focal point of a dendron, (b, c) a linear polymer bearing two or multiple dendrons at the end or backbone, (d–f) multiple polymer chains linked to the focal point or the periphery of a dendron. The HLB of dendrons can be engineered by controlling the ratio of hydrophilic to hydrophobic polymers.

multivalent interactions to be integrated. Poon et al. synthesized a micellar drug delivery system consisting of a PEGylated polyester dendron block and a rigid hydrophobic linear polypeptide core-forming block. Multivalent targeting was achieved through incorporation of differing degrees of FA on the dendron surface. These micelles selectively accumulated in xenografted tumors and delivered paclitaxel with a fourfold enhancement and reduced systemic toxicity [17,39]. Third, the dendritic architecture can substantially enhance the thermodynamic stability of self-assembled structures as measured by the CMC. Linear-dendron based copolymers have been reported to self-assemble at extremely low CMCs and at high HLBs. To investigate this phenomenon, we previously synthesized highly PEGylated dendron coils (PDCs) and compared their self-assembly behaviors to linear-block copolymers at similar HLBs [16]. The results showed that PDCs at similar HLBs had CMCs that were 1–2 orders of magnitude lower than those of the linear-block copolymer counterparts. The decrease in CMC is attributable to the preorganized, conical architecture of PDC, which decreases entropic cost during self-assembly [16]. Recently, Percec et al. synthesized 11 polymer libraries consisting of over 100 amphiphilic Janus dendrimers and found that they could self-assemble into highly monodispersed supramolecular structures denoted as "dendrimersomes" [109]. These dendrimersomes combine the mechanical strength of polymersomes and the biological mimicry of liposomes, and their highly branched structures offers the ease of surface functionalization. Even though a vast number of distinct morphologies have been reported in the current literature, the exact parameters that govern the phase-separation behavior and self-assembly of these dendron-based nanomaterials have not been fully understood.

5.6 FUTURE PERSPECTIVES

Dendrimers and dendron-based materials are one of the most promising platforms used for drug delivery to date. Extensive *in vitro* and *in vivo* research including those presented in this chapter has illuminated the bright future of dendrimers for drug delivery. Drawbacks of dendrimers, such as limited drug payload and toxicity, can be overcome by engineering the dendrimer structure (e.g., PEGylation), modifying the surface groups (e.g., charge neutralization), and incorporation of functional moieties (e.g., targeting, imaging, therapeutic agents). The choice of dendrimer generation (size) is an additional factor to consider when designing new dendrimer-based drug delivery platforms, because it determines the number of surface functional groups available for conjugation of molecules as well as cellular interactions. However, key challenges, such as limited payload and complex synthetic steps, still remain for dendrimers to be clinically implemented.

Dendron-based nanomaterials have emerged for their potential application in drug delivery. Hybridization of dendrons and linear polymers confers many additional benefits that are not seen in either system alone. This combination of dendritic and linear polymer melds high drug loading potential with multifunctional possibilities and multivalent interactions. However, the dendron systems are relatively new and need to undergo extensive validation processes to prove their potential as effective nano-carriers. Nonetheless, the dendritic structures clearly have significant advantages over the linear polymer counterparts and therefore hold a great promise for various delivery applications. With an optimized design considering the backbone structure, size, surface groups, and multifunctionalities, dendritic nanomaterials will undoubtedly play a key role as a versatile delivery vector in the new era of personalized medicine.

REFERENCES

1. Lee, C. C., MacKay, J. A., Frechet, J. M. J., Szoka, F. C. (2005). Designing dendrimers for biological applications. *Nature Biotechnology 23*, 1517–1526.
2. Hong, S. P., Bielinska, A. U., Mecke, A., Keszler, B., Beals, J. L., Shi, X. Y., Balogh, L., Orr, B. G., Baker, J. R., Holl, M. M. B. (2004). Interaction of poly(amidoamine) dendrimers with supported lipid bilayers and cells: hole formation and the relation to transport. *Bioconjugate Chemistry 15*, 774–782.
3. Hong, S., Rattan, R., Majoros, I. J., Mullen, D. G., Peters, J. L., Shi, X., Bielinska, A. U., Blanco, L., Orr, B. G., Baker, Jr., J. R., Holl, M. M. (2009). The role of ganglioside GM1 in cellular internalization mechanisms of poly(amidoamine) dendrimers. *Bioconjugate Chemistry 20*, 1503–1513.
4. Leroueil, P. R., Hong, S., Mecke, A., Baker, Jr., J. R., Orr, B. G., Banaszak Holl, M. M. (2007). Nanoparticle interaction with biological membranes: does nanotechnology present a Janus face? *Accounts of Chemical Research 40*, 335–342.
5. Hong, S., Leroueil, P. R., Majoros, I. J., Orr, B. G., Baker, J. R., Banaszak Holl, M. M. (2007). The binding avidity of a nanoparticle-based multivalent targeted drug delivery platform. *Chemistry & Biology 14*, 107–115.

6. Kukowska-Latallo, J. F., Candido, K. A., Cao, Z., Nigavekar, S. S., Majoros, I. J., Thomas, T. P., Balogh, L. P., Khan, M. K., Baker, J. R. (2005). Nanoparticle targeting of anticancer drug improves therapeutic response in animal model of human epithelial cancer. *Cancer Research 65*, 5317–5324.

7. Myung, J. H., Gajjar, K. A., Saric, J., Eddington, D. T., Hong, S. (2011). Dendrimer-mediated multivalent binding for the enhanced capture of tumor cells. *Angewandte Chemie International Edition 50*, 11769–11772.

8. Wolinsky, J. B., Grinstaff, M. W. (2008). Therapeutic and diagnostic applications of dendrimers for cancer treatment. *Advanced Drug Delivery Reviews 60*, 1037–1055.

9. Tekade, R. K., Kumar, P. V., Jain, N. K. (2009). Dendrimers in oncology: an expanding horizon. *Chemical Reviews 109*, 49–87.

10. Fox, M. E., Szoka, F. C., Frechet, J. M. (2009). Soluble polymer carriers for the treatment of cancer: the importance of molecular architecture. *Accounts of Chemical Research 42*, 1141–1151.

11. Menjoge, A. R., Kannan, R. M., Tomalia, D. A. (2010). Dendrimer-based drug and imaging conjugates: design considerations for nanomedical applications. *Drug Discovery Today 15*, 171–185.

12. Cheng, Y., Zhao, L., Li, Y., Xu, T. (2011). Design of biocompatible dendrimers for cancer diagnosis and therapy: current status and future perspectives. *Chemical Society Reviews 40*, 2673–2703.

13. Patri, A. K., Kukowska-Latallo, J. F., Baker, J. R. (2005). Targeted drug delivery with dendrimers: comparison of the release kinetics of covalently conjugated drug and non-covalent drug inclusion complex. *Advanced Drug Delivery Reviews 57*, 2203–2214.

14. Rosen, B. M., Wilson, C. J., Wilson, D. A., Peterca, M., Imam, M. R., Percec, V. (2009). Dendron-mediated self-assembly, disassembly, and self-organization of complex systems. *Chemical Reviews 109*, 6275–6540.

15. Wurm, F., Frey, H. (2011). Linear-dendritic block copolymers: the state of the art and exciting perspectives. *Progress in Polymer Science 36*, 1–52.

16. Bae, J. W., Pearson, R. M., Patra, N., Sunoqrot, S., Vukovic, L., Kral, P., Hong, S. (2011). Dendron-mediated self-assembly of highly PEGylated block copolymers: a modular nanocarrier platform. *Chemical Communications 47*, 10302–10304.

17. Poon, Z., Chen, S., Engler, A. C., Lee, H. -i., Atas, E., von Maltzahn, G., Bhatia, S. N., Hammond, P. T. (2010). Ligand-clustered "Patchy" nanoparticles for modulated cellular uptake and *in vivo* tumor targeting. *Angewandte Chemie International Edition 49*, 7266–7270.

18. D'Emanuele, A., Attwood, D. (2005). Dendrimer–drug interactions. *Advanced Drug Delivery Reviews 57*, 2147–2162.

19. Whitesides, G. M., Grzybowski, B. (2002). Self-assembly at all scales. *Science 295*, 2418–2421.

20. Buhleier, E., Wehner, W., Vogtle, F. (1978). Cascade-chain-like and nonskid-chain-like syntheses of molecular cavity topologies. *Synthesis-Stuttgart*, 155–158.

21. Tomalia, D. A., Baker, H., Dewald, J., Hall, M., Kallos, G., Martin, S., Roeck, J., Ryder, J., Smith, P. (1985). A new class of polymers: starburst-dendritic macromolecules. *Polymer Journal 17*, 117–132.

22. Bernstein, D. I., Bourne, N., Ayisi, N. K., Ireland, J., Matthews, B., McCarthy, T., Sacks, S. (2003). Evaluation of formulated dendrimer SPL7013 as a microbicide. *Antiviral Research 57*, A66–A66.

23. Herborn, C. U., Barkhausen, J., Paetsch, I., Hunold, P., Mahler, M., Shamsi, K., Nagel, E. (2003). Coronary arteries: contrast-enhanced MR imaging with SH L 643A—experience in 12 volunteers. *Radiology 229*, 217–223.

24. Hayder, M., Poupot, M., Baron, M., Nigon, D., Turrin, C. -O., Caminade, A. -M., Majoral, J. -P., Eisenberg, R. A., Fournie, J. -J., Cantagrel, A., Poupot, R., Davignon, J. -L. (2011). A phosphorus-based dendrimer targets inflammation and osteoclastogenesis in experimental arthritis. *Science Translational Medicine 3*, 81ra35.

25. Duncan, R., Izzo, L. (2005). Dendrimer biocompatibility and toxicity. *Advanced Drug Delivery Reviews 57*, 2215–2237.

26. Malik, N., Wiwattanapatapee, R., Klopsch, R., Lorenz, K., Frey, H., Weener, J. W., Meijer, E. W., Paulus, W., Duncan, R. (2000). Dendrimers: relationship between structure and biocompatibility *in vitro*, and preliminary studies on the biodistribution of ^{125}I-labelled polyamidoamine dendrimers *in vivo*. *Journal of Controlled Release 65*, 133–148.

27. Jain, K., Kesharwani, P., Gupta, U., Jain, N. K. (2010). Dendrimer toxicity: let's meet the challenge. *International Journal of Pharmaceutics 394*, 122–142.

28. Chen, H. T., Neerman, M. F., Parrish, A. R., Simanek, E. E. (2004). Cytotoxicity, hemolysis, and acute *in vivo* toxicity of dendrimers based on melamine, candidate vehicles for drug delivery. *Journal of the American Chemical Society 126*, 10044–10048.

29. Hong, S., Leroueil, P. R., Janus, E. K., Peters, J. L., Kober, M. M., Islam, M. T., Orr, B. G., Baker, Jr., J. R., Banaszak Holl, M. M. (2006). Interaction of polycationic polymers with supported lipid bilayers and cells: nanoscale hole formation and enhanced membrane permeability. *Bioconjugate Chemistry 17*, 728–734.

30. Ihre, H., Padilla De Jesus, O. L., Frechet, J. M. (2001). Fast and convenient divergent synthesis of aliphatic ester dendrimers by anhydride coupling. *Journal of the American Chemical Society 123*, 5908–5917.

31. Ihre, H. R., Padilla De Jesus, O. L., Szoka, Jr., F. C., Frechet, J. M. (2002). Polyester dendritic systems for drug delivery applications: design, synthesis, and characterization. *Bioconjugate Chemistry 13*, 443–452.

32. Padilla De Jesus, O. L., Ihre, H. R., Gagne, L., Frechet, J. M., Szoka, Jr., F. C., (2002). Polyester dendritic systems for drug delivery applications: *in vitro* and *in vivo* evaluation. *Bioconjugate Chemistry 13*, 453–461.

33. Winnicka, K., Bielawski, K., Rusak, M., Bielawska, A. (2009). The effect of generation 2 and 3 poly(amidoamine) dendrimers on viability of human breast cancer cells. *Journal of Health Science 55*, 169–177.

34. Kojima, C. (2010). Design of stimuli-responsive dendrimers. *Expert Opinion on Drug Delivery 7*, 307–319.

35. Kobayashi, H., Brechbiel, M. W. (2003). Dendrimer-based macromolecular MRI contrast agents: characteristics and application. *Molecular Imaging 2*, 1–10.

36. Margerum, L. D., Campion, B. K., Koo, M., Shargill, N., Lai, J. J., Marumoto, A., Sontum, P. C. (1997). Gadolinium(III) DO3A macrocycles and polyethylene glycol coupled to dendrimers—effect of molecular weight on physical and biological properties

of macromolecular magnetic resonance imaging contrast agents. *Journal of Alloys and Compounds 249*, 185–190.

37. Harris, J. M., Chess, R. B. (2003). Effect of pegylation on pharmaceuticals. *Nature Reviews Drug Discovery 2*, 214–221.

38. Kojima, C., Regino, C., Umeda, Y., Kobayashi, H., Kono, K. (2010). Influence of dendrimer generation and polyethylene glycol length on the biodistribution of PEGylated dendrimers. *International Journal of Pharmaceutics 383*, 293–296.

39. Poon, Z., Lee, J. A., Huang, S., Prevost, R. J., Hammond, P. T. (2011). Highly stable, ligand-clustered patchy "micelle" nanocarriers for systemic tumor targeting. *Nanomedicine: Nanotechnology, Biology, and Medicine 7*, 201–209.

40. Gillies, E. R., Dy, E., Frechet, J. M. J., Szoka, F. C. (2005). Biological evaluation of polyester dendrimer: poly(ethylene oxide) "bow-Tie" hybrids with tunable molecular weight and architecture. *Molecular Pharmaceutics 2*, 129–138.

41. Zhao, F., Zhao, Y., Liu, Y., Chang, X., Chen, C., Zhao, Y. (2011). Cellular uptake, intracellular trafficking, and cytotoxicity of nanomaterials. *Small 7*, 1322–1337.

42. Rajendran, L., Knolker, H. J., Simons, K. (2010). Subcellular targeting strategies for drug design and delivery. *Nature Reviews Drug Discovery 9*, 29–42.

43. Kitchens, K. M., Kolhatkar, I. B., Swaan, P. W., Ghandehari, H. (2008). Endocytosis inhibitors prevent poly(amidoamine) dendrimer internalization and permeability across Caco-2 cells. *Molecular Pharmaceutics 5*, 364–369.

44. Albertazzi, L., Serresi, M., Albanese, A., Beltram, F. (2010). Dendrimer internalization and intracellular trafficking in living cells. *Molecular Pharmaceutics 7*, 680–688.

45. Perumal, O. P., Inapagolla, R., Kannan, S., Kannan, R. M. (2008). The effect of surface functionality on cellular trafficking of dendrimers. *Biomaterials 29*, 3469–3476.

46. Mammen, M., Choi, S. K., Whitesides, G. M. (1998). Polyvalent interactions in biological systems: implications for design and use of multivalent ligands and inhibitors. *Angewandte Chemie International Edition 37*, 2755–2794.

47. Badjic, J. D., Nelson, A., Cantrill, S. J., Turnbull, W. B., Stoddart, J. F. (2005). Multivalency and cooperativity in supramolecular chemistry. *Accounts of Chemical Research 38*, 723–732.

48. Kiessling, L. L., Gestwicki, J. E., Strong, L. E. (2006). Synthetic multivalent ligands as probes of signal transduction. *Angewandte Chemie International Edition 45*, 2348–2368.

49. Vance, D., Martin, J., Patke, S., Kane, R. S. (2009). The design of polyvalent scaffolds for targeted delivery. *Advanced Drug Delivery Reviews 61*, 931–939.

50. Kiessling, L. L., Gestwicki, J. E., Strong, L. E. (2000). Synthetic multivalent ligands in the exploration of cell–surface interactions. *Current Opinion in Chemical Biology 4*, 696–703.

51. Wolfenden, M. L., Cloninger, M. J. (2006). Carbohydrate-functionalized dendrimers to investigate the predictable tunability of multivalent interactions. *Bioconjugate Chemistry 17*, 958–966.

52. Liu, S., Maheshwari, R., Kiick, K. L. (2009). Polymer-based therapeutics. *Macromolecules 42*, 3–13.

53. Page, D., Aravind, S., Roy, R. (1996). Synthesis and lectin binding properties of dendritic mannopyranoside. *Chemical Communications* 1913–1914.

54. Thompson, J. P., Schengrund, C. L. (1997). Oligosaccharide-derivatized dendrimers: defined multivalent inhibitors of the adherence of the cholera toxin B subunit and the heat labile enterotoxin of *E. coli* GM1. *Glycoconjugate Journal 14*, 837–845.

55. Reuter, J. D., Myc, A., Hayes, M. M., Gan, Z., Roy, R., Qin, D., Yin, R., Piehler, L. T., Esfand, R., Tomalia, D. A., Baker, Jr., J. R., (1999). Inhibition of viral adhesion and infection by sialic-acid-conjugated dendritic polymers. *Bioconjugate Chemistry 10*, 271–278.

56. Vrasidas, I., de Mol, N. J., Liskamp, R. M. J., Pieters, R. J. (2001). Synthesis of lactose dendrimers and multivalency effects in binding to the cholera toxin B subunit. *European Journal of Organic Chemistry* 4685–4692.

57. Toole, B. P. (2009). Hyaluronan-CD44 interactions in cancer: paradoxes and possibilities. *Clinical Cancer Research 15*, 7462–7468.

58. Lee, Y. C., Lee, R. T. (1995). Carbohydrate–protein interactions—basis of glycobiology. *Accounts of Chemical Research 28*, 321–327.

59. Lis, H., Sharon, N. (1998). Lectins: carbohydrate-specific proteins that mediate cellular recognition. *Chemical Reviews 98*, 637–674.

60. Sears, P., Wong, C. H. (1999). Carbohydrate mimetics: a new strategy for tackling the problem of carbohydrate-mediated biological recognition. *Angewandte Chemie International Edition 38*, 2301–2324.

61. Ashton, P. R., Hounsell, E. F., Jayaraman, N., Nilsen, T. M., Spencer, N., Stoddart, J. F., Young, M. (1998). Synthesis and biological evaluation of alpha-D-mannopyranoside-containing dendrimers. *Journal of Organic Chemistry 63*, 3429–3437.

62. Andre, S., Ortega, P. J. C., Perez, M. A., Roy, R., Gabius, H. J. (1999). Lactose-containing starburst dendrimers: influence of dendrimer generation and binding-site orientation of receptors (plant/animal lectins and immunoglobulins) on binding properties. *Glycobiology 9*, 1253–1261.

63. Baird, E. J., Holowka, D., Coates, G. W., Baird, B. (2003). Highly effective poly (ethylene glycol) architectures for specific inhibition of immune receptor activation. *Biochemistry 42*, 12739–12748.

64. Woller, E. K., Cloninger, M. J. (2002). The lectin-binding properties of six generations of mannose-functionalized dendrimers. *Organic Letters 4*, 7–10.

65. Choudhury, A. K., Kitaoka, M., Hayashi, K. (2003). Synthesis of a cellobiosylated dimer and trimer and of cellobiose-coated polyamidoamine (PAMAM) dendrimers to study accessibility of an enzyme, cellodextrin phosphorylase. *European Journal of Organic Chemistry* 2462–2470.

66. Samuelson, L. E., Sebby, K. B., Walter, E. D., Singel, D. J., Cloninger, M. J. (2004). EPR and affinity studies of mannose-TEMPO functionalized PAMAM dendrimers. *Organic & Biomolecular Chemistry 2*, 3075–3079.

67. Schlick, K. H., Udelhoven, R. A., Strohniever, G. C., Cloninger, M. J. (2005). Binding of mannose-functionalized dendrimers with pea (*Pisum sativum*) lectin. *Molecular Pharmaceutics 2*, 295–301.

68. Kane, R. S. (2010). Thermodynamics of multivalent interactions: influence of the linker. *Langmuir 26*, 8636–8640.

69. Shewmake, T. A., Solis, F. J., Gillies, R. J., Caplan, M. R. (2008). Effects of linker length and flexibility on multivalent targeting. *Biomacromolecules 9*, 3057–3064.

70. Kim, Y., Hechler, B., Gao, Z. G., Gachet, C., Jacobson, K. A. (2009). PEGylated dendritic unimolecular micelles as versatile carriers for ligands of G protein-coupled receptors. *Bioconjugate Chemistry 20*, 1888–1898.

71. Rele, S. M., Cui, W. X., Wang, L. C., Hou, S. J., Barr-Zarse, G., Tatton, D., Gnanou, Y., Esko, J. D., Chaikof, E. L. (2005). Dendrimer-like PEO glycopolymers exhibit anti-inflammatory properties. *Journal of the American Chemical Society 127*, 10132–10133.

72. Gestwicki, J. E., Cairo, C. W., Strong, L. E., Oetjen, K. A., Kiessling, L. L. (2002). Influencing receptor-ligand binding mechanisms with multivalent ligand architecture. *Journal of the American Chemical Society 124*, 14922–14933.

73. Dijkgraaf, I., Rijnders, A. Y., Soede, A., Dechesne, A. C., van Esse, G. W., Brouwer, A. J., Corstens, F. H., Boerman, O. C., Rijkers, D. T., Liskamp, R. M. (2007). Synthesis of DOTA-conjugated multivalent cyclic-RGD peptide dendrimers via 1,3-dipolar cyclo-addition and their biological evaluation: implications for tumor targeting and tumor imaging purposes. *Organic & Biomolecular Chemistry 5*, 935–944.

74. McNerny, D. Q., Kukowska-Latallo, J. F., Mullen, D. G., Wallace, J. M., Desai, A. M., Shukla, R., Huang, B., Banaszak Holl, M. M., Baker, J. R. (2009). RGD Dendron bodies; synthetic avidity agents with defined and potentially interchangeable effector sites that can substitute for antibodies. *Bioconjugate Chemistry 20*, 1853–1859.

75. Martin, A. L., Bernas, L. M., Rutt, B. K., Foster, P. J., Gillies, E. R. (2008). Enhanced cell uptake of superparamagnetic iron oxide nanoparticles functionalized with dendritic guanidines. *Bioconjugate Chemistry 19*, 2375–2384.

76. Vannucci, L., Fiserova, A., Sadalapure, K., Lindhorst, T. K., Kuldova, M., Rossmann, P., Horvath, O., Kren, V., Krist, P., Bezouska, K., Luptovcova, M., Mosca, F., Pospisil, M. (2003). Effects of N-acetyl-glucosamine-coated glycodendrimers as biological modulators in the B16F10 melanoma model *in vivo*. *International Journal of Oncology 23*, 285–296.

77. Zhu, S. J., Hong, M. H., Zhang, L. H., Tang, G. T., Jiang, Y. Y., Pei, Y. Y. (2010). PEGylated PAMAM dendrimer-doxorubicin conjugates: *in vitro* evaluation and *in vivo* tumor accumulation. *Pharmaceutical Research 27*, 161–174.

78. Pan, G. F., Lemmouchi, Y., Akala, E. O., Bakare, O. (2005). Studies on PEGylated and drug-loaded PAMAM dendrimers. *Journal of Bioactive and Compatible Polymers 20*, 113–128.

79. Kojima, C., Kono, K., Maruyama, K., Takagishi, T. (2000). Synthesis of polyamido-amine dendrimers having poly(ethylene glycol) grafts and their ability to encapsulate anticancer drugs. *Bioconjugate Chemistry 11*, 910–917.

80. Beezer, A. E., King, A. S. H., Martin, I. K., Mitchel, J. C., Twyman, L. J., Wain, C. F. (2003). Dendrimers as potential drug carriers; encapsulation of acidic hydrophobes within water soluble PAMAM derivatives. *Tetrahedron 59*, 3873–3880.

81. Morgan, M. T., Carnahan, M. A., Immoos, C. E., Ribeiro, A. A., Finkelstein, S., Lee, S. J., Grinstaff, M. W. (2003). Dendritic molecular capsules for hydrophobic compounds. *Journal of the American Chemical Society 125*, 15485–15489.

82. Majoros, I. J., Myc, A., Thomas, T., Mehta, C. B., Baker, J. R. (2006). PAMAM dendrimer-based multifunctional conjugate for cancer therapy: synthesis, characteriza-tion, and functionality. *Biomacromolecules 7*, 572–579.

83. Khandare, J., Kolhe, P., Pillai, O., Kannan, S., Lieh-Lai, M., Kannan, R. M. (2005). Synthesis, cellular transport, and activity of polyamidoamine dendrimer–methylpredni-solone conjugates. *Bioconjugate Chemistry 16*, 330–337.

84. Lai, P. -S., Lou, P. -J., Peng, C. -L., Pai, C. -L., Yen, W. -N., Huang, M. -Y., Young, T. -H., Shieh, M. -J. (2007). Doxorubicin delivery by polyamidoamine dendrimer conjugation and photochemical internalization for cancer therapy. *Journal of Controlled Release 122*, 39–46.

85. Lim, J., Chouai, A., Lo, S. -T., Liu, W., Sun, X., Simanek, E. E. (2009). Design, synthesis, characterization, and biological evaluation of triazine dendrimers bearing paclitaxel using ester and ester/disulfide linkages. *Bioconjugate Chemistry 20*, 2154–2161.

86. Myc, A., Douce, T. B., Ahuja, N., Kotlyar, A., Kukowska-Latallo, J., Thomas, T. P., Baker, J. R. (2008). Preclinical antitumor efficacy evaluation of dendrimer-based methotrexate conjugates. *Anti-Cancer Drugs 19*, 143–149.

87. Zhang, Y., Thomas, T. P., Desai, A., Zong, H., Leroueil, P. R., Majoros, I. J., Baker, J. R. (2010). Targeted dendrimeric anticancer prodrug: a ethotrexate-folic acid-poly(amido-amine) conjugate and a novel, rapid, "one pot" synthetic approach. *Bioconjugate Chemistry 21*, 489–495.

88. La Thangue, N. B., Kerr, D. J. (2011). Predictive biomarkers: a paradigm shift towards personalized cancer medicine. *Nature Reviews Clinical Oncology 8*, 587–596.

89. van 't Veer, L. J., Dai, H., van de Vijver, M. J., He, Y. D., Hart, A. A., Mao, M., Peterse, H. L., van der Kooy, K., Marton, M. J., Witteveen, A. T., Schreiber, G. J., Kerkhoven, R. M., Roberts, C., Linsley, P. S., Bernards, R., Friend, S. H. (2002). Gene expression profiling predicts clinical outcome of breast cancer. *Nature 415*, 530–536.

90. Glover, D. J., Lipps, H. J., Jans, D. A. (2005). Towards safe, non-viral therapeutic gene expression in humans. *Nature Reviews Genetics 6*, 299–310.

91. Pack, D. W., Hoffman, A. S., Pun, S., Stayton, P. S. (2005). Design and development of polymers for gene delivery. *Nature Reviews Drug Discovery 4*, 581–593.

92. Khalil, I. A., Kogure, K., Akita, H., Harashima, H. (2006). Uptake pathways and subsequent intracellular trafficking in nonviral gene delivery. *Pharmacological Reviews 58*, 32–45.

93. Haensler, J., Szoka, Jr., F. C. (1993). Polyamidoamine cascade polymers mediate efficient transfection of cells in culture. *Bioconjugate Chemistry 4*, 372–379.

94. Yoo, H., Juliano, R. L. (2000). Enhanced delivery of antisense oligonucleotides with fluorophore-conjugated PAMAM dendrimers. *Nucleic Acids Research 28*, 4225–4231.

95. Ren, Y., Kang, C. -S., Yuan, X. -B., Zhou, X., Xu, P., Han, L., Wang, G. X., Jia, Z., Zhong, Y., Yu, S., Sheng, J., Pu, P. -Y. (2010). Co-delivery of as-miR-21 and 5-FU by poly (amidoamine) dendrimer attenuates human glioma cell growth *in vitro*. *Journal of Biomaterials Science, Polymer Edition 21*, 303–314.

96. Zhou, J., Wu, J., Hafdi, N., Behr, J. -P., Erbacher, P., Peng, L. (2006). PAMAM dendrimers for efficient siRNA delivery and potent gene silencing. *Chemical Communications* 2362–2364.

97. Arima, H., Kihara, F., Hirayama, F., Uekama, K. (2001). Enhancement of gene expression by polyamidoamine dendrimer conjugates with alpha-, beta-, and gamma-cyclodextrins. *Bioconjugate Chemistry 12*, 476–484.

98. Tang, M. X., Redemann, C. T., Szoka, Jr., F. C. (1996). *In vitro* gene delivery by degraded polyamidoamine dendrimers. *Bioconjugate Chemistry 7*, 703–714.

99. Kuo, J. -h. S., Liou, M. -j., Chiu, H. -c. (2010). Evaluating the gene-expression profiles of HeLa cancer cells treated with activated and nonactivated poly(amidoamine) den-drimers, and their DNA complexes. *Molecular Pharmaceutics 7*, 805–814.

100. Kim, T. -i., Baek, J. -u., Zhe Bai, C., Park, J. -s. (2007). Arginine-conjugated poly-propylenimine dendrimer as a non-toxic and efficient gene delivery carrier. *Biomaterials 28*, 2061–2067.

101. Patil, M. L., Zhang, M., Betigeri, S., Taratula, O., He, H., Minko, T. (2008). Surface-modified and internally cationic polyamidoamine dendrimers for efficient siRNA delivery. *Bioconjugate Chemistry 19*, 1396–1403.

102. Huang, R. -Q., Qu, Y. -H., Ke, W. -L., Zhu, J. -H., Pei, Y. -Y., Jiang, C. (2007). Efficient gene delivery targeted to the brain using a transferrin-conjugated polyethyleneglycol-modified polyamidoamine dendrimer. *The FASEB Journal 21*, 1117–1125.

103. Steele, T., Shier, W. (2010). Dendrimeric alkylated polyethylenimine nano-carriers with acid-cleavable outer cationic shells mediate improved transfection efficiency without increasing toxicity. *Pharmaceutical Research 27*, 683–698.

104. Jin, S. -E., Bae, J. W., Hong, S. (2010). Multiscale observation of biological interactions of nanocarriers: from nano to macro. *Microscopy Research and Technique 73*, 813–823.

105. Langereis, S., Dirksen, A., Hackeng, T. M., van Genderen, M. H. P., Meijer, E. W. (2007). Dendrimers and magnetic resonance imaging. *New Journal of Chemistry 31*, 1152–1160.

106. Han, L., Li, J., Huang, S., Huang, R., Liu, S., Hu, X., Yi, P., Shan, D., Wang, X., Lei, H., Jiang, C. (2011). Peptide-conjugated polyamidoamine dendrimer as a nanoscale tumor-targeted T1 magnetic resonance imaging contrast agent. *Biomaterials 32*, 2989–2998.

107. Luo, K., Liu, G., He, B., Wu, Y., Gong, Q., Song, B., Ai, H., Gu, Z. (2011). Multifunctional gadolinium-based dendritic macromolecules as liver targeting imaging probes. *Biomaterials 32*, 2575–2585.

108. Becher, P., Schick, M. J., *Nonionic Surfactants Physical Chemistry*. Surfactants Science Series, New York, Marcel Dekker, 1987, pp. 435–491.

109. Percec, V., Wilson, D. A., Leowanawat, P., Wilson, C. J., Hughes, A. D., Kaucher, M. S., Hammer, D. A., Levine, D. H., Kim, A. J., Bates, F. S., Davis, K. P., Lodge, T. P., Klein, M. L., DeVane, R. H., Aqad, E., Rosen, B. M., Argintaru, A. O., Sienkowska, M. J., Rissanen, K., Nummelin, S., Ropponen, J. (2010). Self-assembly of janus dendrimers into uniform dendrimersomes and other complex architectures. *Science 328*, 1009–1014.

110. Jang, W. -D., Kamruzzaman Selim, K. M., Lee, C. -H., Kang, I. -K. (2009). Bioinspired application of dendrimers: from bio-mimicry to biomedical applications. *Progress in Polymer Science 34*, 1–23.

111. Myc, A., Kukowska-Latallo, J., Cao, P., Swanson, B., Battista, J., Dunham, T., Baker, J. R. (2010). Targeting the efficacy of a dendrimer-based nanotherapeutic in heterogeneous xenograft tumors *in vivo*. *Anti-Cancer Drugs 21*, 186–192.

112. Majoros, I. J., Myc, A., Thomas, T., Mehta, C. B., Baker, Jr., J. R. (2006). PAMAM dendrimer-based multifunctional conjugate for cancer therapy: synthesis, characterization, and functionality. *Biomacromolecules 7*, 572–579.

113. Fox, M. E., Guillaudeu, S., Frechet, J. M., Jerger, K., Macaraeg, N., Szoka, F. C. (2009). Synthesis and *in vivo* antitumor efficacy of PEGylated poly(L-lysine) dendrimer-camptothecin conjugates. *Molecular Pharmaceutics 6*, 1562–1572.

114. Singh, P., Gupta, U., Asthana, A., Jain, N. K. (2008). Folate and folate-PEG-PAMAM dendrimers: synthesis, characterization, and targeted anticancer drug delivery potential in tumor bearing mice. *Bioconjugate Chemistry 19*, 2239–2252.

115. Nishiyama, N., Morimoto, Y., Jang, W. D., Kataoka, K. (2009). Design and development of dendrimer photosensitizer-incorporated polymeric micelles for enhanced photodynamic therapy. *Advanced Drug Delivery Reviews 61*, 327–338.

116. Kitchens, K. M., Foraker, A. B., Kolhatkar, R. B., Swaan, P. W., Ghandehari, H. (2007). Endocytosis and interaction of poly (amidoamine) dendrimers with Caco-2 cells. *Pharmaceutical Research 24*, 2138–2145.

117. Goldberg, D. S., Ghandehari, H., Swaan, P. W. (2010). Cellular entry of G3.5 poly (amido amine) dendrimers by clathrin- and dynamin-dependent endocytosis promotes tight junctional opening in intestinal epithelia. *Pharmaceutical Research 27*, 1547–1557.

118. Seib, F. P., Jones, A. T., Duncan, R. (2007). Comparison of the endocytic properties of linear and branched PEIs, and cationic PAMAM dendrimers in B16f10 melanoma cells. *Journal of Controlled Release 117*, 291–300.

119. Saovapakhiran, A., D'Emanuele, A., Attwood, D., Penny, J. (2009). Surface modification of PAMAM dendrimers modulates the mechanism of cellular internalization. *Bioconjugate Chemistry 20*, 693–701.

120. Huang, R. Q., Ke, W. L., Han, L., Liu, Y., Shao, K., Ye, L. Y., Lou, J. N., Jiang, C., Pei, Y. Y. (2009). Brain-targeting mechanisms of lactoferrin-modified DNA-loaded nanoparticles. *Journal of cerebral blood flow and metabolism 29*, 1914–1923.

121. Dhanikula, R. S., Hammady, T., Hildgen, P. (2009). On the mechanism and dynamics of uptake and permeation of polyether-copolyester dendrimers across an *in vitro* blood-brain barrier model. *Journal of Pharmaceutical Sciences 98*, 3748–3760.

122. Yang, W. L., Barth, R. F., Wu, G., Huo, T. Y., Tjarks, W., Ciesielski, M., Fenstermaker, R. A., Ross, B. D., Wikstrand, C. J., Riley, K. J., Binns, P. J. (2009). Convection enhanced delivery of boronated EGF as a molecular targeting agent for neutron capture therapy of brain tumors. *Journal of Neuro-Oncology 95*, 355–365.

123. Waite, C. L., Roth, C. M. (2009). PAMAM-RGD conjugates enhance siRNA delivery through a multicellular spheroid model of malignant glioma. *Bioconjugate Chemistry 20*, 1908–1916.

124. Olson, E. S., Jiang, T., Aguilera, T. A., Nguyen, Q. T., Ellies, L. G., Scadeng, M., Tsien, R. Y. (2010). Activatable cell penetrating peptides linked to nanoparticles as dual probes for *in vivo* fluorescence and MR imaging of proteases. *Proceedings of the National Academy of Sciences of the United States of America 107*, 4311–4316.

125. Cheng, Z., Thorek, D. L., Tsourkas, A. (2010). Gadolinium-conjugated dendrimer nanoclusters as a tumor-targeted T1 magnetic resonance imaging contrast agent. *Angewandte Chemie International Edition 49*, 346–350.

126. Kobayashi, H., Koyama, Y., Barrett, T., Hama, Y., Regino, C. A., Shin, I. S., Jang, B. S., Le, N., Paik, C. H., Choyke, P. L., Urano, Y. (2007). Multimodal nanoprobes for radionuclide and five-color near-infrared optical lymphatic imaging. *ACS Nano 1*, 258–264.

127. Al-Jamal, K. T., Al-Jamal, W. T., Akerman, S., Podesta, J. E., Yilmazer, A., Turton, J. A., Bianco, A., Vargesson, N., Kanthou, C., Florence, A. T., Tozer, G. M., Kostarelos, K. (2010). Systemic antiangiogenic activity of cationic poly-L-lysine dendrimer delays tumor growth. *Proceedings of the National Academy of Sciences of the United States of America 107*, 3966–3971.

128. Kaneshiro, T. L., Jeong, E. K., Morrell, G., Parker, D. L., Lu, Z. R. (2008). Synthesis and evaluation of globular Gd-DOTA-monoamide conjugates with precisely controlled nanosizes for magnetic resonance angiography. *Biomacromolecules 9*, 2742–2748.

129. Taratula, O., Garbuzenko, O. B., Kirkpatrick, P., Pandya, I., Savla, R., Pozharov, V. P., He, H., Minko, T. (2009). Surface-engineered targeted PPI dendrimer for efficient intracellular and intratumoral siRNA delivery. *Journal of Controlled Release 140*, 284–293.

130. Chisholm, E. J., Vassaux, G., Martin-Duque, P., Chevre, R., Lambert, O., Pitard, B., Merron, A., Weeks, M., Burnet, J., Peerlinck, I., Dai, M. S., Alusi, G., Mather, S. J., Bolton, K., Uchegbu, I. F., Schatzlein, A. G., Baril, P. (2009). Cancer-specific transgene expression mediated by systemic injection of nanoparticles. *Cancer Research 69*, 2655–2662.

131. Parrott, M. C., Benhabbour, S. R., Saab, C., Lemon, J. A., Parker, S., Valliant, J. F., Adronov, A. (2009). Synthesis, radiolabeling, and bio-imaging of high-generation polyester dendrimers. *Journal of the American Chemical Society 131*, 2906–2916.

132. Guillaudeu, S. J., Fox, M. E., Haidar, Y. M., Dy, E. E., Szoka, F. C., Frechet, J. M. J. (2008). PEGylated dendrimers with core functionality for biological applications. *Bioconjugate Chemistry 19*, 461–469.

133. Almutairi, A., Rossin, R., Shokeen, M., Hagooly, A., Ananth, A., Capoccia, B., Guillaudeu, S., Abendschein, D., Anderson, C. J., Welch, M. J., Frechet, J. M. J. (2009). Biodegradable dendritic positron-emitting nanoprobes for the noninvasive imaging of angiogenesis. *Proceedings of the National Academy of Sciences of the United States of America 106*, 685–690.

134. Dhanikula, R. S., Argaw, A., Bouchard, J. F., Hildgen, P. (2008). Methotrexate loaded polyether–copolyester dendrimers for the treatment of gliomas: enhanced efficacy and intratumoral transport capability. *Molecular Pharmaceutics 5*, 105–116.

135. Nishiyama, N., Nakagishi, Y., Morimoto, Y., Lai, P. S., Miyazaki, K., Urano, K., Horie, S., Kumagai, M., Fukushima, S., Cheng, Y., Jang, W. D., Kikuchi, M., Kataoka, K. (2009). Enhanced photodynamic cancer treatment by supramolecular nanocarriers charged with dendrimer phthalocyanine. *Journal of Controlled Release 133*, 245–251.

136. van der Poll, D. G., Kieler-Ferguson, H. M., Floyd, W. C., Guillaudeu, S. J., Jerger, K., Szoka, F. C., Frechet, J. M. (2010). Design, synthesis, and biological evaluation of a robust, biodegradable dendrimer. *Bioconjugate Chemistry 21*, 764–773.

6

CHITOSAN-BASED NANOPARTICLES FOR BIOMEDICAL APPLICATIONS

Heebeom Koo, Kuiwon Choi, Ick Chan Kwon, and Kwangmeyung Kim

Center for Theragnosis, Biomedical Research Institute, Korea Institute of Science and Technology, Seongbuk-gu, Seoul, Republic of Korea

6.1 CHITOSAN AS A BIOPOLYMER

In recent years, progress in the field of biotechnology and nanotechnology has made a huge impact on biomedical fields. In addition to the improvement in mechanical devices, nanoparticles have emerged as a multifunctional platform for both diagnosis and therapy of various diseases [1,2]. Researchers have developed various kinds of nanoparticles using polymers, lipids, iron, gold, silica, and so on [3,4]. Among them, polymers are the most widely used components in fabricating nanoparticles [5]. Their major advantage is that their physical or chemical properties can be easily controlled by changing the molecular weights or chemical structures [6]. Therefore, a number of polymer-based nanoparticles have been developed and applied to biomedical fields [7].

Chitosan is one of the representative natural polymers, which is derived from chitin and frequently used for the development of nanoparticles [8,9]. Chitin is the main component of crustacean shell and is composed of β-(1,4)-2-acetamido-D-glucose linked via (1–4) glycosidic bonds. Chitosan is produced by deacetylation of chitin, and the number of amine groups is dependent on the degree of deacetylation. These amine groups endow many useful characteristics with chitosan.

Water solubility of chitosan is mainly dependent on the protonation of amine groups. Because the pK_a value of this amine group is about 6.5, chitosan is generally

Nanoparticulate Drug Delivery Systems: Strategies, Technologies, and Applications, First Edition.
Edited by Yoon Yeo.
© 2013 John Wiley & Sons, Inc. Published 2013 by John Wiley & Sons, Inc.

soluble in solvents with pH below 6.5 but insoluble in solvents with neutral or basic pH [10]. Even though low-molecular-weight chitosans below ~50 kDa are soluble at neutral pH, most chitosans originally have relatively low solubility in water, which is a major hurdle for further development. To enhance the water solubility of chitosan, various chitosan derivatives have been developed by chemical modification. One way to enhance water solubility is to introduce additional cationic or anionic charged groups. Quaternized chitosan (trimethyl chitosan), sulfated chitosan, and carboxymethyl chitosan are representative chitosan derivatives which have charges at a neutral condition [11–13]. Alternatively, conjugation with highly water soluble poly(ethylene glycol) (PEG) or glycol groups (one unit of ethylene oxide) can enhance the solubility of chitosan without additional charges [14–16].

Chitosan is distinguished from many biopolymers by its amine groups providing useful information on its properties. First, this amine group can bind to anionic surface of cellular membrane, enabling efficient cellular adhesion and uptake [17]. This property is very useful in biomedical applications such as drug or gene delivery. Second, chitosan can also bind to anionic DNA or RNA by electronic interactions with these amine groups [18,19]. Consequently, chitosan is a very attractive material for developing gene carriers [20]. Finally, many other molecules can be conjugated to chitosan via these amine groups [21]. This point is particularly important because amide coupling of amine and carboxylic acid groups is the most widely used conjugation method. In addition, a number of functional molecules such as fluorescence dyes are commercially available as activated with N-hydroxysuccinimide ester (NHS) for conjugation to these amine groups [22].

Biodegradability and biocompatibility of chitosan facilitate its use for biomedical applications. The oral LD_{50} value of chitosan is more than 16 g/kg in mice, which is very high and demonstrates the biocompatibility of chitosan [23]. It can be degraded to low-molecular-weight molecules by various enzymes secreted from the kidney. Owing to this biocompatibility, its use as a wound dressing agent was approved for human use by the U.S. Food and Drug Administration (FDA) [24]. On the basis of these points, chitosan is generally believed as a biocompatible biopolymer by many researchers. However, modified chitosans may show more delayed renal excretion due to limited interactions with enzymes [25].

In this review, we focus on chitosan-based nanoparticles for imaging and drug/gene delivery. We introduce many studies on chitosan nanoparticles, which evaluated biological effects in living systems such as cultured cells or animal models. Furthermore, we describe important considerations in developing chitosan-based nanoparticles for diagnosis and therapy of diseases.

6.2 CHITOSAN NANOPARTICLES FOR IMAGING

6.2.1 Optical Imaging

Optical imaging is generally based on fluorescence or luminescence. The advantages are easy handling, high sensitivity, and relatively low cost of imaging equipment [26]. Optical imaging nanoprobes can be prepared by simple conjugation or loading of commercial fluorescence dyes to nanoparticles [27]. Nanoparticles with targeting ability can be used for optical imaging of a specific disease through long-term blood circulation and high accumulation in the target site. For *in vivo* optical imaging, the near-infrared (NIR) region of fluorescence (about 650–900 nm) is frequently used due to the relatively low interference by hemoglobin and water [28].

Ranjan et al. showed that chitosan-based polymeric nanoparticles can be used in *in vivo* optical imaging by simple loading of indocyanine green (ICG), an FDA-approved commercial fluorescence dye [29]. They developed nanoparticles with poly(L-lactide-*co*-epsilon-caprolactone) and poloxamer, and ICG was encapsulated in the nanoparticles. The surface of nanoparticles was further modified by chitosan. The positive charge of chitosan enhanced cellular uptake and tissue accumulation during both *in vitro* and *in vivo* experiments. These results were obtained by NIR optical imaging of the loaded ICG.

Recently, Kim's group demonstrated an excellent tumor-targeting ability of their glycol chitosan nanoparticles by optical imaging [30]. Glycol chitosan is a chitosan derivative with good water solubility. Glycol groups also provide cationic amine groups with PEG-like shielding effects inhibiting interactions of nanoparticles with anionic serum protein. Glycol chitosan nanoparticles were synthesized by conjugating hydrophobic 5β-cholanic acids to hydrophilic glycol chitosan [31] (Figure 6.1a, b). The average size of glycol chitosan nanoparticles was about 250 nm, and NIR fluorescence dye Cy 5.5 was conjugated to it for optical imaging. Owing to the small size, deformability, fast uptake into tumor cells, and the shielding effect of glycol groups, glycol chitosan nanoparticles could be efficiently accumulated within tumors via the enhanced permeability and retention (EPR) effect in newly generated tumor vessels [32–34]. The tumor-targeting ability of glycol chitosan nanoparticles was analyzed in various tumor models. In a flank tumor model, it showed long-term circulation and high accumulation in tumor tissues for a prolonged period (Figure 6.1c). In particular, the EPR effect in tumor tissues and fast penetration of nano-particles were monitored with live optical imaging [35] (Figure 6.1d). Subsequently, a significant correlation between angiogenesis in tumors and *in vivo* behavior of glycol chitosan nanoparticles was shown in a brain tumor model in a time-dependent manner (Figure 6.1e). In a liver tumor model, the high fluorescence signals in tumor tissues than in surrounding normal liver tissues demonstrated the ability of glycol chitosan nanoparticles to escape from the reticuloendothelial system (RES) of the liver (Figure 6.1f). Finally, specific detection of metastatic tumors with them was shown in a lung metastasis tumor model (Figure 6.1g). Furthermore, successful optical imaging of unintended secondary metastatic tumors on forefoot showed the potential of the early detection of metastasis in patients. These results demonstrate the utility of fluorescently labeled chitosan nanoparticles in optical imaging of various cancers.

6.2.2 Magnetic Resonance (MR) Imaging

Nowadays, magnetic resonance (MR) imaging is indispensable for the diagnosis of many diseases in clinical settings. MR imaging mainly provides anatomical imaging

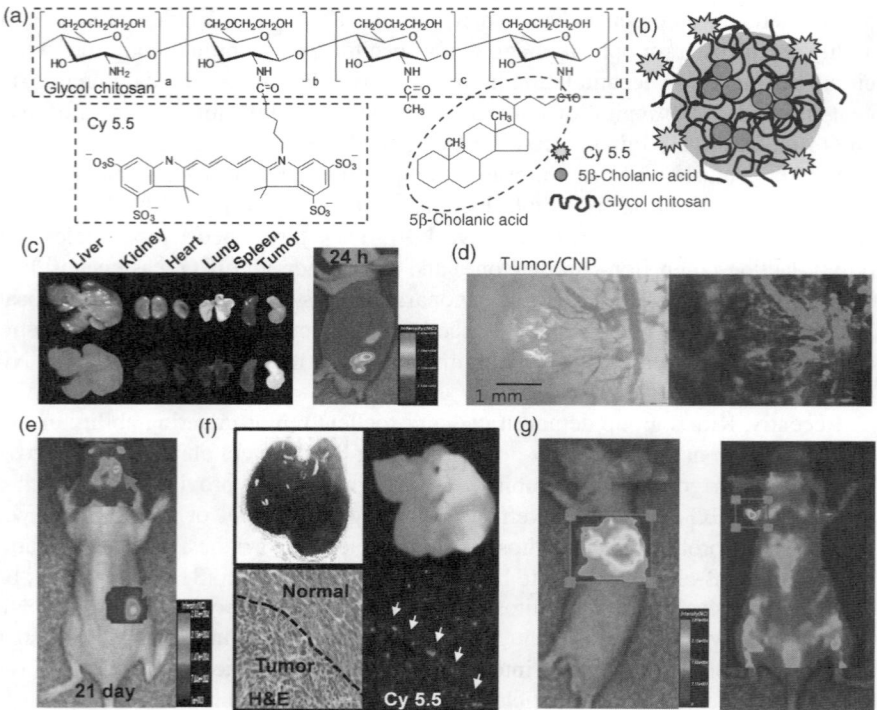

FIGURE 6.1 Chitosan nanoparticles for optical imaging. (a) Chemical structure of Cy5.5 and 5β-cholanic acid conjugated glycol chitosan. (b) Schematic illustration of a self-assembled glycol chitosan nanoparticle. (c) *In vivo* and *ex vivo* tumor imaging with chitosan nanoparticles in left flank tumor model. (d) Vascular penetration imaging in left flank tumor tissue with chitosan nanoparticles. (e) Brain tumor imaging with chitosan nanoparticles. (f) Liver tumor imaging with chitosan nanoparticles (white arrows marks the margin between normal and tumor tissue.). (g) Lung and secondary metastatic tumor imaging with chitosan nanoparticles. (*Source*: Figures reproduced with permission from Reference [30].) (*See insert for color representation of the figure.*)

of organs and tissues, and can visualize internal disorders of the body resulting from diseases. MR imaging is generally categorized into two modes: longitudinal (T1) mode and transverse (T2) mode [36].

In T1 MR imaging, gadolinium (Gd)-based contrast agents can reduce T1 relaxation time by interacting with water and provide bright signals. Gd ions are generally entrapped by chelators such as tetraazacyclododecane tetraacetic acid (DOTA) or diethylenetriaminepentaacetic acid (DTPA) for MR imaging [37]. These chelators can be chemically conjugated to functional nanoparticles for improvement of T1 contrast imaging of target tissues [38]. Since Gd-based T1 MR imaging is based on interactions with water molecules, Gd chelators should be present on the outer environment rather than inside of the nanoparticles. Recently, Kim's group developed a dual imaging probe by conjugation of Gd-DOTA and Cy 5.5 to glycol chitosan backbone of the nanoparticles [39]. In addition to optical imaging by Cy 5.5, intense T1 MR signal of the resulting nanoparticles revealed bright tumor tissues in MR images of tumor-bearing mice.

On the other hand, T2 MR imaging agents, such as iron oxide nanoparticles, shorten the transverse relaxation time and provide dark images of target tissues [40]. Iron oxide nanoparticles are generally synthesized at a high temperature using fatty acids as a surfactant. The iron oxide nanoparticles with such a hydrophobic surface should be hydrophilically modified for applications in aqueous environment of the body [41]. Consequently, many researchers conjugated hydrophilic polymers to them or encapsulated them in hydrophilic nanoparticles to increase water solubility. Chitosan can be employed for varying purposes. For example, Lee et al. developed iron oxide-loaded chitosan nanoparticles with hydrophobic linoleic acid-conjugated water-soluble chitosan and hydrophobic fatty acid-coated iron oxide nanoparticles [42]. These chitosan nanoparticles were stable in an aqueous condition for a long time and generated dark T2 signals in hepatocytes in mice after intravenous injection. The same group also synthesized chitosan nanoparticles conjugated with oleic acid for self-assembly and Cy 5.5 dyes for optical imaging [43] (Figure 6.2a). Subsequently, oleic acid-coated iron oxide nanoparticles were encapsulated in these chitosan nanoparticles for T2 MR imaging (Figure 6.2b). The resulting nanoparticles could provide both optical imaging and T2 MR signals (Figure 6.2c, d). In a U87MG brain-tumor-bearing mouse model, the tumor site was marked by optical fluorescence imaging and T2 MR imaging simultaneously (Figure 6.2e–g).

6.2.3 Computed Tomography (CT) Imaging

Computed tomography (CT) is a powerful imaging technique, which can provide two- or three-dimensional images of living subjects. CT can provide anatomical images of bones and tissues with very high spatial resolution. For enhanced CT imaging, iodine-based contrast agents are used in clinical settings [44]. However, owing to the renal toxicity and high viscosity of iodine solutions, there is an urgent need for alternative CT imaging agents. Recently, many researchers have focused on gold nanoparticles as a new CT imaging agent because gold nanoparticles present a high X-ray absorption coefficient for intense CT signals [45]. Consequently, several groups have tried targeted delivery of gold nanoparticles for *in vivo* CT imaging [46].

Recently, Sun et al. developed a gold nanoparticle-based optical/CT dual imaging probe with glycol chitosan for *in vivo* tumor diagnosis [47]. They used glycol chitosan as a reducing and stabilizing agent during fabrication of gold nanoparticles. Glycol chitosan polymers tightly adsorbed onto the surface of a gold nanoparticle via primary amine groups as reducing agents. Then, matrix metalloproteinase (MMP) activatable peptide probes were conjugated to the glycol chitosan-coated gold nanoparticles for optical imaging [48] (Figure 6.3a). This glycol chitosan-coated gold nanoprobe showed a long circulation time and high accumulation in tumor sites for diagnosis. In a tumor-bearing mouse model, optical and CT imaging of tumor

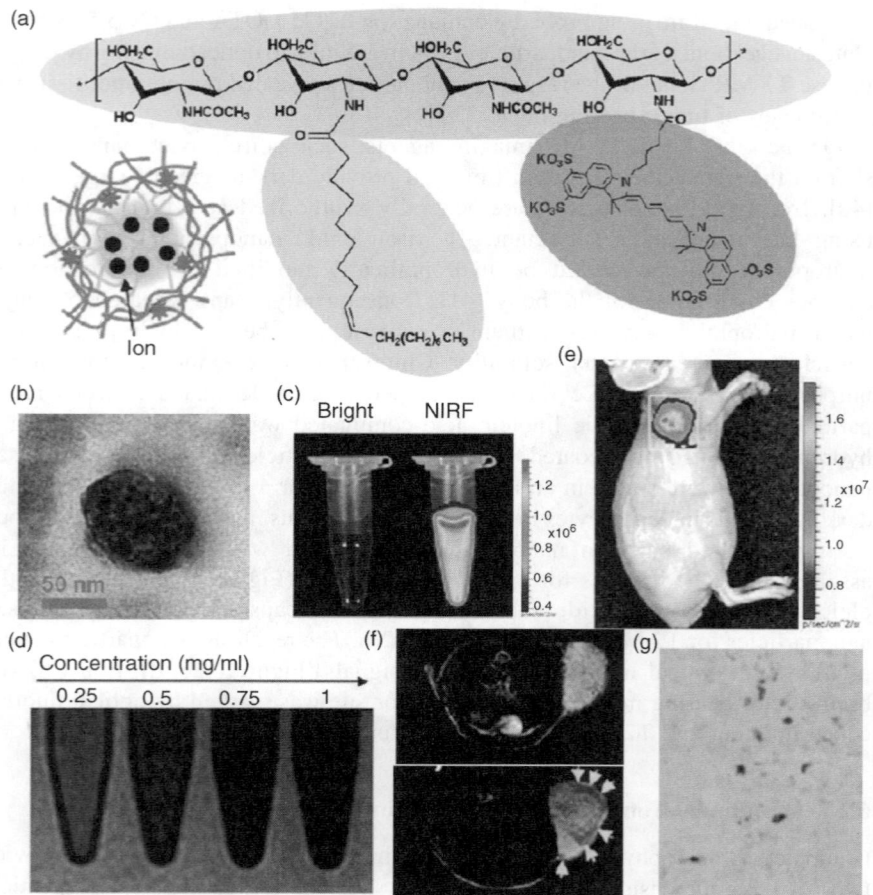

FIGURE 6.2 Chitosan nanoparticles for optical/MR imaging. (a) Chemical structure and illustration of a Cy5.5 and oleic acid conjugated chitosan nanoparticle containing iron oxide nanoparticles (green, yellow, and red color denotes chitosan, oleic acid, and Cy5.5, respectively). (b) Transmission emission microscopy (TEM) image of a chitosan nanoparticle. (c) Bright field and NIR fluorescence images of chitosan nanoparticle solution. (d) MR phantom images of chitosan nanoparticles. (e) NIR fluorescence image of a tumor-bearing mouse after injection of chitosan nanoparticles. (f) MR image of tumor-bearing mouse after injection of chitosan nanoparticles (yellow arrows denote the dark MR signals from chitosan nanoparticles). (g) Prussian blue stained tumor tissue of chitosan nanoparticle-injected mouse. (*Source*: Figures reproduced with permission from Reference [43].) (*See insert for color representation of the figure.*)

tissues was successfully obtained with this nanoprobe (Figure 6.3b, c). Importantly, this research is a good example of a multimodal imaging nanoprobe with synergetic effects [49]. CT imaging can provide three-dimensional anatomical imaging of tumor tissues in the body, while highly sensitive optical imaging can provide

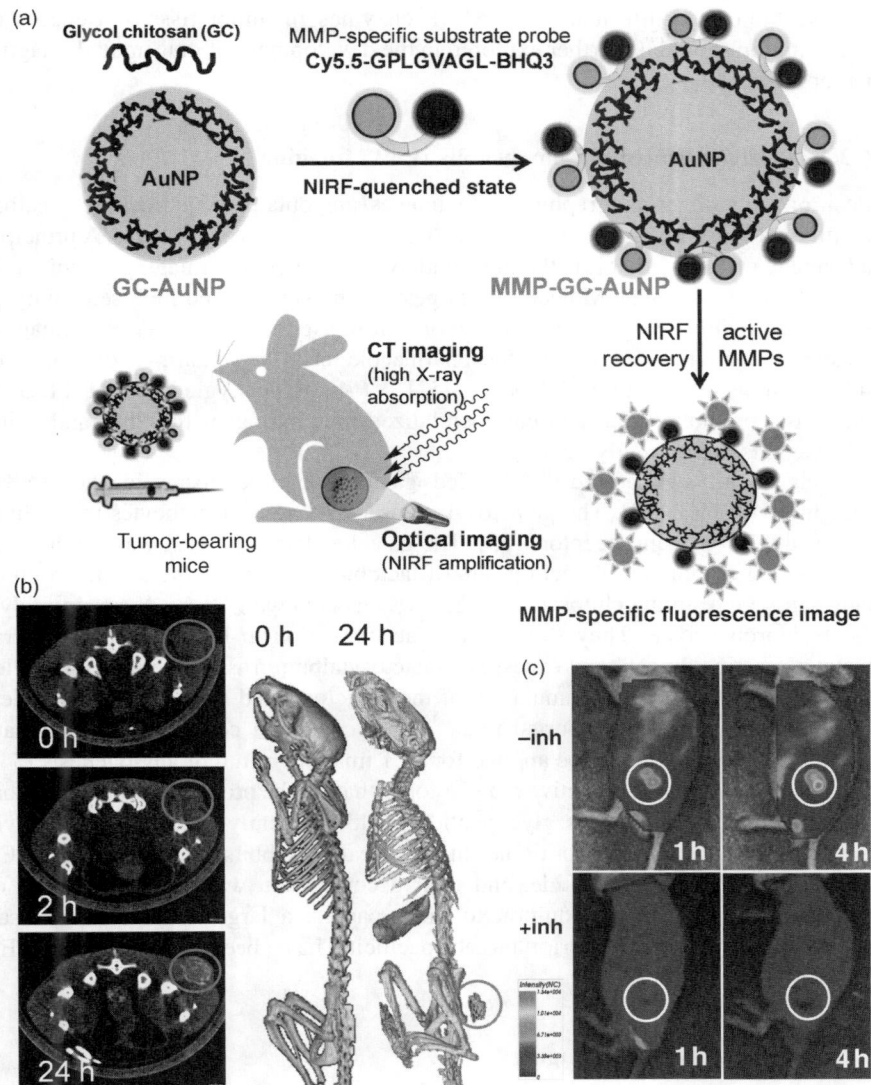

FIGURE 6.3 Chitosan nanoparticles for optical/CT imaging. (a) Schematic illustration of an MMP-specific glycol chitosan-coated gold nanoparticle and their MMP triggered fluorescence activation. (b) CT image of tumor-bearing mouse after injection of chitosan gold nanoparticles (red circles mark the tumor site). (c) NIR fluorescence image of tumor-bearing mouse after injection of chitosan gold nanoparticles (white circles mark the tumor site). (*Source*: Figures reproduced with permission from Reference [47].) (*See insert for color representation of the figure.*)

precise biological information of MMP enzymes in tumor tissues. Benefits of combination should be further explored in the development of multimodal imaging nanoprobes [40].

6.2.4 Positron Emission Tomography (PET) Imaging

Positron emission tomography (PET) images are obtained by tracking positron emitting radioisotopes such as ^{11}C, ^{18}F, ^{64}Cu, and ^{68}Ga in living systems. A principal advantage of PET is that it is the most widely used functional imaging technique in current clinical settings. Molecularly targeted radioisotopes and high sensitivity of PET can provide biological information of the disease [50]. Moreover, PET imaging enables quantitative analysis of the pharmacokinetics and pharmacodynamics of injected molecules with radiolabeling [51]. ^{18}F-fluorodeoxyglucose (^{18}F-FDG) is the most commonly used and commercialized imaging agent for PET, and many other agents are currently developed.

Yang et al. synthesized ^{18}F-labeled galactosylated chitosan for hepatocyte imaging with PET [52]. The galactosyl group can bind to hepatocytes in the liver via asialoglycoprotein receptors [53]. The galactosylated chitosan was synthesized by coupling the carboxylic acid group of lactobionic acid and the amine group of chitosan. The resulting chitosan conjugate was labeled with ^{18}F by N-succinimidyl-4-^{18}F-fluorobenzoate. They showed high accumulation in liver tissue after intravenous injection. Injection of excess neogalactosylalbumin as competitive molecules for receptors prevented accumulation of the galactosylated chitosan nanoparticles.

Recently, Tsao et al. reported that ^{68}Ga-glycopeptide composed of glutamate peptide and chitosan could be applied for PET imaging of tumor angiogenesis [54]. ^{68}Ga is one of the most attractive radioisotopes that can be produced from generators for PET imaging [55]. This glycopeptide could efficiently enter tumor cells via glutamate transporters. Tumor tissues in New Zealand rabbits were observed in PET images with ^{68}Ga-glycopeptide, and their accumulation was higher than that of commercial ^{18}F-FDG. In addition to these papers, a large number of chitosan nanoparticles showing superior target specificity have been developed for PET imaging [9,56].

6.2.5 Ultrasound (US) Imaging

The basic mechanism of ultrasound (US) imaging is the analysis of acoustic signals resulting from reflection or scattering of applied ultrasonic waves inside the body [57]. US imaging has many advantages such as relatively low capital costs, simplicity of the procedure, direct imaging without delay, and low cost of imaging. [58] For the enhancement of US contrast, gas-encapsulated nano- or microbubbles are generally used during imaging [59]. As the core materials of these bubbles, air or inert perfluorocarbon gas has been used. These gas cores are encapsulated by shell materials such as lipids or polymers to enhance the stability and water solubility of bubbles for intravenous injection to patients. Albumin-coated Albunex® and lipid-coated Sonovue® are the most widely used US contrast agents at present.

Liu et al. used chitosan as coating materials of perfluorocarbon gas in developing as an US contrast agent [60]. They synthesized a core-shell structure with silica and chitosan and made hollow nanoparticles by cross-linking of chitosan shell with glutaraldehyde and removing silica cores in hydrofluoride solution. Subsequently, liquid-form perfluoropentane was loaded into the core of nanoparticles. The resulting chitosan/perfluoropentane nanobubbles generated intense US signals in PBS solution similar to Sonovue. Although only a few US contrast agents have been developed with chitosan nanoparticles so far, many chitosan derivatives may be useful for the fabrication of microbubbles for *in vivo* US imaging due to their biocompatibility and wide availability.

6.3   CHITOSAN NANOPARTICLES FOR THERAPY

6.3.1   Drug Delivery

Owing to the positive charge of chitosan, some chitosan nanoparticles have been developed based on electronic interactions with anionic polymers. Sung's group developed pH-responsive nanoparticles with cationic chitosan and anionic poly-γ-glutamic acid [61]. The ionic cross-linking in this particle was further enhanced with tripolyphosphate (TPP) and magnesium sulfate [62] (Figure 6.4a). From pH 2.5 to 6.6, the nanoparticle is relatively stable because both chitosan and poly-γ-glutamic acid are ionized. However, chitosan is deprotonated at pH above 7.0, and nanoparticles become unstable and can release drugs. This study showed that pH-responsive protonation of chitosan can be applied for controlling drug release.

Many previous papers proved that the amine and carboxylic acid groups of chitosan could bind to glycoproteins in mucus by hydrogen bonding [63]. Moreover, recent studies showed that the tight junctions in monolayers of enterocytes could be opened after binding with chitosan molecules [64,65]. Consequently, these propert-ies make chitosan nanoparticles promising drug carriers for oral drug delivery [66,67]. Sung group applied their pH-responsive chitosan nanoparticles to oral delivery of insulin [68,69]. They showed superior transport of insulin across Caco-2 cell monolayers by chitosan nanoparticles using confocal microscopy (Figure 6.4b). The efficient insulin delivery to the duodenum and jejunum by these particles has also been successfully proved in rat models. Furthermore, the time-dependent blood glucose level data in a diabetic rat model demonstrated the great potential of the chitosan nanoparticles as a protein drug carrier for oral delivery.

Because of its cationic charge, chitosan can bind to anionic drugs directly. Heparin is a highly sulfated glycosaminoglycan with intense anionic charge. It is clinically used as a potent anticoagulant, which can form an inhibitory complex with anitithrombin [70]. Sung's group showed that cationic chitosan can bind to anionic heparin and form self-assembled nanoparticles [70]. The chitosan–heparin nano-particles were successfully applied to the oral delivery of heparin. Chitosan enhanced the absorption of heparin in the gastrointestinal tract, resulting in effective anti-coagulation after oral administration in a rat model.

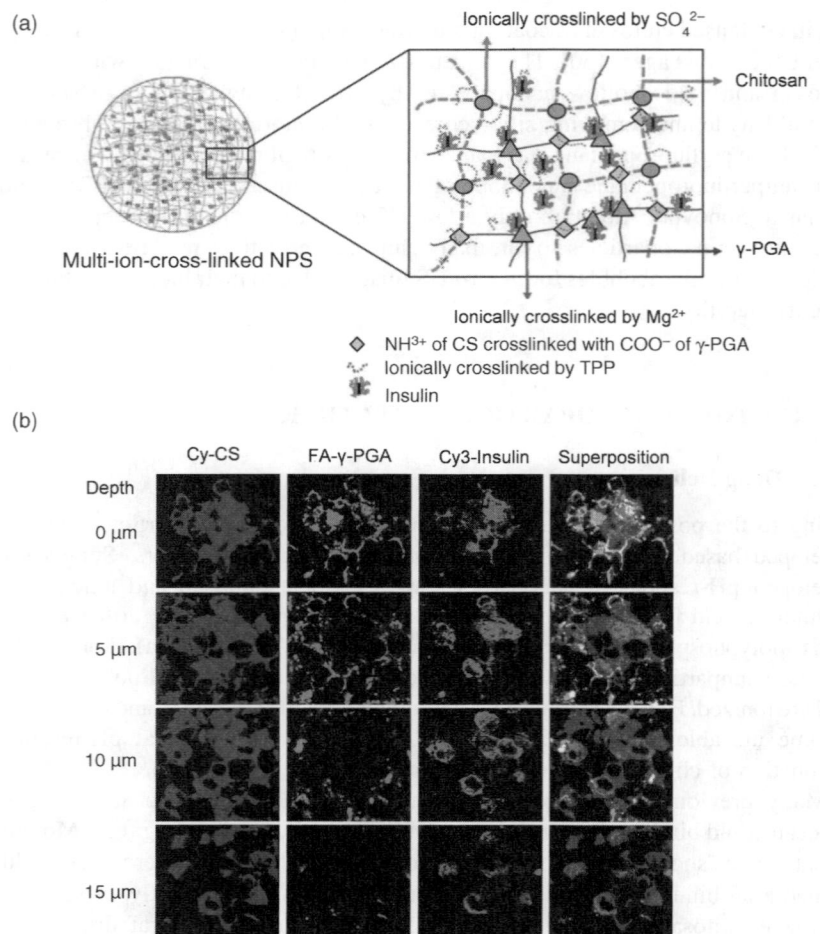

FIGURE 6.4 Chitosan nanoparticles for drug delivery. (a) Schematic illustration of multi-ion-cross-linked chitosan nanoparticles with poly-γ-glutamic acid (γ-PGA). (b) Fluorescence images of Caco-2 cell monolayers with chitosan nanoparticles. (*Source*: Figures reproduced with permission from Reference [61].) (*See insert for color representation of the figure.*)

In case of hydrophobic drugs, amphiphilic chitosan nanoparticles may be a good carrier candidate [71]. Zhao et al. synthesized chitosan nanoparticles by conjugation of hydrophobic linoleic acid and hydrophilic poly(β-maleic acid) to chitosan backbone [72]. Poly(β-maleic acid) could improve the water solubility of original chitosan and facilitate self-assembly of the chitosan conjugate in nanoparticles. A hydrophobic anticancer drug, paclitaxel, was stably loaded into this chitosan nanoparticle, and its therapeutic efficacy has been proved in a tumor-bearing mouse model. Recently, Lee et al. synthesized *N*-acetyl histidine conjugated glycol chitosan nanoparticles and showed successful therapeutic results following an intravenous injection of doxorubicin-loaded nanoparticles in tumor-bearing mice [73]. The biodistribution of nanoparticles and doxorubicin was monitored by optical and radioisotope imaging. In particular, they showed that the ratio of hydrophobic moiety in nanoparticles affected the amount of doxorubicin accumulated in tumor tissues. This study shows that the optimization of nanoparticles for efficient drug delivery is important, and that real-time *in vivo* imaging technique can be a useful tool for optimization [74].

6.3.2 Gene Delivery

Chitosan has long been considered a highly attractive material for developing gene carriers due to its primary amine groups and cationic charges. Chitosan can form a stable complex with nucleotides and increase their cellular uptake. A number of gene delivery carriers were developed with original or modified forms of chitosan, and these efforts are summarized in recent reviews [18,75]. We introduce some representative examples in this section.

Chitosan can be used as a gene carrier in its original form. Leong and colleagues reported that the Factor VIII DNA polyplexes with chitosan could be used in oral gene delivery [76]. Molecular weight and the degree of deacetylation of chitosan have a significant influence on transfection efficiency. In this study, they used chitosan with a molecular weight of 390 kDa and a degree of deacetylation of 84%. Owing to the mucoadhesive character of chitosan, the transfection efficiency with chitosan by oral delivery was more enhanced compared with naked DNA. Moreover, the increased thrombin generation and phenotypic correction in hemophilia A mouse models proved that chitosan was an attractive gene carrier for oral delivery.

Many researchers who are interested in gene therapy of lung diseases have tried several administration routes, including pulmonary administration. Mohri et al. introduced spray–freeze–dried powder of chitosan and DNA for inhalation [77]. They achieved high recovery of dried genes through this method and showed that an optimized amount of chitosan could enhance gene transfection. With dual imaging based on fluorescence and luminescence, the pulmonary administration and gene expression were evaluated in a mouse model. The intense luminescence signal in lung tissues showed that chitosan powder produced with spray–freeze–dried method could be used as pulmonary gene carriers.

To enhance the target cell specificity for gene delivery, some receptor-binding moieties have been conjugated to chitosan nanoparticles [78]. Zheng et al. used folate as a targeting moiety, which can bind to folate receptors abundant in various tumor cells [79]. They synthesized folate conjugated N-trimethyl chitosan with high water solubility and DNA condensing ability [80]. The resulting polymer showed high transfection efficiency *in vitro*, and its cytotoxicity was lower than that of synthetic polymers such as polyethylenimine. For hepatocyte targeting, galactosylated chitosan-polyethylenimine (PEI) was synthesized by Jiang and colleagues [81]. It enhanced the transfection efficiency in liver cells by binding to asialoglycoprotein receptors.

In recent years, small interfering RNA (siRNA) is emerging as a promising therapeutic modality, which can specifically suppress the expression of target protein with no adverse effects [82,83]. Oh group developed poly-L-arginine derivatives of chitosan (CS–PLR), where PEG was also conjugated to inhibit their aggregation under aqueous conditions [84] (Figure 6.5a). This chitosan nanoparticle can form a stable complex with siRNA, and its gene silencing effect was confirmed under cell culture conditions with green fluorescence protein. This PEG–CS–PLR showed enhanced gene silencing efficiency and cellular uptake *in vitro* compared to Lipofectamine™ 2000, commercialized transfection agent and non-PEGylated CS–PLR (Figure 6.5b, c). Finally, they made a tumor-bearing mouse model with B16F10 tumor cells expressing red fluorescence protein (RFP). After intratumoral injection of chitosan nanoparticles containing RFP sequence siRNA, the red fluorescence in tumor site was reduced, proving the *in vivo* gene silencing effect (Figure 6.5d).

Howard and colleagues showed therapeutic results of chitosan/siRNA nanoparticles by intraperitoneal injection [85]. They used an unmodified chitosan with a molecular weight of 114 kDa and a deacetylation degree of 84%. At pH 5.5, chitosan nanoparticles were formed with siRNA of tumor necrosis factor-α (TNF-α)

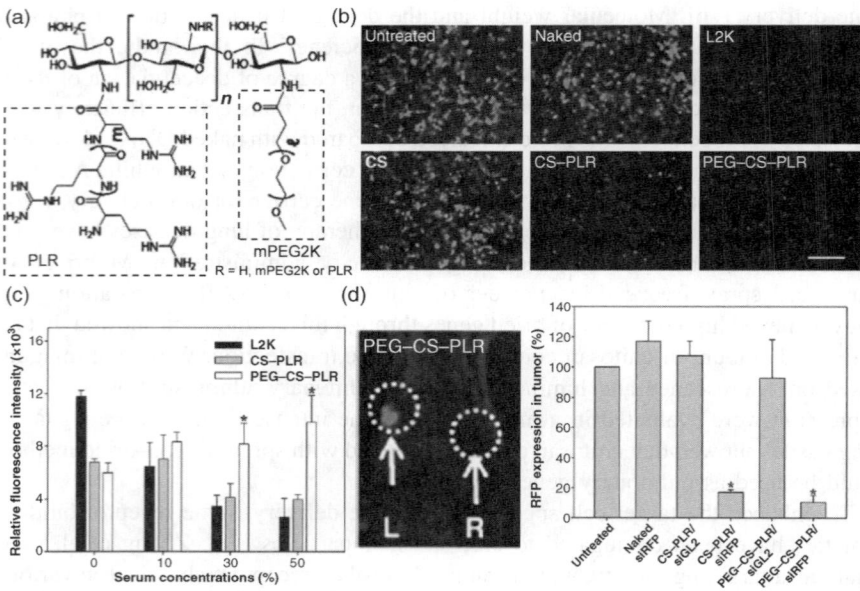

FIGURE 6.5 Chitosan nanoparticles for gene delivery. (a) Chemical structure of PEG–polyarginine–chitosan (PEG–CS–PLR). (b) *In vitro* silencing of green fluorescence protein (GFP) expression with chitosan nanoparticles. (c) Cellular uptake analysis using fluorescently labeled siRNA. *denotes significant difference from other groups (p <0.05) (d) *In vivo* silencing of red fluorescence protein (RFP) expression with chitosan nanoparticles (white circles denote tumor site). (*Source*: Figures reproduced with permission from Reference [84].) (*See insert for color representation of the figure.*)

sequence. During *in vitro* transfection in murine peritoneal macrophages, the chitosan nanoparticles showed efficient knockdown of TNF-α gene. Then, a collagen type II-induced arthritis (CIA) mouse model was established by immunization with an arthrogen-CIA collagen emulsion. After intraperitoneal injection of chitosan nanoparticles, significant anti-inflammatory effect was observed in the CIA model.

For intravenous injection, chitosan nanoparticles containing genes should provide long-term circulation and resist aggregation in blood. Huh et al. reported that glycol chitosan/PEI nanoparticles could be applied to tumor-targeted siRNA delivery via intravenous injection [86]. They made glycol chitosan/PEI nanoparticles by conjugation with 5β-cholanic acid to each polymer. The resulting mixture with two polymer conjugates were self-assembled into a nanoparticle form in an aqueous condition prior to siRNA addition. PEI enhanced the electronic interactions with siRNA, and the glycol group of glycol chitosan enhanced water solubility and inhibited aggregation with serum proteins [87]. In tumor-bearing mice models, this nanoparticle formulation showed approximately fivefold higher accumulation of siRNA compared to the free form of siRNA. *In vivo* gene silencing was also demonstrated in an RFP/B16F10 tumor-bearing mouse model. These results demonstrate the huge potential of chitosan nanoparticles for clinical application as gene carriers.

6.3.3 Photodynamic Therapy

Photodynamic therapy is one of the therapeutic modalities in clinical fields, mainly used in treatments of cancers or skin diseases [88]. It employs photosensitizers, which can generate cytotoxic singlet oxygen when irradiated at an appropriate wavelength. Consequently, targeted delivery of these photosensitizers can result in cell death or tissue destruction in disease site for therapeutic purpose [89]. If they are not delivered to target disease site properly, the therapeutic efficacy reduces and they can result in unintended phototoxicity in normal tissues. Therefore, efficient delivery systems of photosensitizers are needed for successful photodynamic therapy [90].

In 2010, Kim's group applied the amphiphilic glycol chitosan nanoparticles for tumor-targeted delivery of a hydrophobic photosensitizer, protophorphyrin IX (PpIX) [91]. The nanoparticles achieved faster cellular uptake of PpIX than free PpIX, and this uptake rate reflected the efficient death of tumor cells. The *in vivo* data showed high accumulation of PpIX in tumor site during optical imaging and successful tumor suppression in a SCC7 tumor-bearing mouse model upon laser irradiation.

One of the attractive features of photosensitizers is the simultaneous generation of fluorescence with singlet oxygen upon irradiation [92]. This intrinsic fluorescence can provide easy tracking of the photosensitizers in both cell and animal models [93,94]. Kim's group also reported that this fluorescence can be applied to the time-dependent tracking of photosensitizers in a tumor-bearing mouse model by optical imaging [95]. They synthesized two types of glycol chitosan nanoparticles based on physical loading and chemical conjugation of photosensitizers, and imaged biodistribution of the photosensitizers noninvasively by NIR fluorescence (Figure 6.6a).

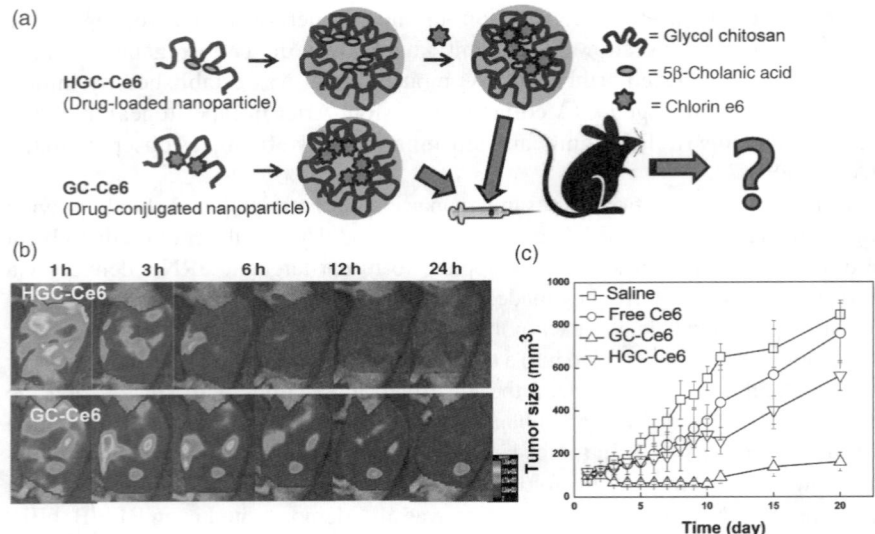

FIGURE 6.6 Chitosan nanoparticles for photodynamic therapy. (a) Schematic illustration of two kinds of glycol chitosan nanoparticles containing photosensitizer, chlorin e6 (Ce6). (b) *In vivo* biodistribution of photosensitizers delivered by chitosan nanoparticles. (c) *In vivo* photodynamic therapy with chitosan nanoparticles. (*Source*: Figures reproduced with permission from Reference [95].) (*See insert for color representation of the figure.*)

The results showed that the loss of photosensitizers by unintended burst release was significantly large in case of physical loading (Figure 6.6b). This loss of photosensitizers could result in lower therapeutic efficacy when compared to photosensitizer-conjugated nanoparticles (Figure 6.6c). This problem of burst release during blood circulation should be overcome by innovative methods to obtain improved therapeutic results [96].

6.4   PROMISES, LIMITATIONS, AND FUTURE PERSPECTIVES OF CHITOSAN NANOPARTICLES

Until now, various forms of nanoparticles have been developed based on chitosan for biomedical applications including imaging, drug delivery, and gene delivery [97]. Chitosan is a biodegradable and biocompatible natural polymer with primary amine groups and cationic charges, and many researchers have explored chitosan due to its unique properties. Consequently, valuable results have been obtained with chitosan nanoparticles in cell and animal models. However, some hurdles remain to be overcome for their clinical applications.

The first challenge is the approval from the Food and Drug Administration (FDA) [98]. For commercialization of chitosan nanoparticles for clinical applications, only an FDA-approved form of chitosan should be used. Various types of modifications or conjugations are performed to improve the functionality of chitosan [99]. However, additional synthetic modification steps are likely to delay the commercialization and clinical translation of nanoparticles.

The second challenge is the quality control of chitosan nanoparticles during manufacture. Physicochemical properties such as size, potential, and molecular weight of chitosan nanoparticles are not as uniform as those of single molecular drugs or imaging agents. This heterogeneity may not be a significant issue at an experimental level, but it may become a potential obstacle from a commercial standpoint. In fact, quality control is a common challenge for most nanoparticle systems for biomedical applications.

While chitosan nanoparticles have a long way to go, there are a number of exciting ongoing studies that employ chitosan in developing nanoparticles for biomedical applications [100]. Investigations are warranted with regard to the potential of chitosan nanoparticles as a system to improve the quality of clinical treatments in the foreseeable future.

REFERENCES

1. Petros, R. A., DeSimone, J. M. (2010). Strategies in the design of nanoparticles for therapeutic applications. *Nature Reviews Drug Discovery 9*, 615–627.

2. Koo, H., Huh, M. S., Ryu, J. H., Lee, D.-E., Sun, I.-C., Choi, K., Kim, K., Kwon, I. C. (2011). Nanoprobes for biomedical imaging in living systems. *Nano Today 6*, 204–220.

3. Park, K., Lee, S., Kang, E., Kim, K., Choi, K., Kwon, I. C. (2009). New generation of multifunctional nanoparticles for cancer imaging and therapy. *Advanced Functional Materials 19*, 1553–1566.

4. Xie, J., Lee, S., Chen, X. (2010). Nanoparticle-based theranostic agents. *Advanced Drug Delivery Reviews 62*, 1064–1079.

5. Kim, J.-H., Park, K., Nam, H. Y., Lee, S., Kim, K., Kwon, I. C. (2007). Polymers for bioimaging. *Progress in Polymer Science 32*, 1031–1053.

6. Fox, M. E., Szoka, F. C., Fréchet, J. M. J. (2009). Soluble polymer carriers for the treatment of cancer: the importance of molecular architecture. *Accounts of Chemical Research 42*, 1141–1151.

7. Bae, Y., Kataoka, K. (2009). Intelligent polymeric micelles from functional poly(ethylene glycol)-poly(amino acid) block copolymers. *Advanced Drug Delivery Reviews 61*, 768–784.

8. Yhee, J., Koo, H., Lee, D., Choi, K., Kwon, I., Kim, K. (2011). Multifunctional chitosan nanoparticles for tumor imaging and therapy. *Advances in Polymer Science 243*, 139–161.

9. Agrawal, P., Strijkers, G. J., Nicolay, K. (2010). Chitosan-based systems for molecular imaging. *Advanced Drug Delivery Reviews 62*, 42–58.

10. Dash, M., Chiellini, F., Ottenbrite, R. M., Chiellini, E. (2011). Chitosan—A versatile semi-synthetic polymer in biomedical applications. *Progress in Polymer Science 36*, 981–1014.

11. Liu, X.-P., Zhou, S.-T., Li, X.-Y., Chen, X.-C., Zhao, X., Qian, Z.-Y., Zhou, L.-N., Li, Z.-Y., Wang, Y.-M., Zhong, Q., Yi, T., Li, Z.-Y., He, X., Wei, Y.-Q. (2010). Anti-tumor activity of *N*-trimethyl chitosan-encapsulated camptothecin in a mouse melanoma model. *Journal of Experimental and Clinical Cancer Research 29*, 76.

12. Bhattacharya, D., Das, M., Mishra, D., Banerjee, I., Sahu, S. K., Maiti, T. K., Pramanik, P. (2011). Folate receptor targeted, carboxymethyl chitosan functionalized iron oxide nanoparticles: a novel ultradispersed nanoconjugates for bimodal imaging. *Nanoscale 3*, 1653–1662.

13. Artan, M., Karadeniz, F., Karagozlu, M. Z., Kim, M.-M., Kim, S.-K. (2010). Anti-HIV-1 activity of low molecular weight sulfated chitooligosaccharides. *Carbohydrate Research 345*, 656–662.

14. Bae, K., Moon, C., Lee, Y., Park, T. (2009). Intracellular delivery of heparin complexed with chitosan-g-poly(ethylene glycol) for inducing apoptosis. *Pharmaceutical Research 26*, 93–100.

15. Lee, S. J., Koo, H., Lee, D.-E., Min, S., Lee, S., Chen, X., Choi, Y., Leary, J. F., Park, K., Jeong, S. Y., Kwon, I. C., Kim, K., Choi, K. (2011). Tumor-homing photosensitizer-conjugated glycol chitosan nanoparticles for synchronous photodynamic imaging and therapy based on cellular on/off system. *Biomaterials 32*, 4021–4029.

16. Hwang, H.-Y., Kim, I.-S., Kwon, I. C., Kim, Y.-H. (2008). Tumor targetability and antitumor effect of docetaxel-loaded hydrophobically modified glycol chitosan nanoparticles. *Journal of Controlled Release 128*, 23–31.

17. Huang, M., Khor, E., Lim, L.-Y. (2004). Uptake and cytotoxicity of chitosan molecules and nanoparticles: effects of molecular weight and degree of deacetylation. *Pharmaceutical Research 21*, 344–353.

18. Mao, S., Sun, W., Kissel, T. (2010). Chitosan-based formulations for delivery of DNA and siRNA. *Advanced Drug Delivery Reviews 62*, 12–27.

19. Noh, S. M., Han, S. E., Shim, G., Lee, K. E., Kim, C.-W., Han, S. S., Choi, Y., Kim, Y. K., Kim, W.-K., Oh, Y.-K. (2011). Tocopheryl oligochitosan-based self assembling oligomersomes for siRNA delivery. *Biomaterials 32*, 849–857.

20. Kim, Y., Jiang, H., Choi, Y., Park, I., Cho, M., Cho, C. (2011). Polymeric nanoparticles of chitosan derivatives as DNA and siRNA carriers. *Advances in Polymer Science 243*, 1–21.

21. Kim, K., Kim, J. H., Park, H., Kim, Y.-S., Park, K., Nam, H., Lee, S., Park, J. H., Park, R.-W., Kim, I.-S., Choi, K., Kim, S. Y., Park, K., Kwon, I. C. (2010). Tumor-homing multifunctional nanoparticles for cancer theragnosis: simultaneous diagnosis, drug delivery, and therapeutic monitoring. *Journal of Controlled Release 146*, 219–227.

22. Park, K., Kim, J.-H., Nam, Y. S., Lee, S., Nam, H. Y., Kim, K., Park, J. H., Kim, I.-S., Choi, K., Kim, S. Y., Kwon, I. C. (2007). Effect of polymer molecular weight on the tumor targeting characteristics of self-assembled glycol chitosan nanoparticles. *Journal of Controlled Release 122*, 305–314.

23. Baldrick, P. (2010) The safety of chitosan as a pharmaceutical excipient. *Regulatory Toxicology and Pharmacology 56*, 290–299.

24. Wedmore, I., McManus, J. G., Pusateri, A. E., Holcomb, J. B. (2006). A special report on the chitosan-based hemostatic dressing: experience in current combat operations. *The Journal of Trauma 60*, 655–658.

25. Kean, T., Thanou, M. (2010). Biodegradation, biodistribution and toxicity of chitosan. *Advanced Drug Delivery Reviews 62*, 3–11.

26. Hilderbrand, S. A., Weissleder, R. (2010). Near-infrared fluorescence: application to *in vivo* molecular imaging. *Current Opinion in Chemical Biology 14*, 71–79.

27. Lee, S., Park, K., Kim, K., Choi, K., Kwon, I. C. (2008). Activatable imaging probes with amplified fluorescent signals. *Chemical Communications (Cambridge, England)* 4250–4260.

28. Weissleder, R. (2001). A clearer vision for *in vivo* imaging. *Nature Biotechnology 19*, 316–317.

29. Ranjan, A. P., Zeglam, K., Mukerjee, A., Thamake, S., Vishwanatha, J. K. (2011). A sustained release formulation of chitosan modified PLCL: poloxamer blend nanoparticles loaded with optical agent for animal imaging. *Nanotechnology 22*, 295104.

30. Na, J. H., Koo, H., Lee, S., Min, K. H., Park, K., Yoo, H., Lee, S. H., Park, J. H., Kwon, I. C., Jeong, S. Y., Kim, K. (2011). Real-time and non-invasive optical imaging of tumor-targeting glycol chitosan nanoparticles in various tumor models. *Biomaterials 32*, 5252–5261.

31. Park, J. H., Kwon, S., Nam, J.-O., Park, R.-W., Chung, H., Seo, S. B., Kim, I.-S., Kwon, I. C., Jeong, S. Y. (2004). Self-assembled nanoparticles based on glycol chitosan bearing 5[beta]-cholanic acid for RGD peptide delivery. *Journal of Controlled Release 95*, 579–588.

32. Torchilin, V. (2011). Tumor delivery of macromolecular drugs based on the EPR effect. *Advanced Drug Delivery Reviews 63*, 131–135.

33. Kim, J.-H., Kim, Y.-S., Park, K., Lee, S., Nam, H. Y., Min, K. H., Jo, H. G., Park, J. H., Choi, K., Jeong, S. Y., Park, R.-W., Kim, I.-S., Kim, K., Kwon, I. C. (2008). Antitumor efficacy of cisplatin-loaded glycol chitosan nanoparticles in tumor-bearing mice. *Journal of Controlled Release 127*, 41–49.

34. Nam, H. Y., Kwon, S. M., Chung, H., Lee, S.-Y., Kwon, S.-H., Jeon, H., Kim, Y., Park, J. H., Kim, J., Her, S., Oh, Y.-K., Kwon, I. C., Kim, K., Jeong, S. Y. (2009). Cellular uptake mechanism and intracellular fate of hydrophobically modified glycol chitosan nanoparticles. *Journal of Controlled Release 135*, 259–267.

35. Maeda, H. (2010). Tumor-selective delivery of macromolecular drugs via the EPR effect: background and future prospects. *Bioconjugate Chemistry 21*, 797–802.

36. Na, H. B., Song, I. C., Hyeon, T. (2009). Inorganic nanoparticles for MRI contrast agents. *Advanced Materials 21*, 2133–2148.

37. Lu, Z.-R., Wang, X., Parker, D. L., Goodrich, K. C., Buswell, H. R. (2003). Poly(l-glutamic acid) Gd(III)-DOTA conjugate with a degradable spacer for magnetic resonance imaging. *Bioconjugate Chemistry 14*, 715–719.

38. Yim, H., Yang, S.-G., Jeon, Y. S., Park, I. S., Kim, M., Lee, D. H., Bae, Y. H., Na, K. (2011). The performance of gadolinium diethylene triamine pentaacetate-pullulan hepatocyte-specific T1 contrast agent for MRI. *Biomaterials 32*, 5187–5194.

39. Nam, T., Park, S., Lee, S.-Y., Park, K., Choi, K., Song, I. C., Han, M. H., Leary, J. J., Yuk, S. A., Kwon, I. C., Kim, K., Jeong, S. Y. (2010). Tumor targeting chitosan nanoparticles for dual-modality optical/MR cancer imaging. *Bioconjugate Chemistry 21*, 578–582.

40. Cheon, J., Lee, J.-H. (2008). Synergistically integrated nanoparticles as multimodal probes for nanobiotechnology. *Accounts of Chemical Research 41*, 1630–1640.

41. Gao, J., Gu, H., Xu, B. (2009). Multifunctional magnetic nanoparticles: design, synthesis, and biomedical applications. *Accounts of Chemical Research 42*, 1097–1107.

42. Lee, C.-M., Jeong, H.-J., Kim, S.-L., Kim, E.-M., Kim, D. W., Lim, S. T., Jang, K. Y., Jeong, Y. Y., Nah, J.-W., Sohn, M.-H. (2009). SPION-loaded chitosan-linoleic acid nanoparticles to target hepatocytes. *International Journal of Pharmaceutics 371*, 163–169.

43. Lee, C.-M., Jang, D., Kim, J., Cheong, S.-J., Kim, E.-M., Jeong, M.-H., Kim, S.-H., Kim, D. W., Lim, S. T., Sohn, M.-H., Jeong, Y. Y., Jeong, H.-J. (2011). Oleyl-chitosan nanoparticles based on a dual probe for optical/MR imaging *in vivo*. *Bioconjugate Chemistry 22*, 186–192.

44. Kweon, S., Lee, H.-J., Hyung, W., Suh, J., Lim, J., Lim, S.-J. (2010). Liposomes coloaded with iopamidol/lipiodol as a RES-targeted contrast agent for computed tomography imaging. *Pharmaceutical Research 27*, 1408–1415.

45. Sun, I.-C., Eun, D.-K., Na, J. H., Lee, S., Kim, I.-J., Youn, I.-C., Ko, C.-Y., Kim, H.-S., Lim, D., Choi, K., Messersmith, P. B., Park, T. G., Kim, S. Y., Kwon, I. C., Kim, K., Ahn, C.-H. (2009). Heparin-coated gold nanoparticles for liver-specific CT imaging. *Chemistry - A European Journal 15*, 13341–13347.

46. Kim, D., Jeong, Y. Y., Jon, S. (2010). A drug-loaded aptamer–gold nanoparticle bioconjugate for combined CT imaging and therapy of prostate cancer. *ACS Nano 4*, 3689–3696.

47. Sun, I.-C., Eun, D.-K., Koo, H., Ko, C.-Y., Kim, H.-S., Yi, D. K., Choi, K., Kwon, I. C., Kim, K., Ahn, C.-H. (2011). Tumor targeting gold particles for CT/optical dual cancer imaging. *Angewandte Chemie International Edition 50*, 9348–9351.

48. Lee, S., Ryu, J. H., Park, K., Lee, A., Lee, S.-Y., Youn, I.-C., Ahn, C.-H., Yoon, S. M., Myung, S.-J., Moon, D. H., Chen, X., Choi, K., Kwon, I. C., Kim, K. (2009). Polymeric nanoparticle-based activatable near-infrared nanosensor for protease determination *in vivo*. *Nano Letters 9*, 4412–4416.

49. Louie, A. (2010). Multimodality imaging probes: design and challenges. *Chemical Reviews 110*, 3146–3195.

50. Chen, K., Conti, P. S. (2010). Target-specific delivery of peptide-based probes for PET imaging. *Advanced Drug Delivery Reviews 62*, 1005–1022.

51. Cho, Y. W., Park, S. A., Han, T. H., Son, D. H., Park, J. S., Oh, S. J., Moon, D. H., Cho, K.-J., Ahn, C.-H., Byun, Y., Kim, I.-S., Kwon, I. C., Kim, S. Y. (2007). *In vivo* tumor targeting and radionuclide imaging with self-assembled nanoparticles: mechanisms, key factors, and their implications. *Biomaterials 28*, 1236–1247.

52. Yang, W., Mou, T., Guo, W., Jing, H., Peng, C., Zhang, X., Ma, Y., Liu, B. (2010). Fluorine-18 labeled galactosylated chitosan for asialoglycoprotein-receptor-mediated hepatocyte imaging. *Bioorganic & Medicinal Chemistry Letters 20*, 4840–4844.

53. Frisch, B., Carriere, M., Largeau, C., Mathey, F., Masson, C., Schuber, F., Scherman, D., Escriou, V. (2004). A new triantennary galactose-targeted PEGylated gene carrier, characterization of its complex with DNA, and transfection of hepatoma cells. *Bioconjugate Chemistry 15*, 754–764.

54. Tsao, N., Wang, C.-H., Her, L.-J., Tzen, K.-Y., Chen, J.-Y., Yu, D.-F., Yang, D. J. (2011). Development of 68Ga-glycopeptide as an imaging probe for tumor angiogenesis. *Journal of Biomedicine and Biotechnology 2011*, 267206.

55. Zhernosekov, K. P., Filosofov, D. V., Baum, R. P., Aschoff, P., Bihl, H., Razbash, A. A., Jahn, M., Jennewein, M., Rösch, F. (2007). Processing of generator-produced 68Ga for medical application. *Journal of Nuclear Medicine 48*, 1741–1748.

56. Min, K. H., Park, K., Kim, Y.-S., Bae, S. M., Lee, S., Jo, H. G., Park, R.-W., Kim, I.-S., Jeong, S. Y., Kim, K., Kwon, I. C. (2008). Hydrophobically modified glycol chitosan nanoparticles-encapsulated camptothecin enhance the drug stability and tumor targeting in cancer therapy. *Journal of Controlled Release 127*, 208–218.

57. Deshpande, N., Needles, A., Willmann, J. K. (2010). Molecular ultrasound imaging: current status and future directions. *Clinical Radiology 65*, 567–581.

58. Voigt, J.-U. (2009). Ultrasound molecular imaging. *Methods 48*, 92–97.

59. Lindner, J. R. (2004). Microbubbles in medical imaging: current applications and future directions. *Nature Reviews Drug Discovery 3*, 527–533.

60. Liu, Y.-L., Wu, Y.-H., Tsai, W.-B., Tsai, C.-C., Chen, W.-S., Wu, C.-S. (2011). Core-shell silica@chitosan nanoparticles and hollow chitosan nanospheres using silica nanoparticles as templates: preparation and ultrasound bubble application. *Carbohydrate Polymers 84*, 770–774.

61. Lin, Y.-H., Sonaje, K., Lin, K. M., Juang, J.-H., Mi, F.-L., Yang, H.-W., Sung, H.-W. (2008). Multi-ion-crosslinked nanoparticles with pH-responsive characteristics for oral delivery of protein drugs. *Journal of Controlled Release 132*, 141–149.

62. Nasti, A., Zaki, N., de Leonardis, P., Ungphaiboon, S., Sansongsak, P., Rimoli, M., Tirelli, N. (2009). Chitosan/TPP and chitosan/TPP-hyaluronic acid nanoparticles: systematic optimisation of the preparative process and preliminary biological evaluation. *Pharmaceutical Research 26*, 1918–1930.

63. Suknuntha, K., Tantishaiyakul, V., Worakul, N., Taweepreda, W. (2011). Characterization of muco- and bioadhesive properties of chitosan, PVP, and chitosan/PVP blends and release of amoxicillin from alginate beads coated with chitosan/PVP. *Drug Development and Industrial Pharmacy 37*, 408–418.

64. Ranaldi, G., Marigliano, I., Vespignani, I., Perozzi, G., Sambuy, Y. (2002). The effect of chitosan and other polycations on tight junction permeability in the human intestinal Caco-2 cell line. *The Journal of Nutritional Biochemistry 13*, 157–167.

65. Lee, E., Kim, H., Lee, I.-H., Jon, S. (2009). *In vivo* antitumor effects of chitosan-conjugated docetaxel after oral administration. *Journal of Controlled Release 140*, 79–85.

66. Lee, E., Lee, J., Lee, I.-H., Yu, M., Kim, H., Chae, S. Y., Jon, S. (2008). Conjugated chitosan as a novel platform for oral delivery of paclitaxel. *Journal of Medicinal Chemistry 51*, 6442–6449.

67. George, M., Abraham, T. E. (2006). Polyionic hydrocolloids for the intestinal delivery of protein drugs: alginate and chitosan—a review. *Journal of Controlled Release 114*, 1–14.

68. Sonaje, K., Lin, Y.-H., Juang, J.-H., Wey, S.-P., Chen, C.-T., Sung, H.-W. (2009). *In vivo* evaluation of safety and efficacy of self-assembled nanoparticles for oral insulin delivery. *Biomaterials 30*, 2329–2339.

69. Sonaje, K., Lin, K.-J., Wey, S.-P., Lin, C.-K., Yeh, T.-H., Nguyen, H.-N., Hsu, C.-W., Yen, T.-C., Juang, J.-H., Sung, H.-W. (2010). Biodistribution, pharmacodynamics and pharmacokinetics of insulin analogues in a rat model: oral delivery using pH-Responsive nanoparticles vs. subcutaneous injection. *Biomaterials 31*, 6849–6858.

70. Chen, M.-C., Wong, H.-S., Lin, K.-J., Chen, H.-L., Wey, S.-P., Sonaje, K., Lin, Y.-H., Chu, C.-Y., Sung, H.-W. (2009). The characteristics, biodistribution and bioavailability of a chitosan-based nanoparticulate system for the oral delivery of heparin. *Biomaterials 30*, 6629–6637.

71. Park, J. H., Saravanakumar, G., Kim, K., Kwon, I. C. (2010). Targeted delivery of low molecular drugs using chitosan and its derivatives. *Advanced Drug Delivery Reviews 62*, 28–41.

72. Zhao, Z., He, M., Yin, L., Bao, J., Shi, L., Wang, B., Tang, C., Yin, C. (2009). Biodegradable nanoparticles based on linoleic acid and poly(β-malic acid) double grafted chitosan derivatives as carriers of anticancer drugs. *Biomacromolecules 10*, 565–572.

73. Lee, B. S., Park, K., Park, S., Kim, G. C., Kim, H. J., Lee, S., Kil, H., Oh, S. J., Chi, D., Kim, K., Choi, K., Kwon, I. C., Kim, S. Y. (2010). Tumor targeting efficiency of bare nanoparticles does not mean the efficacy of loaded anticancer drugs: importance of radionuclide imaging for optimization of highly selective tumor targeting polymeric nanoparticles with or without drug. *Journal of Controlled Release 147*, 253–260.

74. Willmann, J. K., van Bruggen, N., Dinkelborg, L. M., Gambhir, S. S. (2008). Molecular imaging in drug development. *Nature Reviews Drug Discovery 7*, 591–607.

75. Rudzinski, W. E., Aminabhavi, T. M. (2010). Chitosan as a carrier for targeted delivery of small interfering RNA. *International Journal of Pharmaceutics 399*, 1–11.

76. Bowman, K., Sarkar, R., Raut, S., Leong, K. W. (2008). Gene transfer to hemophilia A mice via oral delivery of FVIII-chitosan nanoparticles. *Journal of Controlled Release 132*, 252–259.

77. Mohri, K., Okuda, T., Mori, A., Danjo, K., Okamoto, H. (2010). Optimized pulmonary gene transfection in mice by spray-freeze dried powder inhalation. *Journal of Controlled Release 144*, 221–226.

78. Byrne, J. D., Betancourt, T., Brannon-Peppas, L. (2008). Active targeting schemes for nanoparticle systems in cancer therapeutics. *Advanced Drug Delivery Reviews 60*, 1615–1626.

79. Zheng, Y., Cai, Z., Song, X., Yu, B., Bi, Y., Chen, Q., Zhao, D., Xu, J., Hou, S. (2009). Receptor mediated gene delivery by folate conjugated N-trimethyl chitosan *in vitro*. *International Journal of Pharmaceutics 382*, 262–269.

80. Sahni, J. K., Chopra, S., Ahmad, F. J., Khar, R. K. (2008). Potential prospects of chitosan derivative trimethyl chitosan chloride (TMC) as a polymeric absorption enhancer: synthesis, characterization and applications. *Journal of Pharmacy and Pharmacology 60*, 1111–1119.

81. Jiang, H. L., Kwon, J. T., Kim, Y. K., Kim, E. M., Arote, R., Jeong, H. J., Nah, J. W., Choi, Y. J., Akaike, T., Cho, M. H., Cho, C. S. (2007). Galactosylated chitosan-graft-polyethylenimine as a gene carrier for hepatocyte targeting. *Gene Therapy 14*, 1389–1398.

82. Castanotto, D., Rossi, J. J. (2009). The promises and pitfalls of RNA-interference-based therapeutics. *Nature 457*, 426–433.

83. Oh, Y.-K., Park, T. G. (2009). siRNA delivery systems for cancer treatment. *Advanced Drug Delivery Reviews 61*, 850–862.

84. Noh, S. M., Park, M. O., Shim, G., Han, S. E., Lee, H. Y., Huh, J. H., Kim, M. S., Choi, J. J., Kim, K., Kwon, I. C., Kim, J.-S., Baek, K.-H., Oh, Y.-K. (2010). Pegylated poly-l-arginine derivatives of chitosan for effective delivery of siRNA. *Journal of Controlled Release 145*, 159–164.

85. Howard, K. A., Paludan, S. R., Behlke, M. A., Besenbacher, F., Deleuran, B., Kjems, J. (2008). Chitosan/siRNA nanoparticle-mediated TNF-[alpha] knockdown in peritoneal macrophages for anti-inflammatory treatment in a murine arthritis model. *Molecular Therapy 17*, 162–168.

86. Huh, M. S., Lee, S.-Y., Park, S., Lee, S., Chung, H., Lee, S., Choi, Y., Oh, Y.-K., Park, J. H., Jeong, S. Y., Choi, K., Kim, K., Kwon, I. C. (2010). Tumor-homing glycol chitosan/polyethylenimine nanoparticles for the systemic delivery of siRNA in tumor-bearing mice. *Journal of Controlled Release 144*, 134–143.

87. Saravanakumar, G., Min, K. H., Min, D. S., Kim, A. Y., Lee, C.-M., Cho, Y. W., Lee, S. C., Kim, K., Jeong, S. Y., Park, K., Park, J. H., Kwon, I. C. (2009). Hydrotropic oligomer-conjugated glycol chitosan as a carrier of paclitaxel: synthesis, characterization, and *in vivo* biodistribution. *Journal of Controlled Release 140*, 210–217.

88. Moore, C. M., Pendse, D., Emberton, M. (2009). Photodynamic therapy for prostate cancer—a review of current status and future promise. *Nature Clinical Practice Urology 6*, 18–30.

89. Chatterjee, D. K., Fong, L. S., Zhang, Y. (2008). Nanoparticles in photodynamic therapy: an emerging paradigm. *Advanced Drug Delivery Reviews 60*, 1627–1637.

90. van Nostrum, C. F. (2004). Polymeric micelles to deliver photosensitizers for photodynamic therapy. *Advanced Drug Delivery Reviews 56*, 9–16.

91. Lee, S. J., Park, K., Oh, Y.-K., Kwon, S.-H., Her, S., Kim, I.-S., Choi, K., Lee, S. J., Kim, H., Lee, S. G., Kim, K., Kwon, I. C. (2009). Tumor specificity and therapeutic efficacy of photosensitizer-encapsulated glycol chitosan-based nanoparticles in tumor-bearing mice. *Biomaterials 30*, 2929–2939.

92. Celli, J. P., Spring, B. Q., Rizvi, I., Evans, C. L., Samkoe, K. S., Verma, S., Pogue, B. W., Hasan, T. (2010). Imaging and photodynamic therapy: mechanisms, monitoring, and optimization. *Chemical Reviews 110*, 2795–2838.

93. Koo, H., Lee, H., Lee, S., Min, K. H., Kim, M. S., Lee, D. S., Choi, Y., Kwon, I. C., Kim, K., Jeong, S. Y. (2010). *In vivo* tumor diagnosis and photodynamic therapy via tumoral pH-responsive polymeric micelles. *Chemical Communications 46*, 5668–5670.

94. Park, S., Lee, S. J., Chung, H., Her, S., Choi, Y., Kim, K., Choi, K., Kwon, I. C. (2010). Cellular uptake pathway and drug release characteristics of drug-encapsulated glycol chitosan nanoparticles in live cells. *Microscopy Research and Technique 73*, 857–865.

95. Lee, S. J., Koo, H., Jeong, H., Huh, M. S., Choi, Y., Jeong, S. Y., Byun, Y., Choi, K., Kim, K., Kwon, I. C. (2011). Comparative study of photosensitizer loaded and conjugated glycol chitosan nanoparticles for cancer therapy. *Journal of Controlled Release 152*, 21–29.

96. Lee, H. J., Kim, S. E., Kwon, I. K., Park, C., Kim, C., Yang, J., Lee, S. C. (2010). Spatially mineralized self-assembled polymeric nanocarriers with enhanced robustness and controlled drug-releasing property. *Chemical Communications 46*, 377–379.

97. Peniche, H., Peniche, C. (2011). Chitosan nanoparticles: a contribution to nanomedicine. *Polymer International 60*, 883–889.

98. Eifler, A. C., Thaxton, C. S., Nanoparticle therapeutics: FDA approval, clinical trials, regulatory pathways, and case study, In *Biomedical Nanotechnology*, ed. Hurst, S. J., Humana Press, 2011, pp 325–338.

99. Alves, N. M., Mano, J. F. (2008). Chitosan derivatives obtained by chemical modifications for biomedical and environmental applications. *International Journal of Biological Macromolecules 43*, 401–414.

100. Liu, Z., Jiao, Y., Wang, Y., Zhou, C., Zhang, Z. (2008). Polysaccharides-based nanoparticles as drug delivery systems. *Advanced Drug Delivery Reviews 60*, 1650–1662.

7

POLYMER–DRUG NANOCONJUGATES

RONG TONG

Department of Chemical Engineering, Massachusetts Institute of Technology, Cambridge, MA, USA Laboratory for Biomaterials and Drug Delivery, Department of Anesthesiology, Division of Critical Care Medicine, Children's Hospital Boston, Harvard Medical School, Boston, MA, USA

LI TANG, NATHAN P. GABRIELSON, QIAN YIN, AND JIANJUN CHENG

Department of Materials Science and Engineering, University of Illinois at Urbana–Champaign, Urbana, IL, USA

7.1 INTRODUCTION

Polymeric nanomedicine, an emerging field that involves the use of drug-containing polymeric nanoparticles (NPs) for cancer treatment, is expected to alter the landscape of oncology [1]. The medical application of nanotechnology has been extensive: roughly 40 nanomedicines have already been approved by the Food and Drug Administration (FDA) for clinical use [2–4] and a handful of NPs are currently in preclinical investigations [2,5]. Incorporation of chemotherapeutic agents in NP delivery vehicles can improve water solubility, reduce clearance, reduce drug resistance, and enhance therapeutic effectiveness [6]. Broadly speaking, two approaches have been used to load drugs in NPs for delivery. One is to encapsulate drugs within NPs via noncovalent bonds, that is, distribution of the drug throughout a polymeric matrix during formulation (Figure 7.1a) [7]. The second loading strategy is the formulation of NPs using polymer–drug conjugates, a technique first proposed in 1975 (Figure 7.1b) [8]. In this chapter, we discuss the pros and cons of the two strategies and introduce newly developed polymer-drug conjugates—so-called nanoconjugates—as a formulation strategy which addresses challenges faced by both encapsulates and conjugates

Nanoparticulate Drug Delivery Systems: Strategies, Technologies, and Applications, First Edition.
Edited by Yoon Yeo.
© 2013 John Wiley & Sons, Inc. Published 2013 by John Wiley & Sons, Inc.

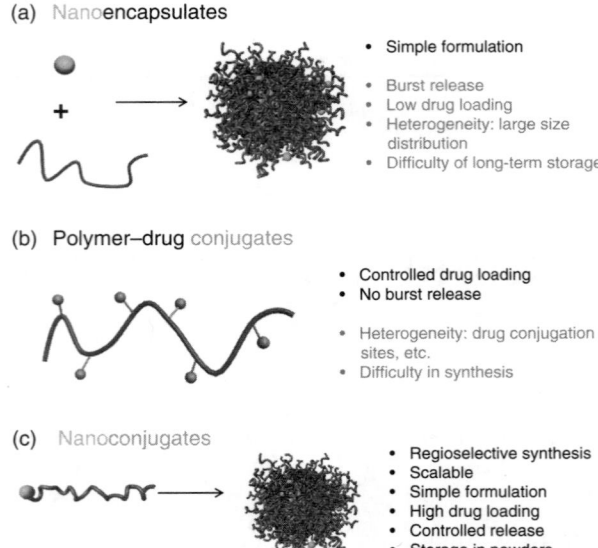

FIGURE 7.1 Pros (black) and cons (gray) of (a) nanoencapsulates (NEs) and (b) polymer–drug conjugates. We propose (c) nanoconjugates (NCs) with reduced heterogeneity to address the challenges in both encapsulated and conjugated systems.

(Figure 7.1c). Along the way, we also discuss topics related to regioselective polymer–drug conjugation chemistry, *in vitro* formulation and characterization, and present preliminary *in vivo* studies that highlight the potential clinical translation of nanoconjugates.

7.2   CURRENT STATUS OF NANOENCAPSULATES AND POLYMER–DRUG CONJUGATES

7.2.1   Nanoencapsulates

Nanoparticles loaded with a drug via encapsulation within a polymer matrix are termed nanoencapsulates. Although nanoencapsulate formulation is straightforward in both concept and fabrication, their clinical use is challenged by a variety of issues (Figure 7.1a). First, nanoencapsulates often show "burst" drug release profiles in aqueous solution, with as much as 80–90% of the encapsulated drug released during the first few to tens of hours [9]. The rapid drug release, also called dose dumping, can prevent the drug from operating within its therapeutic window and cause severe systemic toxicity [10]. Second, drug loading in nanoencapsulates is very low—typically in the range of 1–5% for most NPs studied—and is widely variable depending on the amount of drug being used, the hydrophobicity/hydrophilicity of the drug as well as the compatibility between the drug and the polymer [9,11]. Drug loading, in particular, is a critical measure of the feasibility of nanoencapsulate delivery systems in clinical settings [11]. At low drug loading, large amounts of delivery vehicles are needed to achieve therapeutic concentrations. However, because of the limited body weight and blood volume of animals, administration volumes are fixed. For mice with 20–30 g body weight, the intravenously injected volume must be kept around 100–200 μL [12]. Thus, the intravenous administration of NPs with 1% drug loading in a 100 μL solution at a dose of 50 mg/kg (e.g., docetaxel) would require a 1 g/mL NP solution. In practice, it is impossible to formulate such concentrated solutions and inject them intravenously. Another problem related to drug loading is the lack of a general strategy to achieve quantitative drug encapsulation in many NPs (e.g., polylactide (PLA) and poly(lactic-co-glycolic acid) (PLGA) NPs). In this instance, nonencapsulated drugs may self-aggregate, thus complicating their removal from the NPs [13]. Ultimately, these formulation challenges significantly impact the processability and the clinical transla-tion of nanoencapsulate delivery vehicles for cancer therapy.

7.2.2   Polymer–Drug Conjugates

As difficulties facing nanoencapsulate drug loading began emerging, considerable interest was shifted to drug delivery strategies employing polymer–drug conjugates (Figure 7.1b). In fact, a polymer–protein drug conjugate has already been approved for use in Japan, and a handful of other polymer–drug conjugates are presently transitioning from later clinical trials to the wider market [10,14]. One of the first systems explored, N-(2-hydroxylproply)methacrylamine (HPMA) polymer–drug conjugates developed by Duncan and Kopecek in the late 1970s, has led to various clinical trial candidates [14,15]. Other interesting types of polymer–drug conjugates include cyclodextrin-containing polymers with conjugated camptothecin developed by Davis laboratory (IT-101) [16–20] and dendritic polyester–drug conjugates developed by Frechet and coworkers [21,22].

One of the primary challenges facing polymer–drug conjugate delivery systems is the actual conjugation step. Not only does the conjugation often require extensive post-conjugation purification steps, the heterogeneity of polymer–drug conjugates arising from nonsite-specific coupling reactions may also present bottlenecks to the clinical translation [10]. As shown in Figure 7.2, heterogeneities of polymer–drug conjugates result from (1) molecular weight distributions (MWDs, Mw/Mn) of the polymers (Figure 7.2a), (2) lack of control of the drug conjugation site on the polymer backbone (Figure 7.2b), and (3) lack of regioselectivity with regard to the conjugation site on drugs with multiple conjugation-amenable functional groups (Figure 7.2c). This last point is particularly relevant, as many of the best-selling anticancer small molecule drugs (e.g., paclitaxel, docetaxel, and doxorubicin) contain multiple functional groups (e.g., multihydroxyl groups in all three agents; ketone and amine groups in doxorubicin) for conjugation [23]. In the past several decades, there have been numerous efforts to minimize heterogeneities within polymer–drug conjugates. These include the development of polymers with low MWDs (e.g., dendrimers) [22,24–31], the conjugation of therapeutic agents to specific sites along the polymer backbone (e.g., the termini) [16,32–38], and the activation of specific functional groups on the therapeutic agents by means of

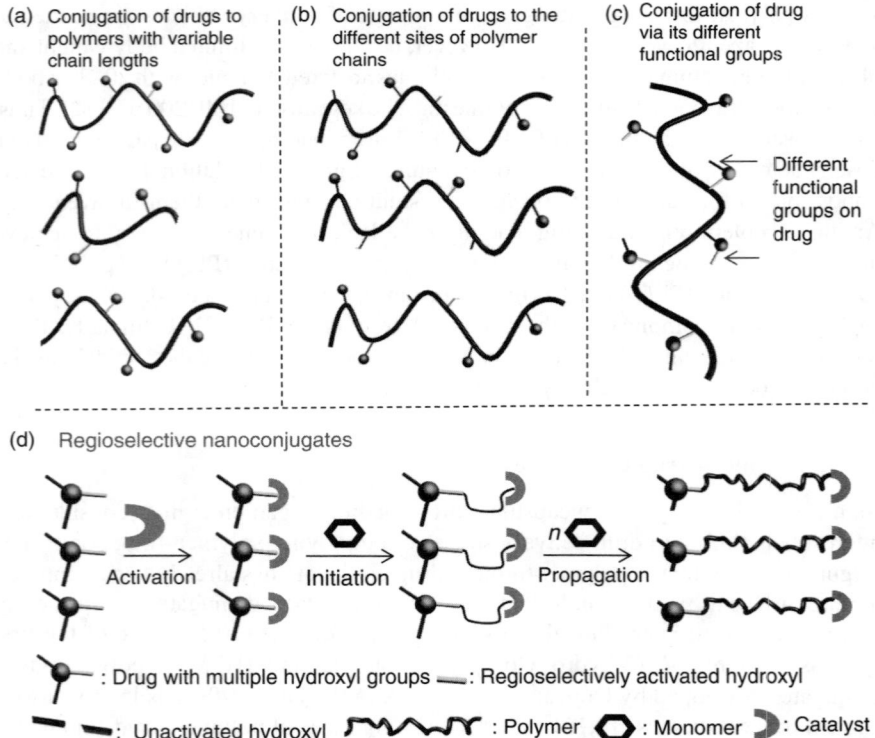

FIGURE 7.2 Heterogeneities in polymer-drug conjugates: (a) polymer–drug conjugates with a broad distribution of polymer chain lengths; (b) polymer–drug conjugates in which the conjugation site on the polymer backbone is uncontrolled; (c) polymer–drug conjugates in which a multifunctional therapeutic agent is conjugated without regioselectivity. Our solution to the heterogeneity problem: (d) drug-initiated, controlled polymerization for the synthesis of polymer–drug conjugates with low MWDs; the drug molecules are conjugated regioselectively to the polymer termini.

protection/deprotection chemistry [39,40]. However, concise synthetic strategies that yield polymer–drug conjugates with minimal heterogeneity for clinical application are still lacking (Figure 7.2a–c).

7.3 NANOCONJUGATES: DESIGN AND SYNTHESIS

7.3.1 Design and General Consideration

We present here a new synthetic strategy to address challenges faced by both nanoencapsulates and polymer–drug conjugates. In particular, we focus on chemistry that allows for site-specific conjugation of a drug to a polymer with a narrow molecular weight distribution (MWD) with quantitative incorporation efficiency.

In designing such a synthetic strategy, we first considered chemical systems which already benefit from controlled and precise chemistry, like the well-established and controlled polymerization methodologies allowing for the preparation of polyesters [41,42], polypeptides [43,44], and hydrocarbon-based synthetic biopolymers [45] with precisely controlled molecular weights (MWs) and narrow MWD. By incorporating these existing strategies in our design, polymer–drug conjugates can be produced with precise and homogeneous chemical structure and, ultimately, improved performance.

Ring-opening polymerization (ROP) for the preparation of PLA polyesters has been investigated extensively [46,47]. The polymerization reaction typically involves lactide (LA) ring opening by a metal–alkoxide (L_nM–OR) to form a RO-terminated LA–metal alkoxide ($ROOCCH(CH_3)O$-ML_n) followed by chain propagation to form RO-terminated PLA [41,46,47]. According to this mechanism, the RO group ultimately is connected to the PLA terminus through an ester bond (Figure 7.3a). This process is well understood and has been used extensively for the

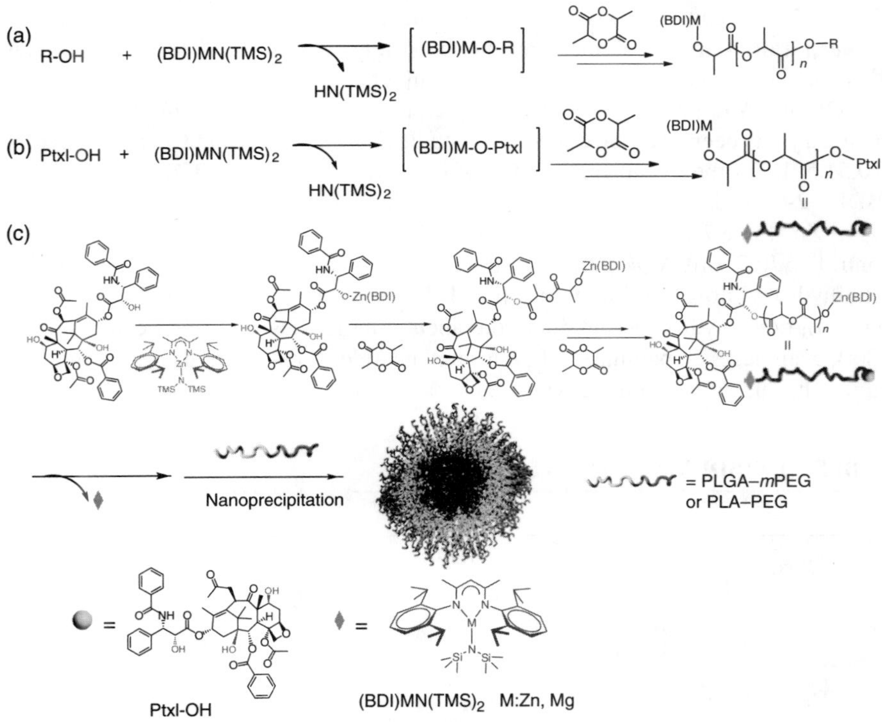

FIGURE 7.3 (a) Mechanism of ROP of PLA initiated by R-OH/BDI-metal catalyst. We propose the mechanism can be applied to (b) complex molecules with dense hydroxyl groups, for example, Ptxl, to initiate ROP. (c) Regioselective coordination of (BDI-II)ZnN(TMS)$_2$ onto 2′-OH of Ptxl to initiate ROP of LA. The polymers can be nanoprecipitated with PLGA–mPEG or PLA–mPEG to formulate sub-120 nm NPs. (*See insert for color representation of the figure.*)

incorporation of hydroxyl-containing small molecules [38], macromolecules [48], and NPs [49] to the termini of PLA.

In the context of delivery technology, hydroxyl groups of drugs can be utilized to achieve controlled living polymerization of LA. The hydroxyl group is the most abundant functional group in natural products, being found in about 65% of the ~78,000 known pharmacophores as well as in several of the best-selling anticancer small molecule drugs. Thus, the active initiator for PLA polymerization can be prepared *in situ* by mixing a hydroxyl(s)-containing drug with an active metal complex, such as a metal-amido [41,50]. If well designed, the *in situ* formed M–ORs can initiate controlled polymerization of LA, resulting in quantitative incorporation of OR with 100% monomer conversion (Figure 7.3b) [41,46]. By judicious control of the metal catalyst ligand, it is also possible to control the regioselective coordination of a structurally complex drug molecule bearing multiple hydroxyl groups (Figure 7.2d). After ROP, the linear PLA will be covalently linked onto the specific hydroxyl group of the drug via an ester bond. PLA–drug conjugates can then be driven to form NPs—called nanoconjugates (NCs)—by nanoprecipitation methods (Figure 7.3c). Unlike nanoencapsulates, the use of a degradable ester linkage between a drug and a polymer prevents burst release from the NPs and facilitates the controlled release of drugs from the conjugates.

ROP of LA to generate polymer–drug conjugates with precise and homogenous chemistry has been examined in several studies performed by the Cheng laboratory [10,51,52]. For these studies, the ROP of LA was catalyzed by (BDI-X)MN(TMS)$_2$ (BDI = 2-((2,6-dialkylphenyl)amino)-4-((2,6-dialkylphenyl)imino)-2-pentene, M = Mg or Zn, Table 7.1), a class of catalysts developed by Coates and coworkers for the controlled ROP of LA [41,50]. The strategy allows for quantitative incorporation of multihydroxyl drugs [51] (e.g., paclitaxel (Ptxl), docetaxel (Dtxl), and doxorubicin (Doxo)), other hydroxyl-containing therapeutic molecules [52] (e.g., camptothecin (Cpt), cyanine, cyclopamine) and even small peptides (goserelin) via ester bonds. By tuning the substituents on the *N*-aryl groups (R$_1$ and R$_2$) and the β-position (R$_3$) of

TABLE 7.1 (BDI-X)ZnN(TMS)$_2$ for NCs Synthesis

	Ligand	R$_1$	R$_2$	R$_3$
	BDI-II	*i*Pr	*i*Pr	H
(BDI-**X**)ZnN(TMS)$_2$				
	BDI-EE	Et	Et	H
	BDI-EI	Et	*i*Pr	H
	BDI-IICN	*i*Pr	*i*Pr	CN

the BDI ligand (Table 7.1) [53], regioselective polymerization on specific hydroxyl groups can be achieved (Table 7.2). The controlled polymerization also can be expanded to other biopolymers, such as poly(δ-valerolactone), poly(trimethylene carbonate), and poly(ε-caprolactone) (Table 7.3 for various polymers loaded with Ptxl and Dtxl). Apart from their controlled chemistry, drug–PLA conjugates can be nanoprecipitated to generate NPs with sub-120 nm diameter, high drug loading, nearly quantitative loading efficiencies, controlled release profiles without burst release, and narrow particle size distributions [51,52,54,55].

The controlled chemistry mentioned earlier was validated in the synthesis of Ptxl–PLA, and Doxo–PLA NCs. In the following discussion of these NCs, we examine the catalyst metal selection to achieve selective activation of hydroxyls in Cpt without disrupting the lactone ring structure [54]. Regioselective polymerization chemistry is also explored in the synthesis of Ptxl–PLA and Doxo–PLA NCs, whose initiator drugs possess multiple hydroxyl groups as well as other functional moieties. Combined, these experiments demonstrate that the proposed NC synthetic strategy is widely applicable for the synthesis of drug–polymer conjugates with various hydroxyl-containing therapeutics and with various polymer backbones.

7.3.2 Synthesis of Cpt–PLA Nanoconjugates

20(S)-Camptothecin (Cpt), a topoisomerase I inhibitor isolated from the Chinese tree *Camptotheca acuminate*, exhibits a broad range of anticancer activity in various animal models [56,57]. In terms of usability, Cpt has low aqueous solubility in its therapeutically active lactone form and is transformed rapidly to its carboxylate analog at physiological pH, producing a highly toxic and therapeutically inactive molecule (Figure 7.4a) [58,59]. While Cpt–polymer conjugates have been prepared to overcome the solubility limitation of the drug, conjugates prepared with conventional coupling chemistry are plagued by various heterogeneities. For example, experiments exploring the conjugation of Cpt via condensation reactions yield MWDs over a range 1.5–2.5 [16]. Meanwhile, direct conjugation of Cpt through its C20-tertiary hydroxyl is a difficult multistep reaction: Cpt must first be converted to a Cpt–amino ester and then conjugated to a polymer containing carboxylate groups via the amine end group of the Cpt–amino ester [16,60]. If the polymer has pendant functional groups, Cpt conjugation is further complicated.

To synthesize Cpt–PLA NCs via ROP, we first explored its ability to form Cpt–metal complexes with various BDI catalysts. When mixed with (BDI-II)MgN (TMS)$_2$, the C20-OH of Cpt formed a (BDI-II)Mg-Cpt alkoxide *in situ*. ROP of LA with this complex resulted in 100% LA conversion. HPLC analysis showed that Cpt was entirely conjugated to PLA with no detectable free Cpt in the polymerization solution. Although the M_n of the resulting polymer was in good agreement with the expected M_n, the MWD of Cpt–LA$_{100}$ was relatively broad (1.31) due in part to chain transfer during polymerization [41]. In order to achieve a better controlled polymerization, we next tested (BDI-II)ZnN(TMS)$_2$. However, the (BDI-II)ZnN (TMS)$_2$/Cpt-mediated ROP resulted in only 61% Cpt incorporation, indicating inefficient formation of Cpt–Zn complexes during the initiation step (Table 7.4)

TABLE 7.2 Formulation of NCs with Hydroxyl-Containing Therapeutic and Dye Molecules[a]

Paclitaxel (Ptxl)
Mitotic inhibitor

Docetaxel (Dtxl)
Mitotic inhibitor

Camptothecin (Cpt)
Topoisomerase I inhibitor

Cyclopamine (Cpa)
Hedgehog pathway inhibitor

Doxorubicin (Doxo)
Intercalating DNA,
topoisomerase II inhibitor

Goserelin (Gos)
Gonadotropin releasing
hormone agonist

Cy5-OH (Cy5)
Fluorescent molecule

Drug	[LA]/[Drug]	NCs	Loading (wt%)	LA Conversion (%)	Loading Efficiency (%)	Size (nm)	PDI	Reference
Cpt	10	Cpt–LA$_{10}$	19.5	>99	>99	72.5 ± 0.7	0.06 ± 0.02	[54]
Ptxl	25	Ptxl–LA$_{25}$	19.2	>99	97	55.6 ± 0.5	0.04 ± 0.01	[51]
Ptxl	15	Ptxl–LA$_{15}$	28.3	>99	95	85.5 ± 1.4	0.09 ± 0.03	[51]
Dtxl	10	Dtxl–LA$_{10}$	35.9	>99	97	77.9 ± 1.5	0.06 ± 0.02	[51]
Doxo	25	Doxo–LA$_{25}$	13.1	>99	98	90.8 ± 0.9	0.09 ± 0.01	[52]
Doxo	10	Doxo–LA$_{10}$	27.4	>99	94	125.2 ± 2.3	0.11 ± 0.01	[52]
Cpa	50	Cpa–LA$_{50}$	5.4	>99	>99	78.0 ± 2.4	0.12 ± 0.01	[52]
Cy5	25	Cy5–LA$_{25}$	12.3	>99	>99	76.3 ± 3.8	0.06 ± 0.01	[52]
Gos	10	Cy5–LA$_{10}$	46.8	>99	>99	120.6 ± 2.7	0.01 ± 0.01	[52]

LA, lactide; NCs, nanoconjugates; PDI, polydispersity.

[a] In molecule structures, the hydroxyl groups are differentiated in dark gray (regioselective conjugated to NCs) and in light gray. The functional group competing with hydroxyl groups (i.e., amine) is labeled in lightest gray.

TABLE 7.3 Ptxl (or Dtxl)/(BDI-II)ZnN(TMS)$_2$ -Mediated ROP of LA, VL, CL, and TMC

Initiator (R)	Monomer	[M]/[R]	Time	Temperature (°C)	Conversion (%)	Incorporation efficiency (%)	M_n/M_{exp} ($\times 10^3$ g/mol)	MWD (M_w/M_n)
Ptxl	LA	200	12	r.t.	>99	>99	29.7/28.1	1.02
Ptxl	VL	100	12	r.t.	>99	>99	15.1/10.8	1.17
Ptxl	CL	200	10	r.t.	>99	>99	20.3/23.7	1.07
Ptxl	TMC	100	6	50	>99	>99	14.7/11.1	1.10
Dtxl	LA	200	12	r.t.	>99	>99	29.6/25.2	1.03
Dtxl	CL	100	10	r.t.	>99	>99	11.2/12.2	1.05
Dtxl	TMC	200	6	50	>99	>99	25.9/21.3	1.19
Dtxl	VL	100	12	r.t.	>99	>99	14.2/10.7	1.15

M, monomer; M_{exp}, expected molecular weight; MWD, molecular weight distribution; r.t., room temperature.

159

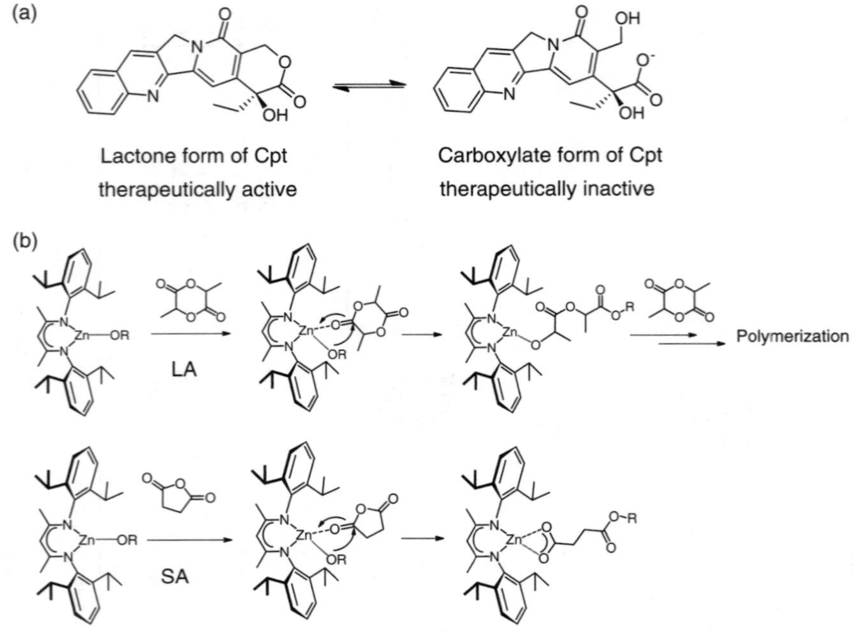

(a)

Lactone form of Cpt
therapeutically active

Carboxylate form of Cpt
therapeutically inactive

(b)

FIGURE 7.4 (a) Equilibrium of Cpt Lactone and carboxylate forms. (b) Suggested insertion-coordination mechanism of (BDI)Zn-OR mediated ring-opening of lactide (LA) and succinic anhydride (SA). R group represents the PLA polymer chain or agents containing hydroxyl group(s) (e.g., Cpt).

[54]. The insufficient activation of the C20-OH of Cpt by (BDI-X)Zn was addressed by subtly tuning the 2- and 6-substituents of the N-aryl groups on the catalyst. Doing so, 100% incorporation efficiencies were observed in both (BDI-EE)ZnN(TMS)$_2$/Cpt- and (BDI-EI)ZnN(TMS)$_2$/Cpt-mediated polymerizations (Table 7.4) [54].

TABLE 7.4   Ring Opening Reaction of SA and Polymerization of LA, Mediated by Cpt– (BDI-X)MN(TMS)$_2$ (M=Mg or Zn)^a

Catalyst	Cpt loading efficiency (%)	Polymer MWD	Cpt–SA (%)	Cpt–Carboxylate^a
(BDI-II)MgN(TMS)$_2$	>99	>1.3	78	Yes
(BDI-II)ZnN(TMS)$_2$	61	<1.1	19	Yes
(BDI-EE)ZnN(TMS)$_2$	>99	>1.2	N.D.	N.D.
(BDI-EI)ZnN(TMS)$_2$	>99	~1.1	89	No

MWD, molecular weight distribution.
^{a}Determined by HPLC analysis of Cpt–SA reaction mediated by different metal catalysts. The Cpt–carboxylate form indicated that the catalyst might have deleterious effect toward Cpt.

In addition to concerns regarding the ROP of LA, we also examined the retention of the lactone ring of Cpt throughout polymerization. As mentioned earlier, the lactone ring of Cpt is unstable and subject to ring opening in the presence of a nucleophile. To ensure that the therapeutically active form of Cpt is released in physiological conditions, the lactone ring of the drug must be maintained throughout the conjugation reaction. As LA is subject to rapid polymerization and the resulting Cpt–PLA conjugate is difficult to be characterized precisely, we used succinic anhydride (SA) as a model monomer to study (BDI-X)ZnN(TMS)$_2$/Cpt-mediated initiation. The ring opening of SA follows the same coordination–insertion mechanism as the initiation step of LA ROP but does not involve the subsequent chain propagation. Thus, the resulting product, Cpt–succinic acid (Cpt–SA), is a small molecule instead of a polymer. As a small molecule, the structure of Cpt–SA can be determined by routine characterization methods (Figure 7.4b). Experiments with SA revealed that Cpt activation by (BDI-II)MgN(TMS)$_2$ resulted in Cpt lactone ring opening (Table 7.4) [54]. The catalyst (BDI-EI)ZnN(TMS)$_2$, however, was able to prevent Cpt carboxylate formation. Using this catalyst, controlled polymerizations were observed over a broad range of LA/Cpt ratios from 75 to 400 in excellent agreement with the expected MWs and with narrow MWDs (1.02–1.18) [54].

With enhanced polymerization chemistry, Cpt–PLA NCs have improved formulation properties. Because both monomer conversion and drug incorporation are quantitative in Cpt–PLA synthesis, drug loadings can be predetermined by adjusting LA/Cpt feeding ratios. At a low monomer/initiator (M/I) ratio of 10, the drug loading can be as high as 19.5% (Cpt–LA$_{10}$, Table 7.2). To our knowledge, this Cpt–PLA NC has one of the highest loadings of Cpt ever reported [54]. Even at this high drug loading, sustained release of Cpt from Cpt–LA$_{10}$ NC was observed through the hydrolysis of the ester linker that connects the Cpt and the PLA without any observed burst release. Furthermore, the released Cpt (in PBS) had an HPLC elution time identical to authentic Cpt and was confirmed to have an identical molecular structure after isolation and characterization by ^{1}H NMR [54]. In contrast, PLA/Cpt NPs prepared by co-precipitation have been previously reported to give low drug loading (0.1–1.5%), low loading efficiency (2.8–38.3%), and poorly controlled release kinetics (20–90% of the encapsulated Cpt released within 1 h in PBS) [61]. Our unique conjugation technique allows for formation of Cpt-containing PLA NCs with superbly controlled formulation parameters, thus making Cpt–PLA NCs potentially useful agents for sustained treatment of cancer *in vivo*.

7.3.3 Synthesis of Ptxl-PLA (Dtxl-PLA) Nanoconjugates

Initially isolated from the bark of the yew tree, Ptxl is a potent anticancer drug that was originally used as a treatment for ovarian and breast cancer. Its use has since expanded and the drug is now also used to treat lung, liver, and other types of cancer. Clinical application of Ptxl is often accompanied with severe, undesirable side effects partially due to the solvent Cremophor EL in the commercial formulation of Taxol^{TM}. To reduce the side effects, various nanoparticulate delivery vehicles have been developed and investigated in the past few decades [62–65], including the FDA-approved albumin-bound Ptxl NPs (Abraxane™). Ptxl's analog, Dtxl (Taxotere™), is currently one of the best-selling chemotherapeutic agents and is used in the treatment of breast, ovarian, prostate, and nonsmall cell lung cancer. Many current NP formulations of both taxanes typically have low drug loadings, uncontrolled encapsulation efficiencies, and significant drug burst release effects when used *in vivo* [10,13,66,67]. Adding to formulation troubles, the densely functionalized structures of both molecules make it difficult to produce homogeneous Ptxl– and Dtxl–polymer conjugates.

Ptxl has three hydroxyl groups at its C2', C1, and C7 positions which can initiate LA polymerization, resulting in Ptxl–PLA conjugates with 1–3 PLA chains attached (Figure 7.5a). The three hydroxyl groups differ in steric hindrance in the order of 2'-OH < 7-OH < 1-OH. The tertiary 1-OH is least accessible and is typically inactive [68]. The 7-OH, however, can potentially compete with 2'-OH [69] for coordination with metal catalysts. In order to differentiate the initiation site between the two hydroxyl groups, we postulated that a metal catalyst with a bulky chelating ligand would preferentially form Ptxl–metal complexes through the 2'-OH for site-specific LA polymerization.

A selective reduction reaction was used to divide Ptxl into two fragments—one with the 2'-OH and the other with the 1- and 7-OHs—in order to study the impact of catalyst on regioselectivity [70]. Tetrabutylammonium borohydride (Bu_4NBH_4) can selectively and quantitatively reduce the C13-ester bond of Ptxl to produce baccatin III (BAC) and (1S,2R)-N-1-(1-phenyl-2,3-dihydroxypropyl)benzamide (PDB; Figure 7.5a) [70]. Thus, Ptxl–LA_5 was generated using a variety of catalysts, reduced with Bu_4NBH_4 and examined using HPLC–MS analysis. Doing so, we found that Mg $(N(TMS)_2)_2$—a catalyst without a chelating ligand—initiates polymerization nonpreferentially at both the 2'-OH and the 7-OH. Its counterpart with a bulky BDI-II ligand (e.g., (BDI-II)MgN(TMS)$_2$) preferentially initiates polymerization at the 2'-OH position. However, the resulting Ptxl–PLAs displayed fairly broad MWD (>1.2). We reasoned that the observation was attributable to fast propagation relative to initiation for the Mg catalysts [41]. Thus, to reduce the MWD of polymers, we utilized a zinc analog, (BDI-II)ZnN(TMS)$_2$, to give a more controlled LA polymerization with a narrow MWD (1.02, Figure 7.5b) [41].

We further examined the effect of the ligand on initiation regioselectivity and LA polymerization by (BDI-X)ZnN(TMS)$_2$/Ptxl by varying the steric bulk of the N-aryl substituents (R_1 and R_2) and the electronic properties of R_3 (Table 7.5) [41,53,71,72]. While both PDB–PLA and BAC–PLA are observed for Ptxl–LA_5 initiated by (BDI-EE)ZnN(TMS)$_2$/Ptxl, analysis of the fragments of the reductive cleavage of (BDI-EI) ZnN(TMS)$_2$/Ptxl showed relatively reduced amounts of BAC–PLA. This indicates that a Zn catalyst with bulky N-aryl substituents will preferentially coordinate with the 2'-OH of Ptxl to initiate PLA polymerization. Of note, various studies suggest that the polymerization process does not lead to deleterious effect on Ptxl—the Ptxl can be loaded and released with intact structure.

SA was again used as a model monomer to study the initiation step of ROP. Such reactions yield a small molecule, Ptxl–succinic acid (Ptxl–SA), whose structure can be determined easily by routine characterization methods (Table 7.5). The results

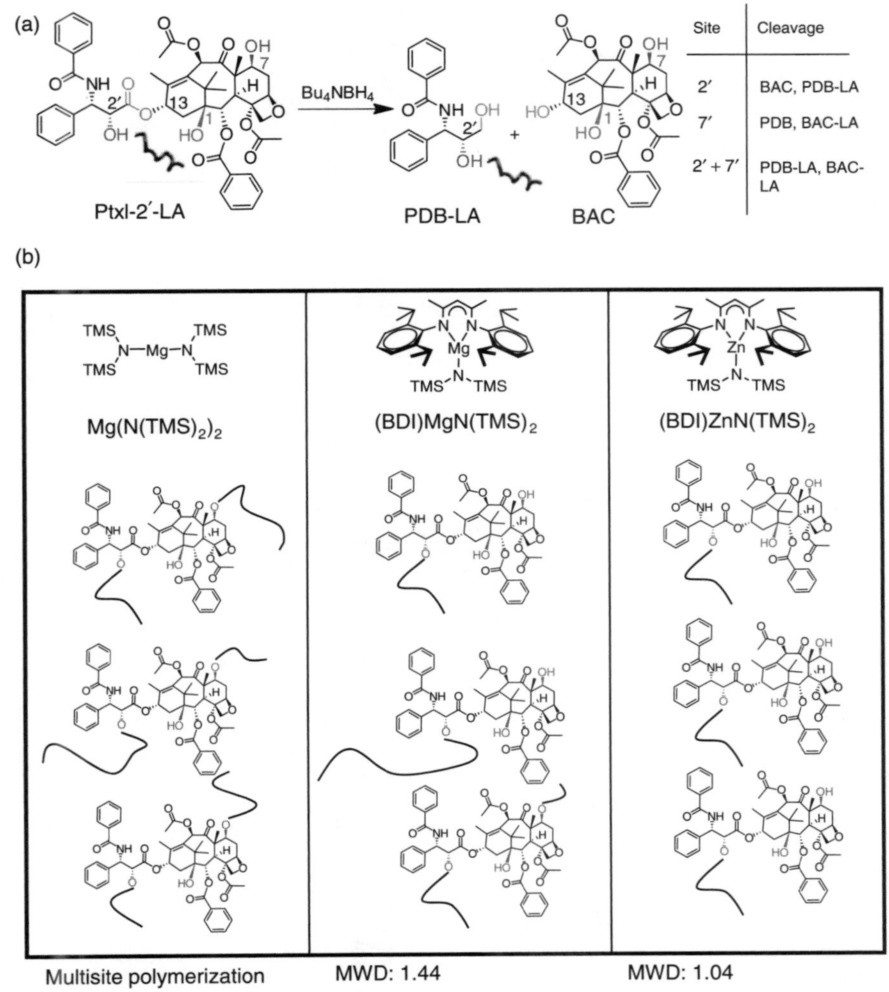

FIGURE 7.5 (a) Bu$_4$NBH$_4$-induced site-specific degradation of Ptxl for the formation of PDB and baccatin (BAC). (b) Scheme of Ptxl–PLA conjugates mediated by different catalysts, with the indication of regioselectivity and molecular weight distributions (MWDs). (*See insert for color representation of the figure.*)

show that the regioselectivity of the Ptxl/SA reaction increases as the sizes of R$_1$ and R$_2$ increase, in good agreement with the reductive reaction study using Bu$_4$NBH$_4$. The catalyst (BDI-II)ZnN(TMS)$_2$ showed the best regioselectivity while (BDI-EE) ZnN(TMS)$_2$ showed the worst. Changing R$_3$ from –H (BDI-II) to the electron-withdrawing –CN group (BDI-IICN) did not change the regioselectivity of the Ptxl/SA reaction. However, the addition of the CN group enhanced the reactivity of the resulting catalyst ((BDI-IICN)ZnN(TMS)$_2$), leading to a higher yield of Ptxl–2′-SA (Table 7.5). Controlled polymerization of PLA was observed when (BDI-II)ZnN

TABLE 7.5 Regioselectivity of the Ptxl/Metal Catalyst-Mediated Ring Opening Reactions

Catalyst	Reduction of Ptxl–LA$_5$[a]	Ptxl–2′-SA (%)	Regioselectivity[a]	Regioselectivity for 2′-OH[b]	Polymer MWD
Mg[N(TMS)$_2$]$_2$	PDB–LA, BAC–LA	N.D.	2′-OH and 7-OH	N.D.	>1.4
(BDI-II)MgN(TMS)$_2$	Majorly PDB–LA, BAC	N.D.	Majorly 2′-OH	N.D.	>1.4
(BDI-II)ZnN(TMS)$_2$	PDB–LA, BAC	40	2′-OH	100	<1.1
(BDI-EE)ZnN(TMS)$_2$	PDB–LA, BAC–LA	30	2′-OH and 7-OH	35	1.3
(BDI-EI)ZnN(TMS)$_2$	Majorly PDB–LA, BAC	45	Majorly 2′-OH	77	~1.2
(BDI-IICN)ZnN(TMS)$_2$	PDB–LA, BAC	54	2′-OH	100	<1.1

[a]Determined by HPLC analysis of Ptxl–LA$_5$ site-specific reduced by Bu$_4$NBH$_4$.
[b]Determined by HPLC analysis of Ptxl–SA ring opening reactions mediated by (BDI-X)Zn catalyst.

(TMS)$_2$/Ptxl was used at various [LA]/[Ptxl] ratios (50/1–300/1), with the obtained MWs in excellent agreement with the expected MWs and the monomodal MWDs in the range of 1.02–1.09.

Dtxl, the Ptxl analog, has four hydroxyl groups at C2', C1, C7, and C10 (Table 7.2). To examine the regioselectivity of Dtxl-initiated ROP of LA, we examined the Dtxl/(BDI-X)ZnN(TMS)$_2$-mediated ring opening reaction of SA. When complexed with (BDI-II)ZnN(TMS)$_2$, Dtxl reacted with SA to yield Dtxl–2'-SA with 100% regioselectivity and 71.5% yield. (BDI-II)ZnN(TMS)$_2$/Dtxl also showed excellent control for the ROP of LA, affording Dtxl–PLA with predictable MWs and very narrow MWDs. NCs formed with either Ptxl–PLA or Dtxl–PLA showed unprecedented high loading—up to 28.3 wt% in Ptxl–LA NCs and up to 35.9 wt% in Dtxl–LA NCs. The detailed formulation methods and efficacy of both conjugates are further discussed in Section 7.4.

Poly(δ-valerolactone) (PVL), poly(trimethylene carbonate) (PTMC), and poly (ε-caprolactone) (PCL) have been used extensively as alternatives to PLA in suturing, drug delivery, and tissue engineering applications. We were therefore interested as to whether the Ptxl (or Dtxl)/(BDI-II)ZnN(TMS)$_2$-mediated ROP of LA could be extended to the synthesis of Ptxl–PCL, Ptxl–PVL, and Ptxl–PTMC conjugates. Ptxl/(BDI-II)ZnN(TMS)$_2$ showed excellent control over the polymerization of δ-valerolactone (VL), trimethylene carbonate (TMC), and ε-caprolactone (CL). All the polymerization reactions gave drug–polymer conjugates with the expected MWs and narrow MWDs ($M_w/M_n < 1.2$) (Table 7.3). Furthermore, the polymerizations of VL and CL proceeded at room temperature, and the monomer conversions were quantitative. The polymerization of TMC, however, required a slightly elevated reaction temperature (6 h at 50°C).

7.3.4 Synthesis of Doxo–PLA Nanoconjugates

Doxo is commonly used in the treatment of a wide range of cancers and leukemia. Clinically, Doxo is administered as Doxil^{TM}, a PEGylated liposome-encapsulated form of the drug [74,75]. Owing to its high hydrophilicity, the encapsulation of Doxo within a hydrophobic polymer matrix can be challenging. For micelles formed by co-precipitating Doxo and PLGA–PEG, reported values indicate Doxo loading and loading efficiency as low as 0.51% and 23%, respectively [76]. Doxo loadings of 0.6–8.7% with loading efficiencies of 11.4–43.6% have also been reported by Hubbell and coworkers in their encapsulation studies using inverse emulsion polymerization of nonprotonated Doxo [77]. However, even with techniques allowing increased loading, burst release of Doxo is reported in many NP delivery systems [40,76,77].

Conjugation of Doxo to polymers can be difficult due to the diverse functional groups presented by the drug. Doxo has three hydroxyl groups, two phenolic hydroxyls, one ketone group, and one amine group. Doxo is also sensitive to pH, heat, metal ions, and light, further complicating its conjugation chemistry [78]. One conjugation strategy is to couple the terminal carboxylate of PLA with Doxo by creating an amide linkage through the 3'-NH$_2$ of Doxo [39,79]. Such Doxo–PLA conjugates, however, cannot release Doxo in its original form by hydrolysis.

Instead, Doxo-3′-lactamide, a prodrug of Doxo, is formed, which does not easily degrade *in vivo* due to the stability of amide bond [80]. Other efforts have been devoted to conjugation between the C13-ketone group of Doxo and hydrazine groups of polymeric carriers by forming an acid-labile hydrazone bond [22,81–83]. However, clinical studies of immunoconjugates with Doxo connected to monoclonal antibodies via hydrazone linkers gave unsatisfactory antitumor effects, ultimately leading to the termination of the clinical development of such immunoconjugates [84].

In previous studies involving the use of metal catalysts for LA polymerization, ROP of LA proceeded predominately by metal–alkoxides (M–ORs) rather than by metal-amides (M-NHRs) [47]. M–OR complexes typically have higher activities for LA ring opening than their amine analogs. For instance, Coates and coworkers reported that $(BDI)ZnOCH(CH_3)_2$ initiated and completed an LA polymerization within 20 min at a M/I ratio of 200 while a similar polymerization mediated by (BDI) $ZnN(TMS)_2$ required 10 h to complete [41]. The chemoselectivity of -OH instead of $-NH_2$ for Zn catalysts has also been confirmed by another molecular pair, 1-pyrenemethanol (Pyr-OH) and 1-pyrenemethylamine ($Pyr-NH_2$) [52].

Doxo has three hydroxyl groups, one at each of its C4′, C9, and C14 positions. Theoretically, LA polymerization can be initiated by any or all of these hydroxyl groups. C9-OH is the most sterically hindered and thus unlikely to initiate polymerization. To evaluate the initiation regioselectivity, we mixed Doxo with the catalyst $(BDI-II)ZnN(TMS)_2$ and succinic anhydride (SA) to mimic the initiation step of LA polymerization. ESI–MS results coupled with NMR analysis revealed that the SA ring was opened by the C14-OH of Doxo rather than by the C4′-OH or C3′-NH_2 of Doxo. When $(BDI-II)ZnN(TMS)_2$ was replaced by $Zn(N(TMS)_2)_2$, a Zn catalyst without ligands, the initiation regioselectivity completely disappeared, with Doxo–4′, 14-bissuccinic ester (Doxo–2SA), and Doxo–4′, 9, 14-trisuccinic ester (Doxo—3SA) being the predominant products (Figure 7.6). Interestingly, the metal activity also had a profound effect on regioselectivity. When the more highly active (BDI-II) $MgN(TMS)_2$ catalyst was used in a similar reaction, the most prevalent product was Doxo–2SA. Thus, by rationally designing ROP metal catalysts, Doxo–PLA conjugates with highly controlled regio- and chemoselectivity are achievable within one step without having to protect the C3′-NH_2 and other competing hydroxyl groups of Doxo [52].

Having completed preliminary investigations with SA, we next studied Doxo-initiated LA polymerization. Doxo/$(BDI-II)ZnN(TMS)_2$-mediated LA polymerization resulted in Doxo–PLA with narrow MWD (<1.2) and the expected MWs. At a low M/I ratio of 10, the drug loading was as high as 27.4% (Doxo–LA_{10}). To our knowledge, this is by far the highest loading ever reported in Doxo-containing polymeric NPs [52]. HPLC analysis revealed that Doxo–LA_{10} conjugates incubated in PBS at 30°C released Doxo in its original, therapeutically active form. In a separate experiment, Doxo–LA_{100} treated with 0.1 M NaOH was shown to release 88–92% of the Doxo in its original form. In this case, the incomplete recovery is likely due to instability of Doxo in NaOH. Combined, these studies suggest that Doxo molecules were linked to PLA through its hydroxyl group(s) by forming ester linker(s) with PLA, which could be hydrolyzed in an alkaline condition. Even at high

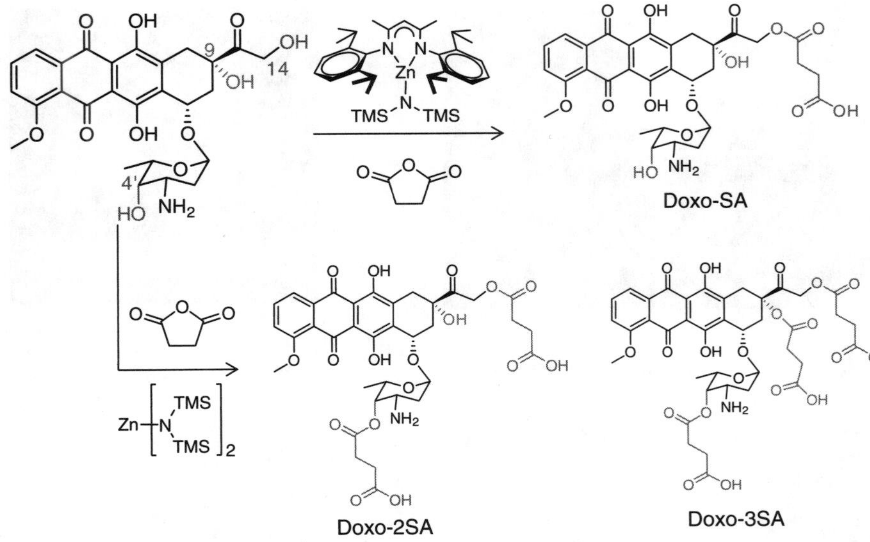

FIGURE 7.6 Scheme of Doxo–SA reaction (1:3 molar ratio) mediated by different Zn catalysts.

drug loading (27.4%), sustained release of Doxo from Doxo–LA$_{10}$ NC was observed without burst release. This is in sharp contrast to the burst release of PLA/Doxo NPs prepared by co-precipitation, in which 90% of the Doxo is released within 3 h [52]. The sustained release of Doxo from NCs with high loading may help alleviate the systemic side effects of Doxil^{TM} without reducing the overall dosage.

7.4 NANOCONJUGATES: FORMULATION AND POTENTIAL APPLICATION

7.4.1 Formulation

Particle size is an important aspect of drug delivery. For cancer therapeutics, formulated particles should have a uniform, monomodal distribution and a diameter <200 nm to take advantage of the enhanced permeability and retention (EPR) effect. As such, particles formed by nanoprecipitation of Ptxl–PLA were evaluated for their size. In general, Ptxl–PLA yielded particles with monomodal distributions while their nanoencapsulate (NEs, see Section 7.2.1) counterparts were polydisperse (Figure 7.7a) [13,85]. As the multimodal distribution of NEs is due in part to the aggregation of nonencapsulated free drug [13], the monomodal distribution observed with NCs is likely related to the unimolecular structure of Ptxl–PLA conjugates. The actual size of Ptxl–PLA NPs prepared by nanoprecipitation can be manipulated by changing the solvent as well as polymer concentration. At a fixed Ptxl–PLA concentration, the size of NCs prepared by precipitating a DMF solution of conjugate

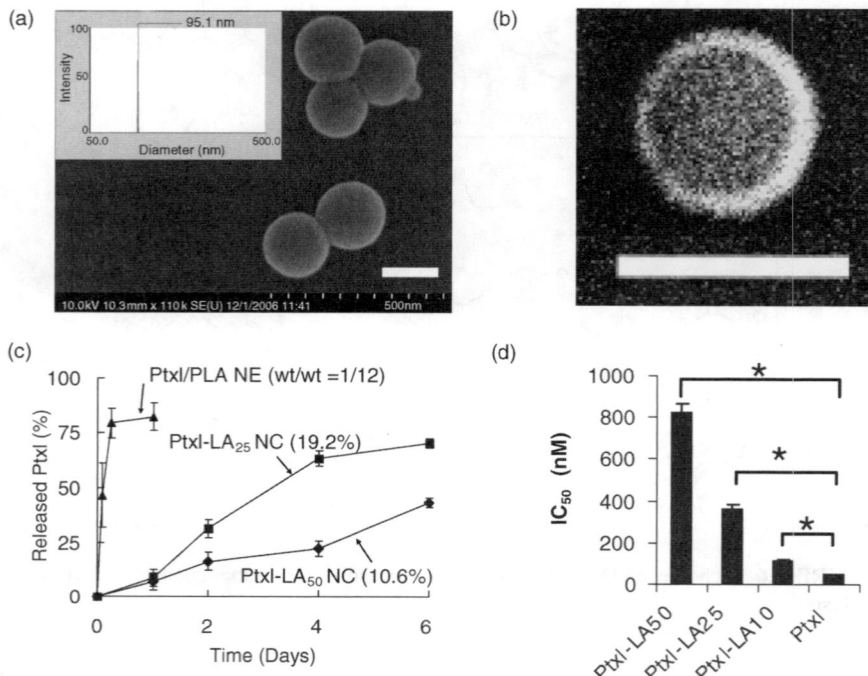

FIGURE 7.7 (a) Ptxl–LA$_{25}$ NCs were analyzed by dynamic light scattering (DLS) and scanning electron microscopy (SEM). (b) SEM image of Ptxl–LA$_{25}$/PLGA–mPEG NCs (Scale bar = 150 nm). The different density of materials were indicated with different gray scale: PEG layer (white color) and hydrophobic PLA part (gray color) (c) Release kinetics of Ptxl from Ptxl–PLA NCs and Ptxl–PLA NE (Prepared by nanoprecipitation of a mixture of Ptxl and PLA (Ptxl–PLA (wt/wt) = 1/12) at 37°C in PBS. (d) IC$_{50}$ values determined by MTT assay of Ptxl–PLA NCs and free Ptxl, which were incubated with PC-3 cells for 24 h. (*Source*: Figures reproduced with permission from Reference [51].)

is typically in a range 60–100 nm, 20–30 nm smaller than those prepared with acetone or THF as a solvent [13]. When nanoprecipitation is carried out using a 1:20 (v:v) mixture of DMF and water, the size of Ptxl–LA NCs shows a linear correlation with the conjugate concentration and can be used to precisely form particles with diameters from 60 to 100 nm.

Surface modification of NPs with PEG is widely used for prolonged systemic circulation and reduced aggregation of NPs in blood [86]. To avoid the removal of unreacted reagents and by-products, poly(lactide)-*b*-methoxylated PEG (PLA–mPEG)—an amphiphilic copolymer that has a 14 kDa PLA and a 5 kDa PEG segment [87]—was used to PEGylate NCs instead of covalently conjugating PEG to NCs [65,85]. The PEGylated structure of the NCs can be seen by scanning transmission electron microscopy (Figure 7.7b). Ptxl–LA NCs with a PEGylated surface have a negative surface zeta potential and thus remain nonaggregated in water and PBS solution by means of surface charge repulsion. In addition, we also developed a one-step co-precipitation method to formulate NCs that could stay nonaggregated in a salt solution with an amphiphilic triblock copolymer PLA–PEG–PLA, a feature that diblock PLA–PEG did not provide [55].

Drug burst release is a long-standing formulation challenge to NEs and often results in undesirable side effects and reduced therapeutic efficacy [67]. Conventional NEs typically show burst release of 60–90% of their payload within a few to tens of hours [9]. Since the Ptxl release kinetics of Ptxl–PLA NCs is determined not simply by diffusion—as is the case with NEs—but by the hydrolysis of the Ptxl–PLA ester linker followed by diffusion out of NCs, the release kinetics of Ptxl from NCs are more controlled and show significantly reduced burst release (Figure 7.7c). For example, Ptxl released from Ptxl–LA$_{25}$ NCs (19.2 wt%) was 8.7% at Day 1 and 70.4% at Day 6. In comparison, 89% of Ptxl was released within 24 h from Ptxl/PLA NE (Figure 7.7c). Release of Ptxl from Ptxl–LA$_{50}$ NCs was slower than from Ptxl–LA$_{25}$ NCs, presumably because of the higher MW of Ptxl–LA$_{50}$ and more compact particle structure. With its controlled drug release, the *in vitro* toxicity of Ptxl–LA NCs correlates with the amount of Ptxl released (Figure 7.7d). For example, Ptxl–LA$_{15}$ NCs have a nearly identical IC$_{50}$ to free Ptxl (87 nM) while the IC$_{50}$ of Ptxl–LA$_{50}$ NCs is an order of magnitude higher. As a result, the toxicity of NCs can be tuned in a wide range simply by controlling NC drug loading.

Aptamers are either single-stranded DNA or RNA that specifically binds to target ligands [88,89]. When used for cancer targeting, aptamers are capable of binding to target antigens with extremely high affinity and specificity in a manner resembling antibody-mediated cancer targeting [90,91]. Unlike antibodies, synthesis of aptamers is an entirely chemical process and thus shows negligible batch-to-batch inconsistency [92,93]. Moreover, aptamers are typically nonimmunogenic and exhibit remarkable stability against pH, temperature, and solvent. The A10 aptamer with 2′-fluoro-modified ribose on all pyrimidines and a 3′-inverted deoxythymidine cap has been identified and utilized to target extracellular prostate-specific membrane antigen (PSMA) [94]. A10 binds to PSMA-positive LNCaP prostate cancer cells but not PSMA-negative PC-3 prostate cancer cells. To demonstrate NC targeting, amine-terminated A10 aptamer was conjugated to PLA–PEG–COOH/Cy5–PLA NCs through carboxylic acid–amine coupling in the presence of EDC and NHS to give aptamer/PLA–PEG–COOH/Cy5–PLA NCs (aptamer–Cy5 NC) [90]. Incubation of aptamer–Cy5 NCs with LNCaP cells for 6 h resulted in substantially increased NC internalization (Figure 7.8) compared to PC-3 cells. These *in vitro* studies demonstrated that NCs conjugated with aptamer-targeting ligand can potentially be used for prostate cancer targeting.

Small-scale NPs that stay nonaggregated in PBS for *in vitro* or *in vivo* laboratory studies are straightforward. However, in order to facilitate clinical translation, NPs need to be prepared in large quantity with well-controlled properties that remain unchanged during the processes of manufacturing, storage, and transport. Because Ptxl is covalently conjugated to PLA through an ester bond that is subject to hydrolysis upon exposure to water, handling of Ptxl–PLA NCs in aqueous solution is undesirable. Thus, NCs have to be formulated in solid form for clinical use. By screening various molecules, we discovered that albumin functions as an excellent

(a)   LNCaP (PSMA+)   (b)   PC-3 (PSMA–)

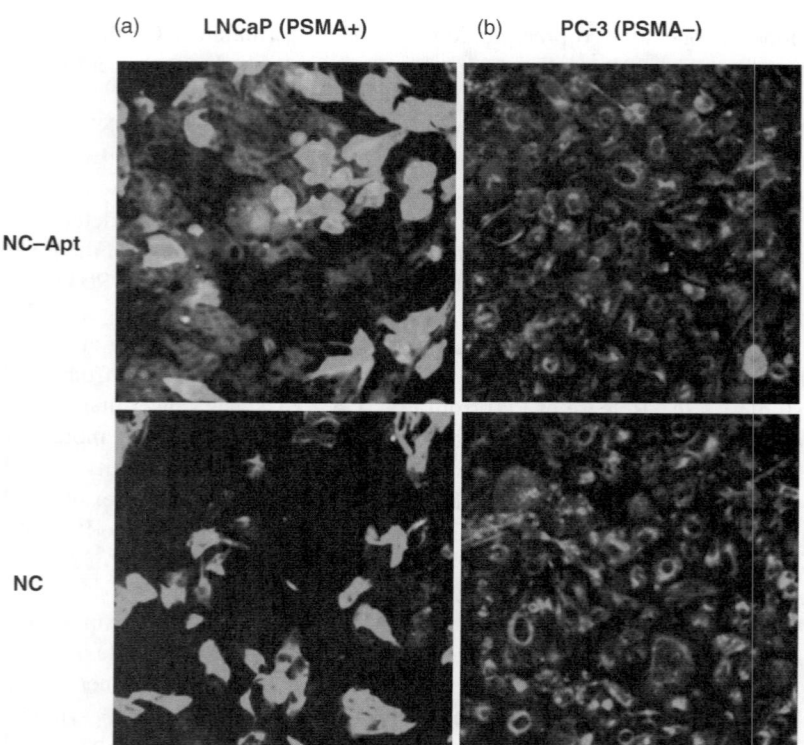

NC–Apt

NC

FIGURE 7.8 Confocal images of LNCaP (a) and PC-3 (b) cells treated with aptamer-functionalized nanoconjugates (NC–Apt, top) and nanoparticles without aptamer (NC, bottom).The Cy-5 incorporated NC or NC–Apt are shown in red. The cells counter stained with Alexa-Flour 488 Phalloidin (binding to cellular actin) are shown in green. (*See insert for color representation of the figure.*)

lyoprotectant for Ptxl–PLA and yields solid NCs that do not aggregate when reconstituted in PBS. By incorporating the albumin-based lyoprotection technique, we demonstrated for the first time that polymer nanoparticles containing a conjugated nucleic acid targeting ligand can be prepared in solid form and still be reconstituted to well-dispersed, nonaggregated particles with functional targeting capability [55].

7.4.2 Theranostic Nanoconjugates

There is growing interest in developing noninvasive, whole-body fluorescent imaging techniques to assess the biodistribution of drug delivery systems or diagnostic agents within patients. To ensure effective measurement of fluorescent signal *in vivo*, it is crucial to use red or near-IR dyes. Quantum dots, a class of inorganic nanocrystals with excellent fluorescent intensity and photostability, can readily be prepared to have a far-red emission band. However, there is a general consensus

Background   Cy5-PLA NP

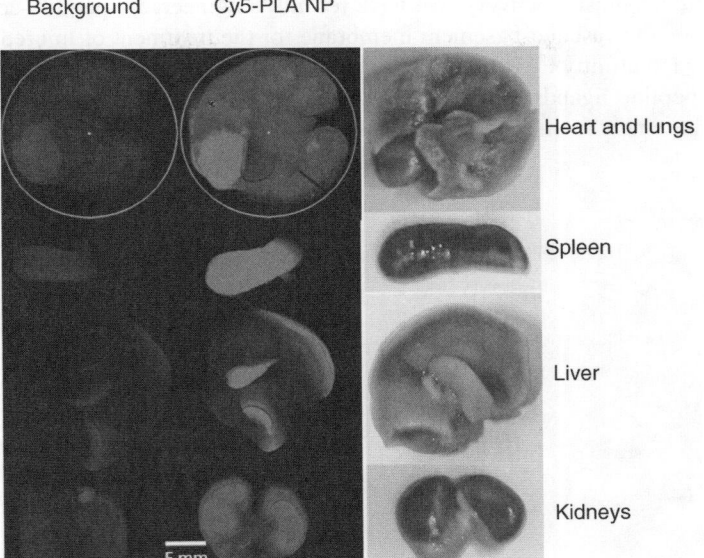

Heart and lungs

Spleen

Liver

Kidneys

FIGURE 7.9 Distribution of Cy5 dye-labeled PLA NCs in various visceral mice organs. Mice were sacrificed 24 h after i.v. injection of Cy5–PLA NCs. (*See insert for color representation of the figure.*)

that quantum dots cannot be used in humans because of their toxicity. Currently, small molecule organic dyes are more promising probes to be coupled with imaging systems for clinical applications as compared to quantum dots. Polymeric nanoparticles can function as good carriers for the delivery of small molecule imaging materials because they can provide prolonged systemic circulation and improved tumor accumulation compared to unformulated drugs. To meet this challenge, it is particularly important that NPs be formulated with stably incorporated fluorescent ligands and controlled formulation parameters (size, surface properties, etc.). Accordingly, our monomodal, narrowly distributed PEGylated Cy5–PLA NPs have the potential to be candidates for whole body *in vivo* imaging. Preliminary *in vivo* studies demonstrated that Cy5–PLA NCs can be easily visualized in various visceral organs with low autofluorescent background ratios (Figure 7.9). Owing to their sub-100 nm size, the Cy5–PLA NCs also can be used to study the lymphatic biodistribution of NPs when coupled with whole-body optical imaging [95]. These promising results support the further development of Cy5–PLA NCs as a model system to assess the *in vivo* pharmacological and pharmacokinetics profiles of NCs.

7.4.3 Nanoconjugates Against Other Diseases

Considering that molecular targeting of cell-based targets may be confounded by inter- or intra-patient heterogeneity in cell surface antigen expression, targeted NPs that can recognize the extracellular matrix have attracted considerable attention for therapeutic/diagnostic delivery. We have recently engineered a peptide-conjugated NP to target the vascular basement membrane for the treatment of injured vasculature. The high affinity C11 peptide was screened from a combinatorial phage library of hepta-peptide ligands against human collagen IV, which represents 50% of the vascular basement membrane (Figure 7.10) [96]. Angioplasty-injured carotid artery

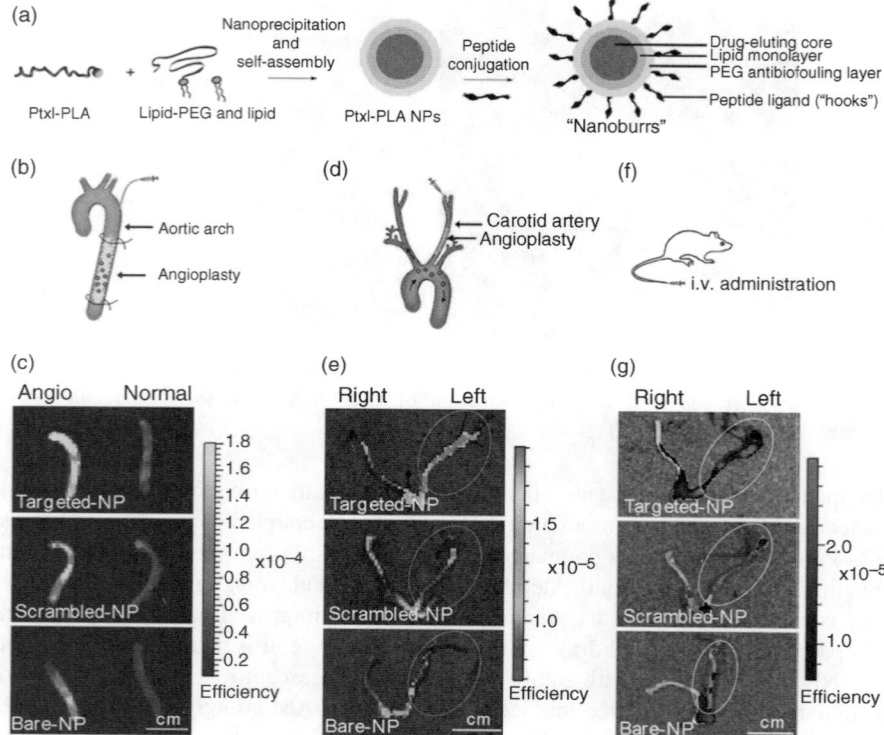

FIGURE 7.10 (a) Schematic of NPs synthesis by nanoprecipitation and self-assembly of Ptxl–LA NCs with lipids and peptide ligands to adhere to the exposed basement membrane during vascular injury. (b) Scheme of *ex vivo* abdominal aorta injury model; samples were delivered into the aorta segment for 5 min. (c) Fluorescence images overlaid on photographs of balloon-injured aortas incubated with NCs with a targeting peptide, compared with scrambled-peptide and nontargeted NPs. (d) *In vivo* intra-aortic administration in a carotid injury model: a catheter was inserted via the external carotid into the common carotid and advanced into the aortic arch. (e) Fluorescence images overlayed on photographs of carotid arteries incubated with NCs with a targeting peptide, compared with scrambled-peptide and nontargeted NPs. (f) *In vivo* systemic administration in a carotid angioplasty model. (g) Fluorescence images overlayed on photographs of carotid arteries incubated with NCs with a targeting peptide, compared with scrambled-peptide and nontargeted NPs. For imaging, Alexa Fluor 647–PLGA dye conjugates were encapsulated in place of Ptxl–PLA drug conjugates. (Scale bar, 1 cm.) *(See insert for color representation of the figure.)*

was used as a model of compromised vasculature to examine the targeting capacity of the C11 peptide-conjugated polymeric NPs. The targeted NPs were delivered via both intra-arterial and i.v. administration and, when compared to nontargeted NPs, showed greater *in vivo* vascular retention at sites of injured vasculature in rats (Figure 7.10) [96]. Although the initial application was for vessel wall targeting in cardiovascular disease, the utility of this peptide-targeted NP system is broad and could be used to diagnose and treat different human diseases where the endothelial lining is compromised.

Immunosuppressive agents have played a pivotal role in ensuring the success of organ transplantation and greatly improved the outcomes of patients with life-threatening, immune-mediated diseases [97]. However, the use of immuno-suppressive agents is hindered by the lack of selectivity as well as major adverse drug reactions. The immunosuppressive agent cyclosporine (CsA), for example, results in a dramatic improvement in short-term allograft survival but also poses a risk of chronic nephrotoxicity because of its narrow therapeutic window [98–100]. We recently developed CsA–PLA NCs to achieve targeted immunosuppression for a wide variety of immune-mediated disorders [101]. CsA–PLA NCs showed superior physicochemical properties with excellent size control, narrow size distribution, and very well controlled release kinetics without a noticeable burst release. CsA–PLA NCs also showed excellent stability in biological media with negligible aggregation. Because of their ability to mediate the sustained release of CsA *in vitro* while suppressing T-cell mediated immune responses, CsA–PLA NCs are an excellent system for immunosuppression of organ rejection. We also developed a novel strategy combining CsA–PLA NCs with dendritic cells (DCs) to efficiently deliver CsA to draining lymph nodes to inhibit T-cell priming in a locally controlled and sustained manner without systemic release. This innovative delivery strategy constitutes a strong basis for future targeted delivery of immunosuppressive drugs with improved efficiency and reduced toxicity.

7.5   CONCLUSIONS AND OUTLOOK

LA ROP-mediated drug conjugation allows for facile regio- and chemoselective incorporation of drugs onto PLA. This in turn allows for the formation of drug delivery vehicles with low polydispersity, predetermined drug loadings (up to ~30%), and quantitative loading efficiencies. The BDI–metal chelating complexes described earlier for Cpt, Ptxl, and Doxo conjugation do not have deleterious effects on drugs and can be easily removed by solvent extraction. Because both Zn and Mg ions are biocompatible and, in fact, are key elements in our dietary mineral supplements, there should be no significant safety concerns regarding the use of these two metal catalysts. Multigram scales of drug-PLA conjugates can be readily prepared within hours using the one-pot polymerization approaches described here. Because drug molecules are covalently conjugated to PLA, the post-reaction formulation process (precipitation, removal of catalysts, nanoprecipitation, sterilization, lyophilization, shipping and handling, etc.) has minimal impact on sample property as compared to drug–polymer NPs prepared via encapsulation methods. This polymerization-mediated conjugation method may be utilized for the formulation of polymer–drug conjugates not only for drug delivery, but also for other controlled release applications (scaffolds, coatings of stents, etc.). Other cyclic esters (e.g., ε-caprolactone and δ-valerolactone) are likely to replace LA and find use as monomers in such drug-initiated polymerizations [102]. The drug/(BDI-X) $ZnN(TMS)_2$-initiated polymerization of these cyclic esters at room temperature has recently been achieved in our laboratory, which will provide further tunability of the release profiles and other physicochemical properties. Given that the lack of a controlled formulation for nanoparticulate drug delivery vehicles is one of the bottlenecks to their clinical development, this unique, ROP-mediated conjugation methodology may contribute to the development of clinically applicable nanomedicines.

Although there has been impressive progress in nanomedicine for cancer treatment in the past few years, enormous tasks remain. The convergence of seemingly disparate scientific fields (e.g., cancer biology, electronics, bioimaging, biomicroelectromechanical systems, computer science, polymer and materials chemistry, biophysics) will only accelerate the development of nanomedicine [103]. Our work will continue in the daunting task of pushing targeted nano-conjugates for clinical evaluation, with the expansion of our technology to other disease-related formulations (e.g., imaging agents or implantable devices). It also will be important to improve intratumoral penetration for enhanced efficacy [104,105]. Given the limitations of spatial and temporal changes in the expression of the target [106], the seemingly pedestrian issues on targeting ligand functionality are also likely to be challenging *in vivo* [107,108]. The prevention and treatment of metastasis will be of particular interest because metastasis is responsible for 90% of cancer deaths [109]. The coordination of different approaches will help optimize the delivery efficiency. A deep understanding of all aspects of the biology of cancers, including tumor microenvironment, will continue to be crucial for design and understanding [110]. We are far from being able to create delivery vehicles to meet the expectation of complete tumor eradication. However, as the field of nanomedicine matures, the objective may be on the horizon.

Acknowledgments

Jianjun Cheng acknowledges support from the NSF (Career Award Program DMR-0748834) and the NIH (NIH Director's New Innovator Award 1DP2OD007246-01, 1R21EB009486A, and 1R21CA139329Z). Rong Tong acknowledges a student fellowship from the Siteman Center for Cancer Nanotechnology Excellence (University of Washington & University of Illinois at Urbana-Champaign) from 2007 to 2010. Li Tang was funded at University of Illinois at Urbana-Champaign from NIH National Cancer Institute Alliance for Nanotechnology in Cancer "Midwest Cancer Nanotechnology Training Center" Grant R25 CA154015A.

REFERENCES

1. Langer, R. (1998). Drug delivery and targeting. *Nature 392*, 5.

2. Davis, M. E., Chen, Z., Shin, D. M. (2008). Nanoparticle therapeutics: an emerging treatment modality for cancer. *Nature Reviews Drug Discovery 7*, 771.

3. Petros, R. A., DeSimone, J. M. (2010). Strategies in the design of nanoparticles for therapeutic applications. *Nature Reviews Drug Discovery 9*, 615.

4. Wagner, V., Dullaart, A., Bock, A. K., Zweck, A. (2006). The emerging nanomedicine landscape. *Nature Biotechnology 24*, 1211.

5. Zhang, L., Gu, F. X., Chan, J. M., Wang, A. Z., Langer, R. S., Farokhzad, O. C. (2008). Nanoparticles in medicine: therapeutic applications and developments. *Clinical Pharmacology and Therapeutics 83*, 761.

6. Farokhzad, O. C., Langer, R. (2009). Impact of nanotechnology on drug delivery. *ACS Nano 3*, 16.

7. Cohen, S., Yoshioka, T., Lucarelli, M., Hwang, L. H., Langer, R. (1991). Controlled delivery systems for proteins based on poly(lactic glycolic acid) microspheres. *Pharmaceutical Research 8*, 713.

8. Ringsdorf, H. (1975). Structure and properties of pharmacologically active polymers. *Journal of Polymer Science Part C-Polymer Symposium 51*, 135.

9. Musumeci, T., Ventura, C. A., Giannone, I., Ruozi, B., Montenegro, L., Pignatello, R., Puglisi, G. (2006). PLA/PLGA nanoparticles for sustained release of docetaxel. *International Journal of Pharmaceutics 325*, 172.

10. Tong, R., Cheng, J. J. (2007). Anticancer polymeric nanomedicines. *Polymer Reviews (Philadelphia, PA, US) 47*, 345.

11. Hamblett, K. J., Senter, P. D., Chace, D. F., Sun, M. M. C., Lenox, J., Cerveny, C. G., Kissler, K. M., Bernhardt, S. X., Kopcha, A. K., Zabinski, R. F., Meyer, D. L., Francisco, J. A. (2004). Effects of drug loading on the antitumor activity of a monoclonal antibody drug conjugate. *Clinical Cancer Research 10*, 7063.

12. Schluep, T., Hwang, J., Cheng, J. J., Heidel, J. D., Bartlett, D. W., Hollister, B., Davis, M. E. (2006). Preclinical efficacy of the camptothecin-polymer conjugate IT-101 in multiple cancer models. *Clinical Cancer Research 12*, 1606.

13. Cheng, J., Teply, B. A., Sherifi, I., Sung, J., Luther, G., Gu, F. X., Levy-Nissenbaum, E., Radovic-Moreno, A. F., Langer, R., Farokhzad, O. C. (2007). Formulation of functionalized PLGA–PEG nanoparticles for *in vivo* targeted drug delivery. *Biomaterials 28*, 869.

14. Duncan, R. (2006). Polymer conjugates as anticancer nanomedicines. *Nature Reviews Cancer 6*, 688.

15. Duncan, R. (2003). The dawning era of polymer therapeutics. *Nature Reviews Drug Discovery 2*, 347.

16. Cheng, J., Khin, K. T., Jensen, G. S., Liu, A. J., Davis, M. E. (2003). Synthesis of linear, beta-cyclodextrin-based polymers and their camptothecin conjugates. *Bioconjugate Chemistry 14*, 1007.

17. Schluep, T., Hwang, J., Hildebrandt, I. J., Czernin, J., Choi, C. H. J., Alabi, C. A., Mack, B. C., Davis, M. E. (2009). Pharmacokinetics and tumor dynamics of the nanoparticle IT-101 from PET imaging and tumor histological measurements. *Proceedings of the National Academy of Sciences of the United States of America 106*, 11394.

18. Davis, M. E. (2009). Design and development of IT-101, a cyclodextrin-containing polymer conjugate of camptothecin. *Advanced Drug Delivery Reviews 61*, 1189.

19. Davis, M. E., Zuckerman, J. E., Choi, C. H. J., Seligson, D., Tolcher, A., Alabi, C. A., Yen, Y., Heidel, J. D., Ribas, A. (2010). Evidence of RNAi in humans from systemically administered siRNA via targeted nanoparticles. *Nature 464*, 1067.

20. Davis, M. E. (2009). The first targeted delivery of siRNA in humans via a self-assembling, cyclodextrin polymer-based nanoparticle: from concept to clinic. *Molecular Pharmaceutics 6*, 659.

21. Fox, M. E., Szoka, F. C., Frechet, J. M. J. (2009). Soluble polymer carriers for the treatment of cancer: the importance of molecular architecture. *Accounts of Chemical Research 42*, 1141.

22. Lee, C. C., Gillies, E. R., Fox, M. E., Guillaudeu, S. J., Frechet, J. M. J., Dy, E. E., Szoka, F. C. (2006). A single dose of doxorubicin-functionalized bow-tie dendrimer cures mice bearing C-26 colon carcinomas. *Proceedings of the National Academy of Sciences of the United States of America 103*, 16649.

23. http://en.wikipedia.org/wiki/List_of_bestselling_drugs.

24. Lee, C. C., MacKay, J. A., Frechet, J. M. J., Szoka, F. C. (2005). Designing dendrimers for biological applications. *Nature Biotechnology 23*, 1517.

25. Srinivasachari, S., Fichter, K. M., Reineke, T. M. (2008). Polycationic beta-cyclodextrin "Click Clusters": monodisperse and versatile scaffolds for nucleic acid delivery. *Journal of the American Chemical Society 130*, 4618.

26. Lu, H., Cheng, J. (2008). Controlled ring-opening polymerization of alpha-amino acid *N*-carboxyanhydrides and facile end group functionalization of polypeptides. *Journal of the American Chemical Society 130*, 12562.

27. Lu, H., Cheng, J. (2007). Hexamethyldisilazane-mediated controlled polymerization of alpha-amino acid *N*-carboxyanhydrides. *Journal of the American Chemical Society 129*, 14114.

28. Cheng, J., Deming, T. J. (2001). Controlled polymerization of beta-lactams using metal-amido complexes: synthesis of block copoly(beta-peptides). *Journal of the American Chemical Society 123*, 9457.

29. Medina, S. H., El-Sayed, M. E. H. (2009). Dendrimers as carriers for delivery of chemotherapeutic agents. *Chemical Reviews 109*, 3141.

30. Gillies, E. R., Frechet, J. M. J. (2002). Designing macromolecules for therapeutic applications: polyester dendrimer-poly(ethylene oxide) "bow-tie" hybrids with tunable molecular weight and architecture. *Journal of the American Chemical Society 124*, 14137.

31. Liu, G., Jia, L. (2004). Design of catalytic carbonylative polymerizations of hetero-cycles. Synthesis of polyesters and amphiphilic poly(amide-block-ester)s. *Journal of the American Chemical Society 126*, 14716.

32. Zhang, X. F., Li, Y. X., Chen, X. S., Wang, X. H., Xu, X. Y., Liang, Q. Z., Hu, J. L., Jing, X. B. (2005). Synthesis and characterization of the paclitaxel/MPEG–PLA block copolymer conjugate. *Biomaterials 26*, 2121.

33. Cheng, J., Khin, K. T., Davis, M. E. (2004). Antitumor activity of beta-cyclodextrin polymer–camptothecin conjugates. *Molecular Pharmaceutics 1*, 183.

34. Shen, Y. Q., Jin, E. L., Zhang, B., Murphy, C. J., Sui, M. H., Zhao, J., Wang, J. Q., Tang, J. B., Fan, M. H., Van Kirk, E., Murdoch, W. J. (2010). Prodrugs forming high drug loading

multifunctional nanocapsules for intracellular cancer drug delivery. *Proceedings of the National Academy of Sciences of the United States of America 132*, 4259.

35. Kim, S. C., Kim, D. W., Shim, Y. H., Bang, J. S., Oh, H. S., Kim, S. W., Seo, M. H. (2001). *In vivo* evaluation of polymeric micellar paclitaxel formulation: toxicity and efficacy. *Journal of Controlled Release 72*, 191.

36. Benny, O., Fainaru, O., Adini, A., Cassiola, F., Bazinet, L., Adini, I., Pravda, E., Nahmias, Y., Koirala, S., Corfas, G., D'Amato, R. J., Folkman, J. (2008). An orally delivered small-molecule formulation with antiangiogenic and anticancer activity. *Nature Biotechnology 26*, 799.

37. Lee, S. H., Zhang, Z. P., Feng, S. S. (2007). Nanoparticles of poly(lactide): tocopheryl polyethylene glycol succinate (PLA–TPGS) copolymers for protein drug delivery. *Biomaterials 28*, 2041.

38. Zhang, Z. P., Feng, S. S. (2006). Nanoparticles of poly(lactide)/vitamin E TPGS copolymer for cancer chemotherapy: synthesis, formulation, characterization and *in vitro* drug release. *Biomaterials 27*, 262.

39. Sengupta, S., Eavarone, D., Capila, I., Zhao, G. L., Watson, N., Kiziltepe, T., Sasisekharan, R. (2005). Temporal targeting of tumour cells and neovasculature with a nanoscale delivery system. *Nature 436*, 568.

40. Yoo, H. S., Lee, K. H., Oh, J. E., Park, T. G. (2000). *In vitro* and *in vivo* anti-tumor activities of nanoparticles based on doxorubicin-PLGA conjugates. *Journal of Controlled Release 68*, 419.

41. Chamberlain, B. M., Cheng, M., Moore, D. R., Ovitt, T. M., Lobkovsky, E. B., Coates, G. W. (2001). Polymerization of lactide with zinc and magnesium beta-diiminate complexes: stereocontrol and mechanism. *Journal of the American Chemical Society 123*, 3229.

42. O'Keefe, B. J., Hillmyer, M. A., Tolman, W. B. (2001). Polymerization of lactide and related cyclic esters by discrete metal complexes. *Journal of the Chemical Society-Dalton Transactions* 2215.

43. Cheng, J., Deming, T. J. (2001). Synthesis and conformational analysis of optically active poly(beta-peptides). *Macromolecules 34*, 5169.

44. Lu, H., Cheng, J. (2008). *N*-trimethylsilyl amines for controlled ring-opening polymerization of amino acid *N*-carboxyanhydrides and facile end group functionalization of polypeptides. *Journal of the American Chemical Society 130*, 12562.

45. Ouchi, M., Terashima, T., Sawamoto, M. (2009). Transition metal-catalyzed living radical polymerization: toward perfection in catalysis and precision polymer synthesis. *Chemical Reviews 109*, 4963.

46. Dechy-Cabaret, O., Martin-Vaca, B., Bourissou, D. (2004). Controlled ring-opening polymerization of lactide and glycolide. *Chemical Reviews 104*, 6147.

47. du Boullay, O. T., Marchal, E., Martin-Vaca, B., Cossio, F. P., Bourissou, D. (2006). An activated equivalent of lactide toward organocatalytic ring-opening polymerization. *Journal of the American Chemical Society 128*, 16442.

48. Hu, Y., Jiang, X., Ding, Y., Zhang, L., Yang, C., Zhang, J., Chen, J., Yang, Y. (2003). Preparation and drug release behaviors of nimodipine-loaded poly(caprolactone)–poly(ethylene oxide)–polylactide amphiphilic copolymer nanoparticles. *Biomaterials 24*, 2395.

49. Chen, F., Gao, Q., Hong, G., Ni, J. (2008). Synthesis of magnetite core–shell nano-particles by surface-initiated ring-opening polymerization of L-lactide. *Journal of Magnetism and Magnetic Materials 320*, 1921.

50. Cheng, M., Attygalle, A. B., Lobkovsky, E. B., Coates, G. W. (1999). Single-site catalysts for ring-opening polymerization: synthesis of heterotactic poly(lactic acid) from rac-lactide. *Journal of the American Chemical Society 121*, 11583.

51. Tong, R., Cheng, J. (2008). Paclitaxel-initiated, controlled polymerization of lactide for the formulation of polymeric nanoparticulate delivery vehicles. *Angewandte Chemie (International edition) 47*, 4830.

52. Tong, R., Cheng, J. (2009). Ring-opening polymerization-mediated controlled formula-tion of polylactide-drug nanoparticles. *Journal of the American Chemical Society 131*, 4744.

53. Moore, D. R., Cheng, M., Lobkovsky, E. B., Coates, G. W. (2002). Electronic and steric effects on catalysts for CO_2/epoxide polymerization: subtle modifications resulting in superior activities. *Angewandte Chemie (International edition) 41*, 2599.

54. Tong, R., Cheng, J. (2010). Controlled synthesis of camptothecin–polylactide conjugates and nanoconjugates. *Bioconjugate Chemistry 21*, 111.

55. Tong, R., Yala, L., Fan, T. M., Cheng, J. (2010). The formulation of aptamer-coated paclitaxel–polylactide nanoconjugates and their targeting to cancer cells. *Biomaterials 31*, 3043.

56. Hertzberg, R. P., Caranfa, M. J., Hecht, S. M. (1989). On the mechanism of topoisom-erase-I inhibition by camptothecin: evidence for binding to an enzyme DNA complex. *Biochemistry 28*, 4629.

57. Muggia, F. M., Dimery, I., Arbuck, S. G., *Camptothecin and its analogs: an overview of their potential in cancer therapeutics*, Academy of Sciences, New York, 1996.

58. Mi, Z. H., Burke, T. G. (1994). Differential interactions of camptothecin lactone and carboxylate forms with human blood components. *Biochemistry 33*, 10325.

59. Mi, Z. H., Burke, T. G. (1994). Marked interspecies variations concerning the inter-actions of camptothecin with serum albumins: a frequency-domain fluorescence spec-troscopic study. *Biochemistry 33*, 12540.

60. Cheng, J., Khin, K. T., Davis, M. E. (2004). Antitumor activity of beta-cyclodextrin polymer: camptothecin conjugates. *Molecular Pharmaceutics 1*, 183.

61. Kunii, R., Onishi, H., Machida, Y. (2007). Preparation and antitumor characteristics of PLA/(PEG–PPG–PEG) nanoparticles loaded with camptothecin. *European Journal of Pharmaceutics and Biopharmaceutics 67*, 9.

62. Li, C. (1998). Complete regression of well-established tumors using a novel water-soluble poly (L-glutamic acid)–paclitaxel conjugate. *Cancer Research 58*, 2404.

63. Singer, J. W. (2005). Paclitaxel poliglumex (XYOTAX, CT-2103): a macromolecular taxane. *Journal of Controlled Release 109*, 120.

64. Fonseca, C., Simoes, S., Gaspar, R. (2002). Paclitaxel-loaded PLGA nanoparticles: preparation, physicochemical characterization and *in vitro* anti-tumoral activity. *Journal of Controlled Release 83*, 273.

65. Gref, R., Minamitake, Y., Peracchia, M., Trubetskoy, V. S., Torchilin, V. P., Langer, R. (1994). Biodegradable long-circulating polymeric nanospheres. *Science 263*, 1600.

66. Panyam, J., Labhasetwar, V. (2003). Biodegradable nanoparticles for drug and gene delivery to cells and tissue. *Advanced Drug Delivery Reviews 55*, 329.

67. Soppimath, K. S., Aminabhavi, T. M., Kulkarni, A. R., Rudzinski, W. E. (2001). Biodegradable polymeric nanoparticles as drug delivery devices. *Journal of Controlled Release 70*, 1.

68. Mastropaolo, D., Camerman, A., Luo, Y. G., Brayer, G. D., Camerman, N. (1995). Crystal and molecular-structure of paclitaxel (taxol). *Proceedings of the National Academy of Sciences of the United States of America 92*, 6920.

69. Mathew, A. E., Mejillano, M. R., Nath, J. P., Himes, R. H., Stella, V. J. (1992). Synthesis and evaluation of some water-soluble prodrugs and derivatives of taxol with antitumor-activity. *Journal of Medicinal Chemistry 35*, 145.

70. Magri, N. F., Kingston, D. G. I., Jitrangsri, C., Piccariello, T. (1986). Modified taxols. 3. Preparation and acylation of baccatin-III. *The Journal of Organic Chemistry 51*, 3239.

71. Moore, D. R., Cheng, M., Lobkovsky, E. B., Coates, G. W. (2003). Mechanism of the alternating copolymerization of epoxides and CO_2 using beta-diiminate zinc catalysts: evidence for a bimetallic epoxide enchainment. *Journal of the American Chemical Society 125*, 11911.

72. Cheng, M., Moore, D. R., Reczek, J. J., Chamberlain, B. M., Lobkovsky, E. B., Coates, G. W. (2001). Single-site beta-diiminate zinc catalysts for the alternating copolymerization of CO_2 and epoxides: catalyst synthesis and unprecedented polymerization activity. *Journal of the American Chemical Society 123*, 8738.

73. http://www.doxil.com/doxil-supply-shortage.

74. http://www.doxil.com/ovarian-cancer/hand-foot-syndrome.

75. Lorusso, D., Di Stefano, A., Carone, V., Fagotti, A., Pisconti, S., Scambia, G. (2007). Pegylated liposomal doxorubicin-related palmar-plantar erythrodysesthesia ('hand-foot' syndrome). *Annals of Oncology 18*, 1159.

76. Yoo, H. S., Park, T. G. (2001). Biodegradable polymeric micelles composed of doxorubicin conjugated PLGA–PEG block copolymer. *Journal of Controlled Release 70*, 63.

77. Missirlis, D., Kawamura, R., Tirelli, N., Hubbell, J. A. (2006). Doxorubicin encapsulation and diffusional release from stable, polymeric, hydrogel nanoparticles. *European Journal of Pharmaceutical Sciences 29*, 120.

78. Altreuter, D. H., Dordick, J. S., Clark, D. S. (2002). Nonaqueous biocatalytic synthesis of new cytotoxic doxorubicin derivatives: exploiting unexpected differences in the regio-selectivity of salt-activated and solubilized subtilisin. *Journal of the American Chemical Society 124*, 1871.

79. Yoo, H. S., Oh, J. E., Lee, K. H., Park, T. G. (1999). Biodegradable nanoparticles containing doxorubicin–PLGA conjugate for sustained release. *Pharmaceutical Research 16*, 1114.

80. Nishiyama, N., Kataoka, K. (2006). Nanostructured devices based on block copolymer assemblies for drug delivery: designing structures for enhanced drug function. *Advances in Polymer Science 193*, 67.

81. Greenfield, R. S., Kaneko, T., Daues, A., Edson, M. A., Fitzgerald, K. A., Olech, L. J., Grattan, J. A., Spitalny, G. L., Braslawsky, G. R. (1990). Evaluation in vitro of adriamycin immunoconjugates synthesized using an acid-sensitive hydrazone linker. *Cancer Research 50*, 6600.

82. Kaneko, T., Willner, D., Monkovic, I., Knipe, J. O., Braslawsky, G. R., Greenfield, R. S., Vyas, D. M. (1991). New hydrazone derivatives of adriamycin and their

immunoconjugates: a correlation between acid stability and cytotoxicity. *Bioconjugate Chemistry 2*, 133.

83. Ulbrich, K., Subr, V. (2004). Polymeric anticancer drugs with pH-controlled activation. *Advanced Drug Delivery Reviews 56*, 1023.

84. Florent, J. C., Monneret, C. (2008). Doxorubicin conjugates for selective delivery to tumors. *Anthracycline Chemistry and Biology II: Mode of Action, Clinical Aspects and New Drugs 283*, 99.

85. Farokhzad, O. C., Cheng, J., Teply, B. A., Sherifi, I., Jon, S., Kantoff, P. W., Richie, J. P., Langer, R. (2006). Targeted nanoparticle–aptamer bioconjugates for cancer chemotherapy *in vivo*. *Proceedings of the National Academy of Sciences of the United States of America 103*, 6315.

86. Caliceti, P., Veronese, F. M. (2003). Pharmacokinetic and biodistribution properties of poly(ethylene glycol)–protein conjugates. *Advanced Drug Delivery Reviews 55*, 1261.

87. Pierri, E., Avgoustakis, K. (2005). Poly(lactide)–poly(ethylene glycol) micelles as a carrier for griseofulvin. *Journal of Biomedical Materials Research. Part A 75A*, 639.

88. Tuerk, C., Gold, L. (1990). Systematic evolution of ligands by exponential enrichment: RNA ligands to bacteriophage-T4 DNA-polymerase. *Science 249*, 505.

89. Ellington, A. D., Szostak, J. W. (1992). Selection *in vitro* of single-stranded-DNA molecules that fold into specific ligand-binding structures. *Nature 355*, 850.

90. Farokhzad, O. C., Jon, S. Y., Khadelmhosseini, A., Tran, T. N. T., LaVan, D. A., Langer, R. (2004). Nanoparticle–aptamer bioconjugates: a new approach for targeting prostate cancer cells. *Cancer Research 64*, 7668.

91. Keefe, A. D., Pai, S., Ellington, A. (2010). Aptamers as therapeutics. *Nature Reviews Drug Discovery 9*, 537.

92. Nimjee, S. M., Rusconi, C. P., Sullenger, B. A. (2005). Aptamers: an emerging class of therapeutics. *Annual Review of Medicine 56*, 555.

93. Cao, Z. H., Tong, R., Mishra, A., Xu, W. C., Wong, G. C. L., Cheng, J. J., Lu, Y. (2009). Reversible cell-specific drug delivery with aptamer-functionalized liposomes. *Angewandte Chemie (International edition) 48*, 6494.

94. Lupold, S. E., Hicke, B. J., Lin, Y., Coffey, D. S. (2002). Identification and characterization of nuclease-stabilized RNA molecules that bind human prostate cancer cells via the prostate-specific membrane antigen. *Cancer Research 62*, 4029.

95. Chaney, E. J., Tang, L., Tong, R., Cheng, J. J., Boppart, S. A. (2010). Lymphatic biodistribution of polylactide nanoparticles. *Molecular Imaging 9*, 153.

96. Chan, J. M., Zhang, L. F., Tong, R., Ghosh, D., Gao, W. W., Liao, G., Yuet, K. P., Gray, D., Rhee, J. W., Cheng, J. J., Golomb, G., Libby, P., Langer, R., Farokhzad, O. C. (2010). Spatiotemporal controlled delivery of nanoparticles to injured vasculature. *Proceedings of the National Academy of Sciences of the United States of America 107*, 2213.

97. Kaplan, B., Meier-Kriesche, H. U. (2004). Renal transplantation: a half century of success and the long road ahead. *Journal of the American Society of Nephrology 15*, 3270.

98. Wong, W., Venetz, J. P., Tolkoff-Rubin, N., Pascual, M. (2005). Immunosuppressive strategies in kidney transplantation: which role for the calcineurin inhibitors? *Transplantation 80*, 289.

99. Chapman, J. R., Nankivell, B. J. (2006). Nephrotoxicity of ciclosporin A: short-term gain, long-term pain? *Nephrology, Dialysis, Transplantation 21*, 2060.

100. Neto, A. B., Haapalainen, E., Ferreira, R., Feo, C. F., Misiako, E. P., Vennarecci, G., Porcu, A., Dib, S. A., Goldenberg, S., Gomes, P. O., Nigro, A. T. (1999). Metabolic and ultrastructural effects of cyclosporin A on pancreatic islets. *Transplant International 12*, 208.

101. Azzi, J., Tang, L., Moore, R., Tong, R., El Haddad, N., Akiyoshi, T., Mfarrej, B., Yang, S. M., Jurewicz, M., Ichimura, T., Lindeman, N., Cheng, J. J., Abdi, R. (2010). Polylactide-cyclosporin A nanoparticles for targeted immunosuppression. *The FASEB Journal 24*, 3927.

102. Rieth, L. R., Moore, D. R., Lobkovsky, E. B., Coates, G. W. (2002). Single-site beta-diiminate zinc catalysts for the ring-opening polymerization of beta-butyrolactone and beta-valerolactone to poly(3-hydroxyalkanoates). *Journal of the American Chemical Society 124*, 15239.

103. Sharp, P. A., Langer, R. (2011). Promoting convergence in biomedical science. *Science 333*, 527.

104. Wong, C., Stylianopoulos, T., Cui, J. A., Martin, J., Chauhan, V. P., Jiang, W., Popovic, Z., Jain, R. K., Bawendi, M. G., Fukumura, D. (2011). Multistage nanoparticle delivery system for deep penetration into tumor tissue. *Proceedings of the National Academy of Sciences of the United States of America 108*, 2426.

105. Sugahara, K. N., Teesalu, T., Karmali, P. P., Kotamraju, V. R., Agemy, L., Greenwald, D. R., Ruoslahti, E. (2010). Coadministration of a tumor-penetrating peptide enhances the efficacy of cancer drugs. *Science 328*, 1031.

106. Jain, R. K., Stylianopoulos, T. (2010). Delivering nanomedicine to solid tumors. *Nature Reviews Clinical Oncology 7*, 653.

107. Torchilin, V. P. (2006). Multifunctional nanocarriers. *Advanced Drug Delivery Reviews 58*, 1532.

108. Sutton, D., Nasongkla, N., Blanco, E., Gao, J. M. (2007). Functionalized micellar systems for cancer targeted drug delivery. *Pharmaceutical Research 24*, 1029.

109. Chaffer, C. L., Weinberg, R. A. (2011). A perspective on cancer cell metastasis. *Science 331*, 1559.

110. Hanahan, D., Weinberg, Robert A. (2011). Hallmarks of cancer: the next generation. *Cell 144*, 646.

8

NANOCRYSTALS PRODUCTION, CHARACTERIZATION, AND APPLICATION FOR CANCER THERAPY

CHRISTIN P. HOLLIS

College of Pharmacy, University of Kentucky, Lexington, KY, USA

TONGLEI LI

Department of Industrial and Physical Pharmacy, Purdue University, West Lafayette, IN, USA

8.1 INTRODUCTION

It is estimated that two-thirds of the newly synthesized drugs are poorly soluble in water [1]. Delivery of these compounds requires significant efforts and, in some cases, becomes extremely difficult. Poor solubility limits the bioavailability when delivered through the oral route; for many compounds, parental delivery, which requires special formulation treatment to overcome the solubility limit, remains to be the only choice. A common practice is to solubilize a poorly soluble drug with specialized organic molecules, which include organic solvents, surfactants, and macromolecules. Unfortunately, adverse effects resulted from the use of such solubilizing molecules, for example, hypersensitivity reactions, nephrotoxicity, neurotoxicity, and neutropenia, have been reported [2,3]. In addition, the individually solubilized drug molecules can also reach healthy organs and tissues indiscriminately and inflict undesired effects. Hence, there is a need for developing formulation that can deliver poorly water-soluble drugs, including anticancer agents.

Nanoparticulate Drug Delivery Systems: Strategies, Technologies, and Applications, First Edition.
Edited by Yoon Yeo.
© 2013 John Wiley & Sons, Inc. Published 2013 by John Wiley & Sons, Inc.

Attempting to target tumors, anticancer drugs are generally formulated as nano-particles in order to take advantage of the so-called enhanced permeability and retention (EPR) effect [4], which characterizes tumors' underdeveloped vasculatures and ineffective lymphatic systems. Various nanosized carrier designs have been developed, including liposomes [5,6], micelles [7], solid lipid nanoparticles [8,9], polymeric nanoparticles [10,11], and dendrimers [12]. In many of these nanoparticle designs, drug molecules are solubilized, dispersed, and encapsulated by specialized chemicals.

Despite a significant amount of efforts on the development of new delivery systems over the last few decades, there are only a few Food and Drug Administration (FDA)-approved cancer therapy products in which poorly soluble drugs are formulated as nanoparticles. Two are liposomal formulations, Doxil® (doxorubicin) and Daunosome® (daunorubicin), and one other is Abraxane®, albumin-bound paclitaxel (PTX) nanoparticles. Impeding the progress in the anticancer drug product development are several factors that lie in the current formulation design strategy. When a drug is solubilized into a nanoconstruct, the multicomponent system is prone to physical instability and structural destruction during storage and/or *in vivo* passage, resulting in drug leakage [13,14]. The drug loading is typically limited [7,15]; high manufacturing cost, multiple steps in preparation and synthesis, and difficulties in scale-up have also limited the potential of the aforementioned vehicles for clinical usage.

Clearly, it has been and will remain a significant challenge to deliver a poorly soluble anticancer compound. Current solubilization approaches face inherent limitations that need to be overcome for the clinical application, particularly when tumor targeting is desired. It is recognized that in order to fully take advantage of the EPR effect, a clinically viable tumor-targeted delivery system has to be stable, should have a high drug loading, and be capable of sustaining the release of drug molecules in the tumor [16]. It is rather futile to develop an intricate vehicle that has little clinical potential. New approaches need to be developed to shift the paradigm of delivering poorly soluble drugs, including antineoplastic agents.

Over the last few years, nanocrystal formulations have been of considerable interest for delivering poorly soluble drugs via oral, parenteral, and other routes [17]. Administered directly as nanosized solid particles, drug nanocrystals require no solubilizing and/or encapsulating chemicals, thus circumventing side effects and instability drawbacks impeding many existing delivery systems. More importantly, production of nanocrystals at the large scale is feasible due to the simplicity of formulation and absence of solubilizing, encapsulating, or conjugating steps. Nano-crystals are expected to target the tumor via the EPR effect and offer less toxicity, improving the anticancer efficacy and pharmacokinetic properties. Nanocrystal formulations may likely rejuvenate the delivery of poorly soluble anticancer compounds, existing and to be discovered.

In this chapter, we first review the production of nanocrystals of pharmaceutical organic compounds. We then focus on the application of nanocrystals for cancer therapy.

8.2 NANOCRYSTAL PRODUCTION

Crystals are regular, periodic supramolecular structures assembled by organic or other types of molecules or atoms. Characteristically, a crystal shows distinct X-ray diffraction patterns, and has a specific melting point, density, and solubility, in addition to other well-defined physicochemical properties. Compared with the amorphous state, in which molecules are randomly packed, the crystalline state is more stable, exemplified by the fact that amorphous materials can readily undergo spontaneous phase transition and recrystallization. As such, using crystalline materials in a formulation assures better physical stability than amorphous formulations.

Nanocrystals of poorly soluble drugs can offer several advantages over solubilized and encapsulated formulations. Besides the physical stability, the dissolution rate of nanocrystals can be significantly enhanced—under sink condition—due to the increase in the surface area. More importantly, manufacturing at the industrial scale is possible. In the following sections, we discuss two major approaches for producing nanocrystals of organic molecules. The top-down approach processes large crystals and physically comminute into smaller pieces; the bottom-up approach yields nanocrystals from solution by crystallization. Several review articles have focused on the techniques, mainly the top-down approach [17–21].

8.2.1 Top-Down Approach

There are two basic diminution techniques: wet milling and high-pressure homogenization. Wet milling involves mechanical attrition, where drug particles are wetted by an aqueous solution of surfactants and subjected to shearing and grinding by mechanical toolings, such as milling balls in a milling container. The particle size is reduced and may reach a few hundred nanometers. The Nanocrystal® technique [22], a wet milling process patented by Élan Co., has yielded several oral products including Rapamune®, Emend®, Tricor®, and Megace ES® [23]. One drawback of the technique is potential contamination from the erosion of metal milling balls or pearls. The amount of contamination is determined by the hardness of drug particles as well as the milling time, which can take up to several days [24]. The use of polymeric beads (e.g., polystyrene derivatives) may be helpful in minimizing the erosion.

High-pressure homogenization is another fragmentation technique. One development is based on jet-stream microfluidization [25], in which two fluid streams of particle suspensions collide under high pressure in a Y-shaped chamber, leading to particle collision and subsequent particle rupture. The marketed product Triglide® developed by SkyePharma Co. is produced by Insoluble Drug Delivery-Particles (IDD-P™) technology, a microfluidization method [26]. Piston-gap homogenizers are also used to produce nanosized solid particles by forcing a suspension of drug particles with a piston through a thin gap under high pressure [27]. The combination of cavitations, high shear forces, and turbulent flow fractures the particles. The outcome of particle fragmentation is decided by several factors, including the power of homogenization, particle hardness, and number of the piston-moving cycles. Compared with wet milling, the homogenization method may yield less contamination during the production process [28]. There are currently two patented technologies for nanocrystal production that employ the piston-gap homogenization method, DissoCubes® and Nanopure®. In the former, developed by SkyePharma Co., drug powders are dispersed in an aqueous solution of surfactants and subjected to homogenization [27]; in the latter, previously owned by PharmaSol GmbH and now acquired by Abbott Co., drug particles are suspended in either nonaqueous dispersion media (e.g., oils and liquefied polyethylene glycol) or aqueous-organic cosolvents (e.g., glycerol–water and ethanol–water mixtures) prior to homogenization [27]. These operations require that one starts with drug particles no larger than 25 μm so that the blockage of the homogenizer can be minimized [29].

Currently, the top-down approach dominates the production of nanocrystals and is used to manufacture several products on the market [21]. Despite its popularity, there are inherent drawbacks in this approach, including the high-energy input, prolonged operation time, possibility of contamination, and decreased crystallinity or stability [30]. The stability concern is particularly critical for a nanocrystal system to be considered for parenteral drug delivery. The amorphous content produced during the high-energy process can lead to phase instability, which in turn causes uncertainty and potential variability in pharmacokinetic behavior, which should be tightly controlled.

8.2.2 Bottom-Up Approach

Nanocrystals can be grown directly from solution. Nucleation and crystal growth thus play the essential role in the bottom-up approach. To induce nucleation, a solution needs to become supersaturated. Cooling, evaporating the solvent, and mixing with an antisolvent are among several ways to create supersaturation [31]. Still, a solution can become supersaturated to some extent without producing any nuclei. This is because nucleation is energetically unfavorable. According to the classical nucleation theory [32–34], the change in free energy when a nucleus starts to grow includes two energy components, one associated with the creation of the new surface, and another with the packing of molecules in the bulk crystal. As depicted in Figure 8.1, the surface-related term, ΔG_S, is always positive (energetically unfavorable) because creation of a new surface of a solid particle requires energy input. On the other hand, the bulk-related term, ΔG_V, is negative (energetically favorable) due to the formation of intermolecular interactions between molecules in the nucleus. When a nucleus is small, ΔG_S dominates, and the overall free energy change is positive so that the development of the nucleus is not spontaneous. As the nucleus grows and surpasses, the so-called critical nucleus size, ΔG_V becomes dominant and the overall term becomes negative. Subsequently, the nucleus can grow freely as long as supersaturation is maintained.

In reality, homogeneous nucleation—nucleation without the aid of substances other than the solute molecules—seldom occurs. Foreign materials, impurities, or even the surface of the crystallization vessel often help to induce nucleation, which is denoted as the heterogeneous nucleation [35]. According to the classical nucleation theory, the presence of a foreign surface lowers the ΔG_S term for a nucleus to form on

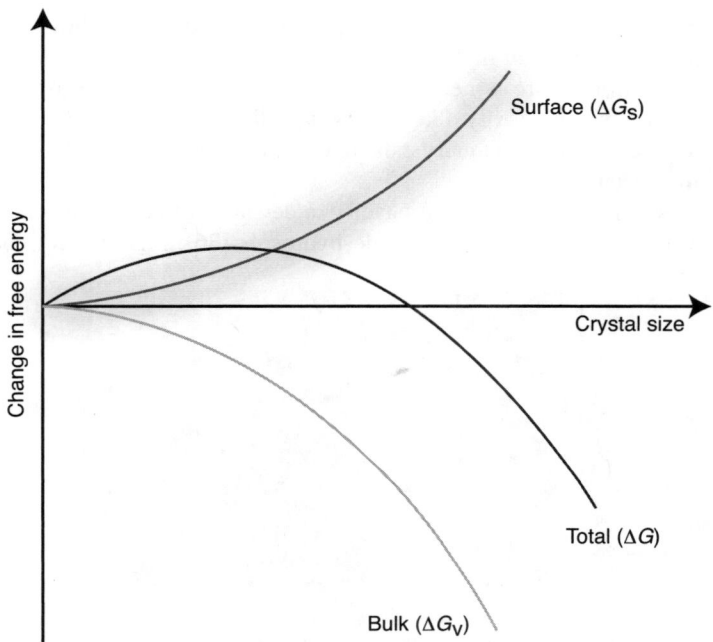

FIGURE 8.1 The change in free energy as a function of crystal size during the nucleation and crystal growth process.

the surface. In other words, the nucleus requires less energy for surface creation when it forms on a pre-existing surface.

In order to produce nanosized crystals from solution, it is thus extremely critical to control the nucleation process. It is important not only for controlling the particle size but also for the size distribution. Ideally, nucleation of a large number of crystalline embryos must occur concurrently from the supersaturated solution. Rapid depletion of the solute molecules from the solution limits further growth of each nucleus so as to achieve a nanosize. Having a large number of nuclei produced at the same time also keeps the particle size distribution narrow. Similar particle sizes are not only preferred for ensuring product quality but also important for maintaining the stability of nanocrystals. A narrow distribution of particle size can greatly minimize the so-called Ostwald ripening phenomenon, in which larger particles grow at the expense of redissolving smaller ones because of their differences in surface energy [36]. For this purpose, the induction of nucleation needs to be abrupt and homogeneous throughout the growth medium. One common strategy is to mix the solution of a drug compound with an antisolvent so that the supersaturation is reached immediately; at the same time, the mixture is subjected to intense sonication and/or mechanical stirring. It is known that air cavities created by sonication may trigger nucleation because of the presence of air–liquid interfaces and locally concentrated liquid pockets. Nevertheless, because nucleation is such a difficult process to control [37], subtle changes in growth conditions can result in drastic variations in product quality. For producing nanocrystals, extreme care is needed for identifying a suitable set of peripheral conditions [38], which mostly likely vary for distinct compounds. Figure 8.2 shows a few samples of nanocrystals that were produced by the bottom-up approach in our laboratory.

There are several approaches that have been developed to create supersaturation and induce nucleation [20]. They include hydrosols [36], high-gravity controlled

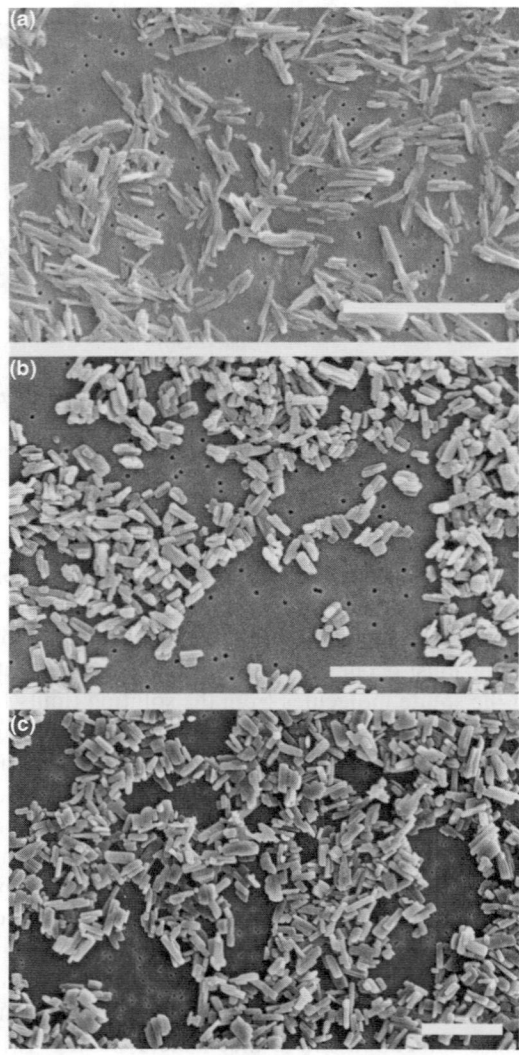

FIGURE 8.2   SEM images of nanocrystals grown in our laboratory: camptothecin (a), paclitaxel (b), and ZSTK474 (c). Scale bar: 2 μm.

precipitation technology (HGCP) [39,40], confined impinging jets [41], supercritical fluid (SCF) technology [42], and sonoprecipitation [43], as discussed further.

8.2.2.1 Hydrosols. Considered to be a more mature method, hydrosols or aqueous nanosuspensions of poorly water-soluble drugs have been developed by Sandoz (now Novartis) [44]. The active ingredient is dissolved in aqueous-organic miscible cosolvents (e.g., ethanol, isopropanol, and acetone), while various peptisers and/or stabilizing agents (e.g., citric acid, gelatin, ethyl cellulose N7, lecithin) may be added to the aqueous phase to prevent aggregation. A matrix former, generally sugar or sugar alcohol (e.g., mannitol), is also added to the solution to prevent agglomeration and aid in the resuspension process. In the laboratory setting, solid nanoparticles can be produced by mechanically mixing the various components in a beaker while removing the organic solvent through evaporation. For the industrial production, the organic and aqueous phases are pumped into a static mixer and forced through an atomizing nozzle, leading to the formation and drying of nanoparticles [44–47]. One study reported that spray-dried cyclosporin A contained 38 μm particle agglomerates that were then reduced to 120 nm upon redispersion in water [45]; the nanoparticles produced by this method were nonetheless amorphous [45,46].

8.2.2.2 High-Gravity Controlled Precipitation (HGCP). HGCP is considered to be one of the most promising techniques to produce nanoprecipitations at the commercial scale [20]. The key component in HGCP is a rotating packed bed (RPB), where two liquid streams are pumped in and mixed together vigorously by centrifugation. Subsequently, the mixture is spread into thin films and ultimately broken into tiny droplets. In addition to using antisolvents, mixing two streams of reactants can also produce precipitates chemically in HGCP. Depending on the compound and conditions, a product may be amorphous or crystalline. Nanoparticles of a few pharmaceutical compounds, including cephradine [48], danazol [49], and salbutamol sulfate [50,51], have been successfully produced by this method.

8.2.2.3 Flash Nanoprecipitation by Confined Liquid Impinging Jets (CLIJ). In this technique, two respective jet streams of a drug solution and an antisolvent are forced together by opposing nozzles that are mounted in a small chamber to create localized, intensified mixing. The mixing causes precipitation of nanoparticles within the residence time of mixing. The velocity, drug concentration, and volume ratio of the two streams are considered to be the factors determining the particle size distribution of a product. CLIJ has been utilized in the production of salbutamol sulfate, ibuprofen, cyclosporin A, and amphotericin B [52–54]. GRAS (generally regarded as safe) additives may be added to assist the production.

8.2.2.4 Supercritical Fluid (SCF) Technology. This method takes advantage of the unique physical properties of supercritical fluids, including low density and viscosity as well as high diffusivity of solute molecules to attain rapid mixing. In addition, quick and easy removal of a SCF without excessive drying can greatly facilitate the precipitation of nanoparticles. Supercritical carbon dioxide (SCO_2) is mostly used and considered green or environmentally friendly for processing most pharmaceuticals. Depending on the solubility of a compound in SCO_2, production can be accomplished by allowing the drug–SCO_2 mixture to expand under ambient conditions (method called rapid expansion of supercritical solution or RESS [55]) or by using SCO_2 as the antisolvent when the drug is poorly soluble in SCO_2 (a process is called supercritical antisolvent or SAS [55]). In both the methods, it is critical that other solvents used are miscible with SCO_2; water is actually immiscible. Amoxicillin, ampicillin, and rifampicin nanoparticles were successfully produced by the SCF approach [56]. 100–500 nm protein nanoparticles of insulin, rhDNase, lysosome, and albumin were precipitated from aqueous ethanol solution by using ethanol as the co-solvent [57].

8.2.2.5 Sonoprecipitation. Ultrasonic waves can create cavitations that subsequently collapse, releasing shock waves that may assist in rapid and more uniform nucleation [58]. Ultrasound can also reduce particle agglomeration by minimizing particle contacts [43]. Drug microparticles were reportedly produced by the sonocrystallization method [59–61]; it is also shown that the method could produce nanoparticles of amorphous cefuroxime axetil [62]. Factors that influence the quality of a final product include (i) ultrasound frequency, intensity, and power, (ii) sonication probe size and immersion depth, (iii) solution volume, and (iv) duration. Although the experimental setup appears to be straightforward, this technology has not been adopted for commercial production.

The quality of nanocrystals produced by the bottom-up approach is expected to be superior to that by the top-down method. Particle size and morphology can be better controlled, and it is possible to produce pure nanocrystals with little or no amorphous content. Shorter production time, less energy consumption, and reduced contamination also make the precipitation techniques desirable. Clearly, the bottom-up approach holds great promise for developing nanocrystals for parenteral delivery, which demands uncompromised quality and stability. Although no such product is currently in the market, there is certainly a growing interest and need in this field. Challenges remain in controlling nucleation and subsequent crystal growth so that batch-to-batch reproducibility can be achieved. It is certain that among various methods that are being developed, ones that are simple, reproducible, and yet cost-effective for production scale-up should have great potential in commercial production [36].

8.2.3 Combined Approach

There have been efforts to combine the two aforementioned approaches. The NanoEdge™ patent by Baxter claims that nanocrystals, ranging from 400 to 2 μm, can be prepared by crystallization followed by high-pressure homogenization [63]. Our research group used an antisolvent approach to grow nanocrystals by rapidly precipitating anticancer drugs under intense stirring and sonication [64,65]. To further reduce the size, the nanocrystal suspension can subsequently be homogenized. Nanocrystals with an average particle size ranging from 200 to 800 nm have been successfully produced in our laboratory. The second step of homogenization breaks down larger crystals and, particularly, aggregates, achieving a much narrower particle size distribution.

8.3 PARTICLE STABILIZATION

Nanocrystals and nanoparticles of drugs, whether prepared by the top-down or bottom-up approach, often require surfactants to be physically attached to the surface of each particle in order to prevent the particles from aggregation during their preparation, storage, and administration. Particle aggregation mutually facilitates Ostwald's ripening and eventually becomes irreversible, leading to larger particles and precipitates. Surfactant usage is particularly necessary when the surface of nanocrystals is uncharged. In general, polymeric surfactants are commonly used to introduce steric repulsion between particles; electrostatic repulsion also promotes particle segregation when ionic surfactants are used. Povidones, pluronics, and cellulosics are the most commonly used surfactants. Natural biological molecules, such as lecithins and cholic acid derivatives, are utilized as well [66].

Finding a suitable surfactant for a particular drug is experimentally demanding [67]. A stabilizer obviously needs to have sufficient affinity for the particle surface and the amount used should provide sufficient steric and/or electrostatic repulsion between the particles [68]. Excess amounts of surfactants do not always lead to better coverage due to the tendency of surfactants to form micelles when the concentration surpasses the critical micelle concentration (CMC) [69,70]. A recent study of the effects of 13 stabilizers during wet milling of 9 drug compounds [71] concluded that the semisynthetic polymers, including hydroxypropylmethylcellulose (HPMC), methylcellulose (MC), hydroxyethylcellulose (HEC), hydroxypropylcellulose (HPC), carboxymethylcellulose sodium salt (NaCMC), and alginic acid sodium salt (NaAlg), yielded rather poor stabilizing performance, likely because of their high viscosity. Less viscous stabilizers, including linear synthetic polymers (PVP K30 and K90) and synthetic copolymers (poloxamer 188 and Kollicoat IR®), showed better stabilizing capabilities, especially at higher concentrations. Tween® 80 and TPGS (D-α-tocopherol polyethylene glycol 1000 succinate) gave the best stabilizing performance. Moreover, the timing of surfactant addition may affect particle properties. Stabilizers are commonly added along with the starting micro-crystalline materials for wet milling or homogenization. However, the addition of a stabilizer prior to crystallization may alter the nucleation and subsequent crystal growth outcome. For example, the morphology of the nanocrystals of ZSTK474, a novel PI3K inhibitor [72], that were prepared in our laboratory by the bottom-up approach varied from an elongated to a plate-like shape due to the addition of 1% (w/v) Pluronic® F127 (Figure 8.3).

Because nanocrystals undergo consistent dynamic changes, especially in solution or in contact with moisture—dissolution and surface recrystallization—albeit at a slow rate, surface adsorption by surfactants may not be able to sustain stability of nanocrystals for the long-term storage. Alternatively, nanocrystal products may be

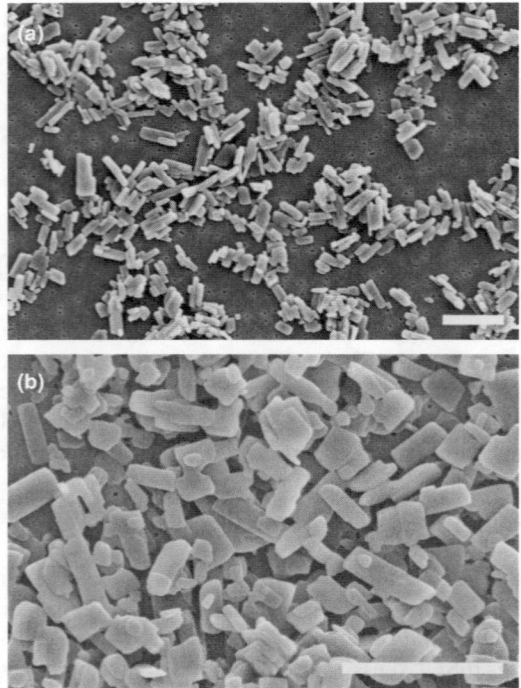

FIGURE 8.3   SEM images of ZSTK474 nanocrystals grown in the absence (a) and presence (b) of Pluronic F127. Scale bar: 2 μm.

further processed by freeze drying, spray drying, or pelletization [19]; water-soluble sugars, such as mannitol, trehalose, and dextran, are commonly added as matrix formers prior to drying [73]. In the absence of a solvent, the surface dynamics and mobility are greatly minimized. For further sterilization, nanosuspensions for intravenous use are often subjected to terminal heat and gamma irradiation [74] as well as filtration [75], and stored as liquid suspension or lyophilized solid form.

8.4   PARTICLE CHARACTERIZATION

8.4.1   Particle Size

Nanocrystals need to be fully evaluated to determine their (i) particle size and size distribution, (ii) surface charge, (iii) crystallinity, and (iv) dissolution rate [18,29,73,76]. For rapid and noninvasive determination of the mean particle size and size distribution, photon correlation spectroscopy (PCS) is mostly used for particles ranging from 3 nm to 3 μm. The size distribution is indicated by polydispersity index (PDI), which is indicative of the width of particle size distribution. A PDI value of 0.1–0.25 suggests that the size distribution is fairly narrow, while a PDI value greater than 0.5 indicates a very broad distribution. To achieve long-term stability, it is important to maintain the PDI parameter as low as possible. When the particle size is >3 μm, laser diffractometry (LD) can be used. Note that the particle size data obtained by LD are volume based, and those by PCS are light intensity weighted. Also note that these techniques work well for spherical particles; for nanocrystals, which tend to have large aspect ratios (i.e., length/width), the size results often deviate significantly from those determined by electron microscope imaging. For nonspherical particles, LD may overestimate the particle size distribution [77,78]. Over-prediction of the median diameters is also observed [79]. Thus, caution is needed when measuring and reporting particle size of irregular shape particles [80]. A coulter counter, which yields the absolute number of particles per volume unit of different size classes, can be used as an additional method, for example, to quantify any contamination from microparticles. When a formulation is intended for intravenous delivery, it is essential to ensure that the particle size is smaller than the smallest blood capillary (a few micrometers) so that potential capillary blockade or emboli formation can be avoided [29].

8.4.2 Surface Charge

Surface charging is important for the stability of nanocrystals in solution. According to the Derjaguin, Verwey, Landau, and Overbeek (DLVO) theory, the stability of a colloidal system depends on the balance between two counteracting forces, attractive (van der Waals) and repulsive (electrical double layer) [81]. The electrical double layer surrounding a particle consists of an inner region (Stern layer), where counterions are strongly bound to the particle, and an outer region (diffuse layer), where counterions are loosely associated. Within the diffuse layer, there exists a notional boundary, named the slipping plane, which separates counterions that move along with the particle and those that stay with the bulk fluid. The electric potential difference at the slipping plane versus a point in the bulk medium is defined as the zeta potential. It can be measured by using light scattering to monitor the electrophoretic mobility when the colloidal system is subjected to an electric field. In addition to the sample's surface chemistry, pH of the medium, conductivity of the fluid, and concentration of additives can influence the distributions of counterions surrounding the interface and thereby the zeta potential. To resist flocculation and aggregation, particles need to have sufficiently high repulsion forces. It is generally recognized that a specific value of zeta potential can be useful in predicting the physical stability of a nanosuspension formulation. ±30 mV or larger (absolute value) is desired for an electrostatically stabilized suspension. For particles with smaller zeta potentials, surfactants may be needed to provide further steric and/or electrostatic repulsions [82]. Numerous studies have shown that surface potential can have a drastic influence on cellular uptake or even tumor targeting of nanoparticles [83–87]. One study of cationic and anionic polymeric nanoparticles with similar particle size (90–100 nm) reported that anionic nanoparticles were transited through the degradative lysosomal pathway within the cell, while the cationic particles were transcytosed and accumulated at the basolateral membrane [84]. On the other hand, positively charged micelles caused higher hemolytic activity—due to strong electrostatic interactions between the particles and the anionic erythrocyte membrane—and accumulated in the liver proportionally to the positive charge density; in contrast, negatively charged micelles were not hemolytic [87]. Highly charged nanoparticles had a much higher opsonization rate than neutral or slightly charged nanoparticles of the same size [87,88].

8.4.3   Crystallinity

Crystallinity of a product is often assessed by X-ray diffraction (XRD), complemented by differential scanning calorimetry (DSC). When a compound is known to form hydrates or solvates, thermogravimetric analysis (TGA) is used as well to determine any weight loss at elevated temperature. These analyses also play a key role in characterizing polymorphic forms of a crystalline formulation. In one study to formulate SN 30191, an anticancer drug (PI3K inhibitor), by high-pressure homogenization, the premilled form II particles were converted to form I when the sample was subjected to a pressure between 500 and 1500 bar [89]. Further increase in pressure to 1750 bar resulted in the formation of a hydrate form. Polymorphic transition can considerably change the solubility and dissolution rate, which may alter a compound's *in vivo* performance, such as its bioavailability.

8.4.4   Dissolution

Dissolution behavior of nanocrystals can be determined by the USP dissolution apparatus 2 (paddle) or similar methods. A large volume of dissolution medium (e.g., more than 500 mL) may be needed to minimize the influence of withdrawing samples during the measurement. Careful attention is needed when carrying out the solubility measurement of nanocrystals. It is reported that solubility results determined by centrifugation, ultracentrifugation, or filtration may be questionable because of poor separation of solid particles from the solution [90]. In the presence of high-concentration stabilizers or surfactants, solubility of a drug may be apparently increased, likely due to micellar solubilization of the drug [90]. It must be noted that although the dissolution rate of a nanocrystal formulation can be enhanced considerably because of the increase in surface area, solubility increase is minimal. For instance, the solubility of loviride in a nanosuspension was reportedly increased by 15% when the particle size was about 150 nm [90]. For a poorly soluble drug, the absolute increase in the dissolution rate of its nanocrystal formulation is still marginal.

8.5   NANOCRYSTALS FOR CANCER THERAPY

Combinatorial chemistry and high throughput screening have promoted a steady increase in the number of drug candidates over the years but the majority have been poorly soluble [91–93]. Among various formulation strategies, reducing the size of solid particles of a drug compound has gained considerable attention as a delivery method [73]. Although the marketed nanocrystal-based products are mainly for oral delivery, formulated mostly by the top-down approach, nanocrystals have been exploited for other routes as well, including parenteral, dermal, mucosal, ocular, and pulmonary [17]. Relative to the conventional delivery approaches, nanocrystal formulations are free of organic solvents or other solubilizing chemicals, which may cause adverse effects. The administration of escalated doses of nanocrystals, which can be up to 3–10-fold higher than the maximum tolerated dose of a conventional formulation, may result in improved patient tolerability and enhanced therapeutic efficacy [94,95]. More importantly, nanocrystals offer a much stable and well-integrated platform for drug delivery, circumventing many drawbacks associated with the existing delivery designs, such as drug leakage and systemic breakdown during the storage or even administration stage. Such merits become even more desirable for the intravenous delivery of poorly soluble chemotherapeutic agents. The high drug loading and sustained release of drug molecules allow the nanocrystals to fully take advantage of the EPR effect for targeting tumors. To date, the number of studies reported to use nanocrystals for cancer therapy is limited (Table 8.1). Because of the promising values, more studies are expected to emerge. Below, we briefly review some recent studies that characterize the behaviors and performance of nanocrystal formulations of antineoplastic agents *in vitro* and *in vivo*.

Free drug molecules dissolved from nanocrystals enter cells through passive diffusion. Additionally, nanocrystals can be taken up by cells via active transport process, including the clathrin- and caveolae-mediated endocytosis. Particles of a size of 200 nm or less are preferentially internalized by clathrin-coated pits, while particles between 200 and 500 nm are taken up by the caveolae-mediated pathway [96,97]. The maximum size of particles that can internalized through cell membrane is 500 nm [98]. In one of our studies, the addition of chlorpromazine (CPZ), which disrupts clathrin and hence clathrin-mediated endocytosis, reduced the cellular uptake of camptothecin (CPT) nanocrystals by about 20% compared with that of CPT nanocrystals without the inhibitor [64]. Moreover, nanocrystals are often observed to exert greater cytotoxicity than a solution formulation, particularly when the incubation time is relatively short. One study shows that, against PC-3 prostate cancer cells, the IC_{50} value of deacetymycoepoxydiene (DM) nanocrystals was much lower than that of the drug solution after 12 h incubation; the IC_{50} value of the nanosuspension and solution converged when observed at 48 h [99]. Similar observations were made in our laboratory when IC_{50} values of paclitaxel nanocrystals and its solution against the breast cancer cell MDA-MB-231 were compared (not published).

One of the early *in vivo* studies of nanocrystals was conducted in a murine adenocarcinoma model by testing nanocrystals of piposulfan, camptothecin, etoposide, and paclitaxel that were prepared by ball milling [94]. Solubility values of these drugs range from 200 to <4 μg/ml. It was found that, when given intravenously using a multiple-dose regimen at the maximum tolerated dose (MTD), the nanocrystal formulations of the four chemotherapeutic drugs were able to suppress the tumor burden with no death reported, while control formulations using organic solvents and surfactants to solubilize the drugs caused 20% or more mortality. The significant

TABLE 8.1 Nanocrystal Formulations of Anticancer Drugs

Compound	Method	Stabilizer(s)	Average Size (nm)	Zeta Potential (mV)	Reference
Piposulfan	MM	Tween 80, Span80	210.2 ± 38.9	N/A	[94]
Camptothecin	MM	Pluronic F108	202.3 ± 30.5	N/A	[94]
Etoposide	MM	Pluronic F127	256.2 ± 53.0	N/A	[94]
Paclitaxel	MM	Pluronic F127	279.2 ± 29.60	N/A	[94]
Asulacrine	HPH	Poloxamer 188 (Pluronic F68)	702 ± 0.02(d90)	N/A	[105]
SN30191	HPH	Poloxamer 407, Solutol HS15, Mannitol	1200–1300 (d90)	N/A	[89]
Paclitaxel	3PNET	Pluronic F127	122 ± 35	0.82	[95,118]
Oridonin	HPH	Pluronic F68, lecithin, HPMC, PVP	322.7; 103.3 ± 1.5 and 897.2 ± 14.2; 912.5 ± 17.6	−26.74 ± 2.68; −20.3 ± 0.4, −21.8 ± 0.8; N/A	[70,106,119]
DeacetyMycoepoxydiene	HPH	Lecithin, Pluronic F68, HPMC, PVP	423 ± 11	−23.1 ± 3.5	[99]
Quercetin	HPH, EPAS	Pluronic F68, Lecithin	213.6 ± 29.3 (HPH); 282.6 ± 0.02 (EPAS)	−22.48 ± 4.6 (HPH); −21.12 ± 2.7 (EPAS)	[120]
Paclitaxel	SLbL, US	Polyelectrolytes, BSA	100 ± 20 nm (SLbL); 220 ± 20 (US)	−45 ± 3	[121]
Camptothecin	Anti-solvent	Without stabilizer	234.4 ± 35.6	−41.8 ± 1.3	[64]
Paclitaxel	Anti-solvent	Without stabilizer	438.5 ± 8.3	−15–22	[65]

HPH, high pressure homogenization; MM, media milling; EPAS, evaporative precipitation into aqueous solution; 3PNET, three-phase nanoparticle engineering technology; SLbL, sonication assisted layer-by-layer polyelectrolyte coating; US, ultrasonication.

196

incre~se in the MTD of nanocrystals is also reported by a recent study in which paclitaxel nanocrystals were formulated by a three-phase nanoparticle engineering technique and tested in two murine tumor models [95]. The nanocrystals were given at a dose equivalent to the MTD of a Cremophor-EL solubilized dosage form (20 mg/kg), and both formulations showed similar treatment efficacy. The nanocrystals showed much improved therapeutic outcome when administered at 60 mg/kg. The PTX nanosuspensions were also tested by oral administration but found to be less effective than the parenteral dosage form.

Intravenously administered nanocrystals are expected to reach tumors via the EPR effect. We found that camptothecin nanocrystals, administered in a MCF-7 xenografted murine tumor model, yielded significantly better antitumor activity ($p < 0.01$) than the drug salt solution [64]. Biodistribution results at 24 h after the injection further revealed that CPT nanocrystals accumulated in the tumor at a much higher level than the salt solution. Note that a majority of nanocrystals (or any nanoparticle drug delivery systems) are still being taken up by the reticuloendothelial system (RES), mainly liver and spleen [100]. A common practice to reduce the RES uptake and enhance biocompatibility is to modify the surface of a nanosystem is commonly modified, physically or chemically, with biocompatible polymers, such as polyethylene glycol (PEG), that is known to prolong the systemic circulation [101]. Because nanocrystals continue to dissolve after their administration, surface coating by polymeric surfactants onto the nanocrystals may not work. Thus, there is a critical need to investigate new ways to incorporate biocompatible molecules (and other functional molecules) to nanocrystals. Nonetheless, relative to solution-based delivery systems, nanocrystals are expected to accumulate in the tumor by at least an order of magnitude more (from less than one to a few percent) at the same dose level. Given the stable integrity, nanocrystals hold a great promise in targeted cancer therapy.

The particle size of a nanoparticle delivery system has a fundamental influence on the pharmacokinetics (PK) and biodistribution of the nanoparticles. It has been suggested that particles between 50 and 300 nm can achieve prolonged blood circulation [100,102], which helps their accumulation in solid tumors via the EPR effect [100,103,104]. With the same surface charge of -25 mV, polymeric nanoparticles with the size of 150 nm were reported to have significantly longer circulation time due to the low uptake by the RES than particles of 500 nm [85]. Compared with a solution formulation, optimally designed nanovehicles are expected to bear improved PK behaviors, including increased half-life ($t_{1/2}$), reduced clearance (Cl), increased area under the curve (AUC), and increased mean residence time (MRT) [104]. The PK of intravenously injected nanocrystals of chemotherapeutics (e.g., paclitaxel, asulacrine, and oridonin) was reported to be quite distinctive, best described by a two-compartmental model, where the elimination was non-zero order [105–107]. In a rabbit model, the pharmacokinetic profile of 103 nm oridonin nanocrystals, which were shown to dissolve rapidly, was similar to that of the drug solution [70]. On the contrary, the 300 nm paclitaxel [107] and 700 nm asulacrine nanocrystals [105] were rapidly taken up by the RES organs, such as liver, spleen, and lungs, resulting in shortened distribution half-lives ($t_{1/2}\alpha$) and decreased AUC. The decreased and yet prolonged AUC in plasma was found to be beneficial to minimize the toxic drug exposure. Interestingly, the 900 nm oridonin nanocrystals were found to achieve decreased Cl and $t_{1/2}\alpha$ and increased AUC as compared with the dissolved drug [70]. Overall, prolonged elimination half-life ($t_{1/2}\beta$) and MRT were observed for nanocrystals between 300 and 900 nm [105–107]. The sustained blood concentrations likely resulted from slow release of drug molecules from nanocrystals trapped by the RES organs [105,106]. Size has also been found to affect particle accumulation and retention in the tumor. Small particles may accumulate faster in tumors, but larger ones can be retained longer [108]. Moreover, particle accumulation in the tumor relies on the leakiness of the tumor vasculature, which is further determined by the tumor type [109,110], histological grade, and malignant severity [111]. The cutoff sizes of pores in the tumor vasculature range from 200 nm to as large as 1.2 μm, with the majority between 380 and 780 nm [110]. Collectively, it has been suggested that in order to take advantage of the EPR effect and target tumors, nanoparticles should be smaller than 400 nm [108]. The particle size is nonetheless one of many factors that determine the pharmacokinetics and biodistribution of nanoparticles. Surface chemistry plays an important role as well. Because their surface is decorated with hydroxyl groups, amorphous silica particles with sizes of 50 and 250 nm, respectively, exhibited significantly higher blood to liver ratios compared with 180 nm PEG-coated solid lipid nanoparticles in a Wistar rat model [102]. Despite their intrinsic hydrophilicity, which resulted in decreased RES clearance, detailed toxicological studies of amorphous silica particles are still limited [112].

A recent clinical study of Panzem®, a nanocrystal dispersion of 2-methoxyestradiol (2ME2), an endogenous estradiol-17β metabolite that bears antimitotic and antioangiogenic activities, demonstrated an improved pharmacokinetic performance and antitumor activity relative to its capsule formulation when given orally to 16 patients with advanced solid malignancies and 14 patients with taxane-refractory, metastatic castrate-resistant prostate cancer [113,114]. Despite the fact that the 2ME2 nanocrystal formulation was well-tolerated and showed evidence of biologic activity, it did not appear to have sufficient clinical activity and the study was terminated at phase II. No other clinical studies to test parenteral nanocrystal formulations for cancer therapy have been reported.

8.6 FUTURE PERSPECTIVES

Nanocrystals have shown promising potentials for drug delivery, particularly for delivering poorly water-soluble anticancer drugs. Major challenges still remain and must be overcome before nanocrystal-based formulation can be fully embraced for treating patients. One such difficulty is production of high-quality nanocrystals with a uniform particle size and morphology. These properties are especially desired for parenteral dosage forms. The top-down approach lacks a well-defined mechanism to control the particle size; the high-energy process may damage the crystallinity of fragmented particles, which presents a concern for the physical stability of a final product. On the other hand, the bottom-up approach stems from solution crystallization, which has been extensively investigated over decades with regard to mechanism and kinetics. The broad understanding will eventually lead to technical breakthroughs in controlling the particle size, size distribution, and shape of nano-crystals produced from solution. More studies nevertheless remain to be done, particularly for fine-tuning the nucleation process at the industrial scale to ensure uniformity and reproducibility.

Another challenge of using nanocrystals for cancer therapy is seemingly a lack of feasible means to enhance biocompatibility and improve cancer targeting. Tumor-targeting moieties to augment cellular uptake cannot be chemically attached to the crystal surface. Physical adsorption directly or conjugated with surface-coated surfactants may not be suitable either because of continuous dissolution of nano-crystals. Also because of the dynamic nature of dissolution, any surfactants used to prevent nanocrystals from agglomeration and/or to enhance systemic circulation can be easily detached from the nanocrystal surface during blood circulation and extravasation. These limitations seem to drastically dampen the application of nanocrystals for cancer therapy. Similarly, it seems to be difficult, if not impossible, to develop theranostic nanocrystal systems that integrate imaging or diagnostic agents with a drug substance.

So comes the hybrid nanocrystal concept. Over the last few years, our laboratory has developed nanosized hybrid crystals, through which a functional material is physically integrated. The concept was inspired by a well-known phenomenon in solid-state chemistry wherein impurities are entrapped in the crystal lattice of a host particle. The existence of foreign substances in a host is typically insignificant, having little effect on the structure and integrity of the host crystal structure but capable of casting drastic influence on the optical appearance and other physical properties of the host. Examples are abundant. Colored diamonds are one famous example; synthetic hybrid crystals include alloys and semiconductors. The most relevant examples are dyeing crystals, where guest dye molecules are integrated in organic crystals [115]. Hybrid nano-crystals of paclitaxel were developed in our laboratory to integrate fluorescent probes and tested in a murine tumor model [65]. The results indicated that the nanocrystals were capable of not only exerting antitumor effect but permitting real-time imaging of the delivery system in animals. Figure 8.4 shows the hybrid paclitaxel nanocrystals and

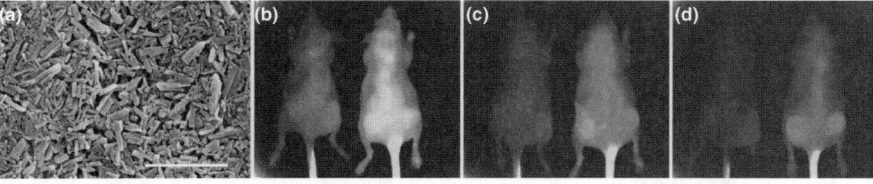

FIGURE 8.4 SEM image of paclitaxel/fluorophore (FPR-749) nanocrystals (a, scale bar: 2 μm), and fluorescence images obtained by IVIS of two mice bearing MCF-7 xenograft tumors in rear flanks. FPR-749 solution (left) and the hybrid nanocrystals (right) were intravenously injected via tail vein and images were captured at 25 min (b), 4 h (c) and 48 h (d), respectively. Exposure time was 0.25 s, F-stop 2, and binning medium. (*See insert for color representation of the figure.*)

whole-body fluorescence images of mice intravenously injected with either dissolved or integrated fluorophore. We are currently expanding the concept to integrate other types of functional compounds, including radioisotopes, ligands, and polymers. Because the guest substance is integrated within a nanocrystal host, hybrid nanocrystals are expected to advance the usage of nanocrystals for cancer therapy.

Finally, although the focus of this chapter is organic nanocrystals, we cannot discount the development and application of inorganic nanocrystals in drug delivery and cancer treatment. Quantum dots (QDs) have been widely used in cellular imaging [116,117]. Other inorganic nanosystems, including iron oxide, silica, gold, and even carbon nanotubes have been explored as drug delivery systems.

Acknowledgments

The project described was supported by Grant Number R25CA153954 from the National Cancer Institute. The content is solely the responsibility of the authors and does not necessarily represent the official views of the National Cancer Institute or the National Institutes of Health.

REFERENCES

1. Hauss, D. J. (2007). Oral lipid-based formulations. *Advanced Drug Delivery Reviews 59*, 667–676.

2. Singla, A. K., Garg, A., Aggarwal, D. (2002). Paclitaxel and its formulations. *International Journal of Pharmaceutics 235*, 179–192.

3. Rowinsky, E. K., Eisenhauer, E. A., Chaudhry, V., Arbuck, S. G., Donehower, R. C. (1993). Clinical toxicities encountered with paclitaxel (Taxol(R)). *Seminars in Oncology 20*, 1–15.

4. Maeda, H., The enhanced permeability and retention (EPR) effect in tumor vasculature: the key role of tumor-selective macromolecular drug targeting, In *Advances in Enzyme Regulation*, Vol. 41, ed. Weber, G., Pergamon, Great Britain, 2001, 189–207.

5. Maruyama, K. (2011). Intracellular targeting delivery of liposomal drugs to solid tumors based on EPR effects. *Advanced Drug Delivery Reviews 63*, 161–169.

6. Andresen, T. L., Jensen, S. S., Jorgensen, K. (2005). Advanced strategies in liposomal cancer therapy: problems and prospects of active and tumor specific drug release. *Progress in Lipid Research 44*, 68–97.

7. Torchilin, V. P. (2007). Micellar nanocarriers: pharmaceutical perspectives. *Pharmaceutical Research 24*, 1–16.

8. Wong, H. L., Bendayan, R., Rauth, A. M., Li, Y. Q., Wu, X. Y. (2007). Chemotherapy with anticancer drugs encapsulated in solid lipid nanoparticles. *Advanced Drug Delivery Reviews 59*, 491–504.

9. Bummer, P. M. (2004). Physical chemical considerations of lipid-based oral drug delivery—solid lipid nanoparticles. *Critical Reviews in Therapeutic Drug Carrier Systems 21*, 1–19.

10. Tong, R., Cheng, J. J. (2007). Anticancer polymeric nanomedicines. *Polymer Reviews 47*, 345–381.

11. Pridgen, E. M., Langer, R., Farokhzad, O. C. (2007). Biodegradable, polymeric nanoparticle delivery systems for cancer therapy. *Nanomedicine 2*, 669–680.

12. Wolinsky, J. B., Grinstaff, M. W. (2008). Therapeutic and diagnostic applications of dendrimers for cancer treatment. *Advanced Drug Delivery Reviews 60*, 1037–1055.

13. Gulati, M., Grover, M., Singh, S., Singh, M. (1998). Lipophilic drug derivatives in liposomes. *International Journal of Pharmaceutics 165*, 129–168.

14. Mayer, L. D., Krishna, R., Webb, M., Bally, M. (2000). Designing liposomal anticancer drug formulations for specific therapeutic applications. *Journal of Liposome Research 10*, 99–115.

15. Garcia-Fuentes, M., Alonso, M. J., Torres, D. (2005). Design and characterization of a new drug nanocarrier made from solid–liquid lipid mixtures. *Journal of Colloid and Interface Science 285*, 590–598.

16. Ruenraroengsak, P., Cook, J. M., Florence, A. T. (2010). Nanosystem drug targeting: facing up to complex realities. *Journal of Controlled Release 141*, 265–276.

17. Shegokar, R., Müller, R. H. (2010). Nanocrystals: industrially feasible multifunctional formulation technology for poorly soluble actives. *International Journal of Pharmaceutics 399*, 129–139.

18. Müller, R. H., Jacobs, C., Kayser, O. (2001). Nanosuspensions as particulate drug formulations in therapy rationale for development and what we can expect for the future. *Advanced Drug Delivery Reviews 47*, 3–19.

19. Müller, R. H., Moschwitzer, J., Bushrab, F. N., Manufacturing of nanoparticles by milling and homogenization techniques, In *Nanoparticle Technology for Drug Delivery, Drugs, and Pharmaceutical Sciences*, eds. Gupta, R. B., Kompella, U. B., Taylor & Francis Group, LLC, New York, 2006, pp. 21–51.

20. Chan, H. K., Kwok, P. C. L. (2011). Production methods for nanodrug particles using the bottom-up approach. *Advanced Drug Delivery Reviews 63*, 406–416.

21. Rabinow, B. E. (2004). Nanosuspensions in drug delivery. *Nature Reviews Drug Discovery 3*, 785–796.

22. Liversidge, G. G., Cundy, K. C., Bishop, J. F., Czekai, D. A. Surface modified drug nanoparticles. U.S. Patent 5,145,684. 1991.

23. Junghanns, J., Müller, R. H. (2008). Nanocrystal technology, drug delivery and clinical applications. *International Journal of Nanomedicine 3*, 295–309.

24. Merisko-Liversidge, E., Liversidge, G. G., Cooper, E. R. (2003). Nanosizing: a formulation approach for poorly-water-soluble compounds. *European Journal of Pharmaceutical Sciences 18*, 113–120.

25. Haynes, D. H. Phospholipid-coated microcrystals: injectable formulations of water insoluble drugs. U.S. Patent 5,091,187. 1992.

26. Keck, C. M., Müller, R. H. (2006). Drug nanocrystals of poorly soluble drugs produced by high pressure homogenisation. *European Journal of Pharmaceutics and Biopharmaceutics 62*, 3–16.

27. Müller, R. H., Becker, R., Kruss, K. P., Peters, K. Pharmaceutical nanosuspensions for medicament administration as systems with increased saturation solubility and rate of solution. U.S. Patent 5,858,410. 1999.

28. Illig, K. J., Mueller, R. L., Ostrander, K. D., Swanson, J. R. (1996). Use of microfluidizer processing for preparation of pharmaceutical suspensions. *Pharmaceutical Technology 20*, 78–88.

29. Patravale, V. B., Date, A. A., Kulkarni, R. M. (2004). Nanosuspensions: a promising drug delivery strategy. *Journal of Pharmacy and Pharmacology 56*, 827–840.

30. Begat, P., Young, P. M., Edge, S., Kaerger, J. S., Price, R. (2003). The effect of mechanical processing on surface stability of pharmaceutical powders: visualization by atomic force microscopy. *Journal of Pharmaceutical Sciences 92*, 611–620.

31. Panagiotou, T., Fisher, R. J. (2008). Form nanoparticles via controlled crystallization. *Chemical Engineering Progress 104*, 33–39.

32. Becker, R., Döring, W. (1935). Kinetische behandlung der keimbildung in übersättingten dämpfen. *Annals of Physics 24*, 719–752.

33. Kashchiev, D., van Rosmalen, G. M. (2003). Review: nucleation in solutions revisited. *Crystal Research and Technology 38*, 555–574.

34. Volmer, M., *Kinetik der Phasenbildung*, Steinkopf., Dresden, 1939.

35. Mullin, J. W., *Crystallization*, Butterworth-Heinemann, Oxford, 2001.

36. Horn, D., Rieger, J. (2001). Organic nanoparticles in the aqueous phase—theory, experiment, and use. *Angewandte Chemie-International Edition 40*, 4331–4361.

37. Oxtoby, D. W. (1998). Nucleation of first-order phase transitions. *Accounts of Chemical Research 31*, 91–97.

38. Dirksen, J. A., Ring, T. A. (1991). Fundamentals of crystallization—kinetic effects on particle-size distributions and morphology. *Chemical Engineering Science 46*, 2389–2427.

39. Chen, J. F., Wang, Y. H., Guo, F., Wang, X. M., Zheng, C. (2000). Synthesis of nanoparticles with novel technology: high-gravity reactive precipitation. *Industrial & Engineering Chemistry Research 39*, 948–954.

40. Hu, T. T., Wang, J. X., Shen, Z. G., Chen, J. F. (2008). Engineering of drug nanoparticles by HGCP for pharmaceutical applications. *Particuology 6*, 239–251.

41. Panagiotou, T., Mesite, S. V., Fisher, R. J. (2009). Production of norfloxacin nano-suspensions using microfluidics reaction technology through solvent/antisolvent crystallization. *Industrial & Engineering Chemistry Research 48*, 1761–1771.

42. Byrappa, K., Ohara, S., Adschiri, T. (2008). Nanoparticles synthesis using supercritical fluid technology—towards biomedical applications. *Advanced Drug Delivery Reviews 60*, 299–327.

43. Luque de Castro, M. D., Priego-Capote, F. (2007). Ultrasound-assisted crystallization (sonocrystallization). *Ultrasonics Sonochemistry 14*, 717–724.

44. List, M., Sucker, H. Pharmaceutical colloidal hydrosols for injection. G.B. Patent 2,200,048. 1988.

45. Gassmann, P., Sucker, H. Improvements in pharmaceutical compositions. WO 018105. 1992.

46. List, M., Sucker, H. Hydrosols of pharmacologically active agents and their pharmaceutical compositions comprising them. U.S. Patent 5,389,382. 1995.

47. Gassmann, P. S., H. Pharmaceutical compositions comprised of stabilized peptide particles. U.S. Patent 6,447,806. 2002.

48. Zhong, J., Shen, Z. G., Yang, Y., Chen, J. F. (2005). Preparation and characterization of uniform nanosized cephradine by combination of reactive precipitation and liquid anti-solvent precipitation under high gravity environment. *International Journal of Pharmaceutics 301*, 286–293.

49. Zhao, H., Wang, J. X., Wang, Q. A., Chen, J. F., Yun, J. (2007). Controlled liquid antisolvent precipitation of hydrophobic pharmaceutical nanoparticles in a microchannel reactor. *Industrial & Engineering Chemistry Research 46*, 8229–8235.

50. Chiou, H., Li, L., Hu, T. T., Chan, H. K., Chen, J. F., Yun, J. (2007). Production of salbutamol sulfate for inhalation by high-gravity controlled antisolvent precipitation. *International Journal of Pharmaceutics 331*, 93–98.

51. Hu, T., Chiou, H., Chan, H. K., Chen, J. F., Yun, J. (2008). Preparation of inhalable salbutamol sulphate using reactive high gravity controlled precipitation. *Journal of Pharmaceutical Sciences 97*, 944–949.

52. Chiou, H., Chan, H. K., Heng, D., Prud'homme, R. K., Raper, J. A. (2008). A novel production method for inhalable cyclosporine A powders by confined liquid impinging jet precipitation. *Journal of Aerosol Science 39*, 500–509.

53. Chiou, H., Chan, H. K., Prud'homme, R. K., Raper, J. A. (2008). Evaluation on the use of confined liquid impinging jets for the synthesis of nanodrug particles. *Drug Development and Industrial Pharmacy 34*, 59–64.

54. Kumar, V., Adamson, D. H., Prud'homme, R. K. (2010). Fluorescent polymeric nano-particles: aggregation and phase behavior of pyrene and amphotericin B molecules in nanoparticle cores. *Small 6*, 2907–2914.

55. Rogers, T. L., Johnston, K. P., Williams, R. O. (2001). Solution-based particle formation of pharmaceutical powders by supercritical or compressed fluid CO_2 and cryogenic spray-freezing technologies. *Drug Development and Industrial Pharmacy 27*, 1003–1015.

56. Reverchon, E., De Marco, I., Torino, E. (2007). Nanoparticles production by super-critical antisolvent precipitation: a general interpretation. *Journal of Supercritical Fluids 43*, 126–138.

57. Bustami, R. T., Chan, H. K., Dehghani, F., Foster, N. R. (2000). Generation of micro-particles of proteins for aerosol delivery using high pressure modified carbon dioxide. *Pharmaceutical Research 17*, 1360–1366.

58. Lancaster, R. W., Singh, H., Theophilus, A. L. Apparatus and process for preparing crystalline particles. WO 0038811. 2000.

59. Abbas, A., Srour, M., Tang, P., Chiouc, H., Chan, H. K., Romagnoli, J. A. (2007). Sonocrystallisation of sodium chloride particles for inhalation. *Chemical Engineering Science 62*, 2445–2453.

60. Dhumal, R. S., Biradar, S. V., Paradkar, A. R., York, P. (2009). Particle engineering using sonocrystallization: salbutamol sulphate for pulmonary delivery. *International Journal of Pharmaceutics 368*, 129–137.

61. Kaerger, J. S., Price, R. (2004). Processing of spherical crystalline particles via a novel solution atomization and crystallization by sonication (SAXS) technique. *Pharmaceutical Research 21*, 372–381.

62. Dhumal, R. S., Biradar, S. V., Yamamura, S., Paradkar, A. R., York, P. (2008). Preparation of amorphous cefuroxime axetil nanoparticles by sonoprecipitation for enhancement of bioavailability. *European Journal of Pharmaceutics and Biopharmaceutics 70*, 109–115.

63. Kipp, J. E., Wong, J. C. T., Doty, M. J., Rebbeck, C. L. Microprecipitation method for preparing submicron suspensions. U.S. Patent 6,869,617. 2001.

64. Zhang, H., Hollis, C. P., Zhang, Q., Li, T. L. (2011). Preparation and antitumor study of camptothecin nanocrystals. *International Journal of Pharmaceutics 415*, 293–300.

65. Zhao, R. S., Hollis, C. P., Zhang, H., Sun, L. L., Gemeinhart, R. A., Li, T. L. (2011). Hybrid nanocrystals: achieving concurrent therapeutic and bioimaging functionalities toward solid tumors. *Molecular Pharmaceutics 8*, 1985–1991.

66. Van Eerdenbrugh, B., Van den Mooter, G., Augustijns, P. (2008). Top-down production of drug nanocrystals: nanosuspension stabilization, miniaturization and transformation into solid products. *International Journal of Pharmaceutics 364*, 64–75.

67. Kocbek, P., Baumgartner, S., Kristl, J. (2006). Preparation and evaluation of nano-suspensions for enhancing the dissolution of poorly soluble drugs. *International Journal of Pharmaceutics 312*, 179–186.

68. Lee, J., Lee, S. J., Choi, J. Y., Yoo, J. Y., Ahn, C. H. (2005). Amphiphilic amino acid copolymers as stabilizers for the preparation of nanocrystal dispersion. *European Journal of Pharmaceutical Sciences 24*, 441–449.

69. Deng, Z., Xu, S., Li, S. (2008). Understanding a relaxation behavior in a nanoparticle suspension for drug delivery applications. *International Journal of Pharmaceutics 351*, 236–243.

70. Gao, L., Zhang, D. R., Chen, M. H., Zheng, T. T., Wang, S. M. (2007). Preparation and characterization of an oridonin nanosuspension for solubility and dissolution velocity enhancement. *Drug Development and Industrial Pharmacy 33*, 1332–1339.

71. Van Eerdenbrugh, B., Vermant, J., Martens, J. A., Froyen, L., Van Humbeeck, J., Augustijns, P., Van Den Mooter, G. (2009). A screening study of surface stabilization during the production of drug nanocrystals. *Journal of Pharmaceutical Sciences 98*, 2091–2103.

72. Yaguchi, S. I., Fukui, Y., Koshimizu, K., Yoshimi, H., Matsuno, T., Gouda, H., Hirono, S., Yamazaki, K., Yamori, T. (2006). Antitumor activity of ZSTK474, a new phosphatidylitiositol 3-kinase inhibitor. *Journal of the National Cancer Institute 98*, 545–556.

73. Kesisoglou, F., Panmai, S., Wu, Y. H. (2007). Nanosizing—oral formulation development and biopharmaceutical evaluation. *Advanced Drug Delivery Reviews 59*, 631–644.

74. Na, G. C., Stevens, H. J., Yuan, B. O., Rajagopalan, N. (1999). Physical stability of ethyl diatrizoate nanocrystalline suspension in steam sterilization. *Pharmaceutical Research 16*, 569–574.

75. Zheng, J. Y., Bosch, H. W. (1997). Sterile filtration of NanoCrystal(TM) drug formulations. *Drug Development and Industrial Pharmacy 23*, 1087–1093.

76. Wu, L. B., Zhang, J., Watanabe, W. (2011). Physical and chemical stability of drug nanoparticles. *Advanced Drug Delivery Reviews 63*, 456–469.

77. Matsuyama, T., Yamamoto, H. (2004). Particle shape and laser diffraction: a discussion of the particle shape problem. *Journal of Dispersion Science and Technology 25*, 409–416.

78. Muhlenweg, H., Hirleman, E. D. (1998). Laser diffraction spectroscopy: influence of particle shape and a shape adaptation technique. *Particle & Particle Systems Characterization 15*, 163–169.

79. Endoh, S., Kuga, Y., Ohya, H., Ikeda, C., Iwata, H. (1998). Shape estimation of anisometric particles using size measurement techniques. *Particle & Particle Systems Characterization 15*, 145–149.

80. Stevens, N., Shrimpton, J., Palmer, M., Prime, D., Johal, B. (2007). Accuracy assessments for laser diffraction measurements of pharmaceutical lactose. *Measurement Science & Technology 18*, 3697–3706.

81. Buckton, G., *Interfacial Phenomena in Drug Delivery and Targeting*, Harwood Academic Publishers, Switzerland, 1995, 133–134.

82. Jacobs, C., Müller, R. H. (2002). Production and characterization of a budesonide nanosuspension for pulmonary administration. *Pharmaceutical Research 19*, 189–194.

83. Chung, T. H., Wu, S. H., Yao, M., Lu, C. W., Lin, Y. S., Hung, Y., Mou, C. Y., Chen, Y. C., Huang, D. M. (2007). The effect of surface charge on the uptake and biological function of mesoporous silica nanoparticles 3T3-L1 cells and human mesenchymal stem cells. *Biomaterials 28*, 2959–2966.

84. Harush-Frenkel, O., Rozentur, E., Benita, S., Altschuler, Y. (2008). Surface charge of nanoparticles determines their endocytic and transcytotic pathway in polarized MDCK cells. *Biomacromolecules 9*, 435–443.

85. He, C. B., Hu, Y. P., Yin, L. C., Tang, C., Yin, C. H. (2010). Effects of particle size and surface charge on cellular uptake and biodistribution of polymeric nanoparticles. *Biomaterials 31*, 3657–3666.

86. Juliano, R. L., Stamp, D. (1975). Effect of particle-size and charge on clearance rates of liposomes and liposome encapsulated drugs. *Biochemical and Biophysical Research Communications 63*, 651–658.

87. Xiao, K., Li, Y. P., Luo, J. T., Lee, J. S., Xiao, W. W., Gonik, A. M., Agarwal, R. G., Lam, K. S. (2011). The effect of surface charge on in vivo biodistribution of PEG-oligocholic acid based micellar nanoparticles. *Biomaterials 32*, 3435–3446.

88. Dobrovolskaia, M. A., Aggarwal, P., Hall, J. B., McNeil, S. E. (2008). Preclinical studies to understand nanoparticle interaction with the immune system and its potential effects on nanoparticle biodistribution. *Molecular Pharmaceutics 5*, 487–495.

89. Sharma, P., Zujovic, Z. D., Bowmaker, G. A., Denny, W. A., Garg, S. (2011). Evaluation of a crystalline nanosuspension: polymorphism, process induced transformation and in vivo studies. *International Journal of Pharmaceutics 408*, 138–151.

90. Van Eerdenbrugh, B., Vermant, J., Martens, J. A., Froyen, L., Van Humbeeck, J., Van den Monter, G., Augustijns, P. (2010). Solubility increases associated with crystalline drug nanoparticles: methodologies and significance. *Molecular Pharmaceutics 7*, 1858–1870.

91. Gardner, C. R., Walsh, C. T., Almarsson, O. (2004). Drugs as materials: valuing physical form in drug discovery. *Nature Reviews Drug Discovery 3*, 926–934.

92. Lipinski, C. A. (2000). Drug-like properties and the causes of poor solubility and poor permeability. *Journal of Pharmacological and Toxicological Methods 44*, 235–249.

93. Lipinski, C. A., Lombardo, F., Dominy, B. W., Feeney, P. J. (1997). Experimental and computational approaches to estimate solubility and permeability in drug discovery and development settings. *Advanced Drug Delivery Reviews 23*, 3–25.

94. MeriskoLiversidge, E., Sarpotdar, P., Bruno, J., Hajj, S., Wei, L., Peltier, N., Rake, J., Shaw, J. M., Pugh, S., Polin, L., Jones, J., Corbett, T., Cooper, E., Liversidge, G. G.

(1996). Formulation and antitumor activity evaluation of nanocrystalline suspensions of poorly soluble anticancer drugs. *Pharmaceutical Research 13*, 272–278.

95. Liu, F., Park, J. Y., Zhang, Y., Conwell, C., Liu, Y., Bathula, S. R., Huang, L. (2010). Targeted cancer therapy with novel high drug-loading nanocrystals. *Journal of Pharmaceutical Sciences 99*, 3542–3551.

96. Rejman, J., Oberle, V., Zuhorn, I. S., Hoekstra, D. (2004). Size-dependent internalization of particles via the pathways of clathrin-and caveolae-mediated endocytosis. *Biochemical Journal 377*, 159–169.

97. Zuhorn, I. S., Kalicharan, R., Hoekstra, D. (2002). Lipoplex-mediated transfection of mammalian cells occurs through the cholesterol-dependent clathrin-mediated pathway of endocytosis. *Journal of Biological Chemistry 277*, 18021–18028.

98. Yu, X., Valmikinathan, C. M., Rogers, A., Wang, J., Nanotechnology and drug delivery, In *Biomedical Nanostructures*, eds. Gonsalves, K. E., Halberstadt, C. R., Laurencin, C. T., Wiley, Hoboken, NJ, 2007, 93–113.

99. Wang, Y. C., Liu, Z. P., Zhang, D. R., Gao, X. H., Zhang, X. Y., Duan, C. X., Jia, L. J., Feng, F. F., Huang, Y. J., Shen, Y. M., Zhang, Q. A. (2011). Development and in vitro evaluation of deacety mycoepoxydiene nanosuspension. *Colloids and Surfaces B-Biointerfaces 83*, 189–197.

100. Moghimi, S. M., Hunter, A. C., Murray, J. C. (2001). Long-circulating and target-specific nanoparticles: theory to practice. *Pharmacological Reviews 53*, 283–318.

101. Shenoy, D., Little, S., Langer, R., Amiji, M. (2005). Poly(ethylene oxide)-modified poly (beta-amino ester) nanoparticles as a pH-sensitive system for tumor-targeted delivery of hydrophobic drugs: Part 2. In vivo distribution and tumor localization studies. *Pharmaceutical Research 22*, 2107–2114.

102. Barbe, C., Bartlett, J., Kong, L. G., Finnie, K., Lin, H. Q., Larkin, M., Calleja, S., Bush, A., Calleja, G. (2004). Silica particles: a novel drug-delivery system. *Advanced Materials 16*, 1959–1966.

103. Alexis, F., Pridgen, E., Molnar, L. K., Farokhzad, O. C. (2008). Factors affecting the clearance and biodistribution of polymeric nanoparticles. *Molecular Pharmaceutics 5*, 505–515.

104. Li, S. D., Huang, L. (2008). Pharmacokinetics and biodistribution of nanoparticles. *Molecular Pharmaceutics 5*, 496–504.

105. Ganta, S., Paxton, J. W., Baguley, B. C., Garg, S. (2009). Formulation and pharmacokinetic evaluation of an asulacrine nanocrystalline suspension for intravenous delivery. *International Journal of Pharmaceutics 367*, 179–186.

106. Gao, L., Zhang, D. R., Chen, M. H., Duan, C. X., Dai, W. T., Jia, L. J., Zhao, W. F. (2008). Studies on pharmacokinetics and tissue distribution of oridonin nanosuspensions. *International Journal of Pharmaceutics 355*, 321–327.

107. Wang, Y. L., Li, X. M., Wang, L. Y., Xu, Y. L., Cheng, X. D., Wei, P. (2011). Formulation and pharmacokinetic evaluation of a paclitaxel nanosuspension for intravenous delivery. *International Journal of Nanomedicine 6*, 1497–1507.

108. Bae, Y. H., Park, K. (2011). Targeted drug delivery to tumors: myths, reality and possibility. *Journal of Controlled Release 153*, 198–205.

109. Hashizume, H., Baluk, P., Morikawa, S., McLean, J. W., Thurston, G., Roberge, S., Jain, R. K., McDonald, D. M. (2000). Openings between defective endothelial cells explain tumor vessel leakiness. *American Journal of Pathology 156*, 1363–1380.

110. Hobbs, S. K., Monsky, W. L., Yuan, F., Roberts, W. G., Griffith, L., Torchilin, V. P., Jain, R. K. (1998). Regulation of transport pathways in tumor vessels: role of tumor type and microenvironment. *Proceedings of the National Academy of Sciences of the United States of America 95*, 4607–4612.

111. Daldrup, H., Shames, D. M., Wendland, M., Okuhata, Y., Link, T. M., Rosenau, W., Lu, Y., Brasch, R. C. (1998). Correlation of dynamic contrast-enhanced MR imaging with histologic tumor grade: comparison of macromolecular and small-molecular contrast media. *American Journal of Roentgenology 171*, 941–949.

112. Fadeel, B., Garcia-Bennett, A. E. (2010). Better safe than sorry: understanding the toxicological properties of inorganic nanoparticles manufactured for biomedical applications. *Advanced Drug Delivery Reviews 62*, 362–374.

113. Tevaarwerk, A. J., Holen, K. D., Alberti, D. B., Sidor, C., Arnott, J., Quon, C., Wilding, G., Liu, G. (2009). Phase I trial of 2-methoxyestradiol NanoCrystal dispersion in advanced solid malignancies. *Clinical Cancer Research 15*, 1460–1465.

114. Harrison, M. R., Hahn, N. M., Pili, R., Oh, W. K., Hammers, H., Sweeney, C., Kim, K., Perlman, S., Arnott, J., Sidor, C., Wilding, G., Liu, G. (2011). A phase II study of 2-methoxyestradiol (2ME2) NanoCrystalA (R) dispersion (NCD) in patients with taxane-refractory, metastatic castrate-resistant prostate cancer (CRPC). *Investigational New Drugs 29*, 1465–1474.

115. Kahr, B., Gurney, R. W. (2001). Dyeing crystals. *Chemical Reviews 101*, 893–951.

116. Smith, A. M., Gao, X. H., Nie, S. M. (2004). Quantum dot nanocrystals for in vivo molecular and cellular imaging. *Photochemistry and Photobiology 80*, 377–385.

117. Smith, A. M., Nie, S. M. (2010). Semiconductor nanocrystals: structure, properties, and band gap engineering. *Accounts of Chemical Research 43*, 190–200.

118. Deng, J. X., Huang, L., Liu, F. (2010). Understanding the structure and stability of paclitaxel nanocrystals. *International Journal of Pharmaceutics 390*, 242–249.

119. Zhang, Z., Zhang, X. M., Xue, W., YangYang, Y. N., Xu, D. R., Zhao, Y. X., Lou, H. Y. (2010). Effects of oridonin nanosuspension on cell proliferation and apoptosis of human prostatic carcinoma PC-3 cell line. *International Journal of Nanomedicine 5*, 735–742.

120. Gao, L., Liu, G. Y., Wang, X. Q., Liu, F., Xu, Y. F., Ma, J. (2011). Preparation of a chemically stable quercetin formulation using nanosuspension technology. *International Journal of Pharmaceutics 404*, 231–237.

121. Pattekari, P., Zheng, Z., Zhang, X., Levchenko, T., Torchilin, V., Lvov, Y. (2011). Top-down and bottom-up approaches in production of aqueous nanocolloids of low solubility drug paclitaxel. *Physical Chemistry Chemical Physics 13*, 9014–9019.

9

CLEARANCE OF NANOPARTICLES DURING CIRCULATION

SEUNG-YOUNG LEE AND JI-XIN CHENG

Weldon School of Biomedical Engineering, Purdue University, West Lafayette, IN, USA

9.1 INTRODUCTION

Administration of conventional therapeutic agents for cancer therapy is often limited due to their chemical properties, such as low solubility and poor stability in blood. These properties cause a short circulation half-life and low therapeutic efficacy, thus necessitating frequent administrations. In addition, toxic side effects of these agents due to nonspecific biodistribution have been additional issues in cancer chemotherapy. To overcome the limitations of conventional therapeutic agents, researchers have explored drug delivery systems that can safely and efficiently carry drugs and deliver them to the tumor site without affecting normal tissues [1–3]. In this regard, nanoparticle systems have gained a lot of interest because of the drug loading capacity and the ability to easily pass through anatomical barriers. Delivery of nanoparticles to solid tumors is facilitated by the leakiness of tumor blood vessels [4,5], and the discovery of various tumor-specific ligands has enabled more efficient delivery of nanoparticles to cancer cells [6,7].

One of the key factors to determine the targeting efficacy of nanoparticles is their blood-circulation time because long-circulating nanoparticles have more chances to transport into tumor tissues [8,9]. The circulation half-life of the nanoparticles primarily depends on the rate of the biological clearance [10,11]. The clearance mechanisms can be classified into three categories: (i) disintegration of nanoparticles by protein adsorption, (ii) opsonization-mediated nanoparticle removal by immune

Nanoparticulate Drug Delivery Systems: Strategies, Technologies, and Applications, First Edition.
Edited by Yoon Yeo.
© 2013 John Wiley & Sons, Inc. Published 2013 by John Wiley & Sons, Inc.

cells, and (iii) filtration by organs with fenestrated vasculature. Biological clearance of nanoparticles is influenced by physiochemical properties of the nanoparticles, such as surface charge, composition, size, shape, and deformability [12–14]. To extend the circulation time of nanoparticles in the blood stream, nanoparticles are modified in various ways.

9.2 NANOPARTICLE-BASED DRUG CARRIERS IN CANCER THERAPY

9.2.1 Nanoparticle-Based Drug Carriers

Nanoparticles with a size ranging from 1 to 1000 nm have been considered promising drug carriers. Their distinct advantages include flexible control of particle size and surface properties, the compatibility with various routes of administration (e.g., topical, enteral, and parenteral), and the ability to protect anticancer drugs against clearance in the blood and to provide controlled drug release such as sustained or pulsatile drug release [15–18].

Nanoparticle-based drug carriers were first introduced in 1950–1960s as liposomes and polymer–drug conjugates [19,20]. Since then, various nanoparticle-based drug carriers have been developed for the enhancement of chemotherapy [21,22]. In particular, advancement in nanotechnology and biomaterials has contributed to the improvement of existing nanoparticle systems and the development of new drug delivery systems (Table 9.1) [11,23–43]. These nanoparticle-based drug carriers help enhance drug localization in tumor tissues and reduce cytotoxicity for normal tissues compared to free drug.

9.2.2 Tumor Targeting Strategies

An ideal nanoparticle-based drug carrier for chemotherapy should first guide a drug to the targeted tumor tissues while preserving therapeutic activity of the drug during circulation. Secondly, after reaching tumor tissues, nanoparticles should release the drug to kill tumor tissues selectively and site specifically, reducing collateral damages to normal tissues [44–46]. Currently, passive and active tumor targeting strategies are used to deliver nanoparticles to the tumor sites.

9.2.2.1 Passive Tumor Targeting. Most solid tumors possess leaky vasculature induced by vascular permeability factors such as bradykinin, nitric oxide (NO), peroxynitrite ($ONOO^-$), prostaglandins, vascular endothelial growth factor (VEGF), matrix metalloproteinases (MMPs), kallikrein, and tumor necrosis factor (TNF) [4,47]. High concentrations of these secreted factors around tumor tissues accelerate the formation of new blood vessels (neovascularization or angiogenesis) to supply oxygen and nutrients for tumor growth. This results in hyperpermeable vasculature, which allows macromolecules or nanoparticles to access tumor tissue. In addition, many solid tumors are known to develop a defective lymphatic drainage system, causing the molecules and nanoparticles to be retained there [48]. The cut-off size of

TABLE 9.1   Nanoparticle-Based Drug Delivery Systems

System	Structure	Structure Illustration	Example of Compounds	Reference
Polymeric nanoparticles	Hydrophobic moieties or polymers are conjugated to hydrophilic backbone polymer and form self-aggregation		PEG–PLA/PLGA LMWC–PLGA GC–CA HA–CA	[23] [24] [11] [25]
Polymersome	Amphiphilic block copolymers assemble and form a polymer bilayer shell		PEO–PEE PEG–PLA PEO–PBO	[26] [27] [28]
Polymeric micelles	Amphiphilic block copolymers assemble into small spherical particles of a hydrophobic core and hydrophilic shell		PEG–PLA PEG–PCL PEG–PLGA PEO–PPO–PEO	[29] [30] [31] [32]
Dendrimer	Well-defined branched synthetic polymer with a spherical three-dimensional shape		PEG–PAMAM PGLSA Polypeptide	[33] [34] [35]
Liposome	Amphiphilic lipids assemble into spherical particles with lipid bilayer shell		PC–cholesterol PEG–DSPE	[36] [37]
Inorganic nanoparticles	Metal-based nanoparticles with polymer surface modification		Gold Iron oxide	[38] [39]
Protein nanoparticles	Protein-based formulation using the interaction with drugs		Albumin	[40]
Hybrid nanoparticles	At least two different materials are used to form spherical particles composed of a core and shell		Albumin–PMMA PLGA–QD–siRNA PLGA-dendrimer	[41] [42] [43]

Hydrophobic polymer; ⌇⌇⌇⌇ hydrophilic polymer. ▨▨▨▨ Lipid; ● drug.
PEG, poly(ethylene glycol); PLA, poly(lactic acid); PLGA, poly(lactic-*co*-glycolic acid); LMWC, low molecular-weight chitosan; GC, glycol chitosan; CA, cholanic acid; HA, hyaluronic acid; PEE, poly(ethyl ethylene); PEO, poly(ethylene oxide); PBO, poly(butylene oxide); PCL, poly(caprolactone); PPO, poly(propylene oxide); PAMAM, poly(amido amine); PGLSA, poly(glycerol succinic acid); PC, phosphatidylcholine; DSPE, distearoylphosphatidylethanolamine; PMMA, poly(methyl methacrylate); Dox, doxorubicin; QD, quantum dot.

gap junctions between endothelial cells in permeable tumor blood vessels is around 200–800 nm [5], which enables accumulation of nanoparticles having the similar size. This phenomenon is known as the enhanced permeability and retention (EPR) effect, a critical principle for the passive tumor targeting of nanoparticles [49]. After accumulation in the tumor tissues, the drug is released into extracellular matrices, where it diffuses into intracellular sites.

Most nanoparticle-based drug carriers rely entirely on this EPR mechanism to facilitate delivery to the tumor. Despite a wealth of documentations on the EPR effect, tumor targeting efficacy through the EPR effect remains controversial. This controversy stems from the fact that there is no major driving force facilitating the passive extravasation of nanoparticles through the leaky vasculature [50]. It is currently believed that passive nanoparticle extravasation via the EPR effect occurs by diffusion and convection in blood flow. However, these forces actually play a minor role in the transport of nanoparticles to the interstitium of tumors compared to bulk fluid flow [51]. Interstitial fluid pressure in tumor is higher than intravascular pressure due to the defective lymphatic drainage [52]. Furthermore, blood pressure in solid tumors tends to increase toward the tumor center, which drives overall mass transport of fluid away from the center [53,54]. As a result, nanoparticle transport to central tumor regions is remarkably limited. A recent report describes that this problem was experimentally addressed by increasing systemic blood pressure with angiotensin II (AT-II) [55] or facilitating tumor blood flow using nitroglycerin (NG) [56]. This approach resulted in temporary augmentation of the EPR effect and drug delivery to tumor tissues [57].

9.2.2.2 Active Tumor Targeting. Another strategy for tumor targeted drug delivery is active targeting using tumor-binding ligands [58,59]. Here nanoparticles are decorated with a ligand as a targeting moiety on the surface, which helps their entry into the cells. Cellular internalization of these nanoparticles occurs via receptor-mediated endocytosis [60,61] and is especially important in biotherapeutic delivery such as DNA and small interfering RNA (siRNA) [62]. Various targeting ligands for cancer therapy, such as RGD peptides for targeting $\alpha_v\beta_3$ integrins in tumor vasculature and folic acid for targeting folate receptors on cancer cells, have been identified and used for targeted drug delivery (Table 9.2) [63–71]. However, the mechanism of active targeting nanoparticles is not entirely clear [72], and the effectiveness of ligands remains controversial. Several recent reports have shown only a modest improvement in tumor accumulation of the ligand-decorated nanoparticles compared to nanoparticles with no ligands [73,74]. The results of these studies suggest that the tumor localization of nanoparticles is independent of the ligand and primarily relies on the EPR effect.

9.3 BIOLOGICAL CLEARANCE OF NANOPARTICLES

When the nanoparticle delivery to tumor tissues largely depends on the EPR effect, circulation time of nanoparticles in the blood is one of the most important properties [8]. The long circulation time provides nanoparticles with a great opportunity to transport

TABLE 9.2 Tumor Molecules and Ligands for Active Targeting

Target Site	Target Molecule (or Receptor)	Ligand	Reference
Tumor vascular targeting	VEGF receptor	CBO-P11	[63]
	$\alpha_v\beta_3$	cRGD peptide	[64]
	VCAM-1	Anti-VCAM-1 antibody	[65]
	MT1-MMP	Anti-MT1-MMP antibody	[66]
Tumor cell targeting	EGF receptor	EGF	[67]
	Transferrin receptor	Transferrin	[68]
	Folate receptor	Folic acid	[69]
	PSMA	Anti-PSMA aptamer	[70]
	ASGP receptor	Carbohydrate	[71]

VEGF, vascular endothelial growth factor; cRGD, cyclic arginine–glycine–aspartic acid; VCAM-1, vascular cell adhesion molecule-1; EGF, epidermal growth factor; PSMA, prostate specific membrane antigen; ASGP, asialoglycoprotein.

into tumor tissues through the leaky vasculature and interact with target receptors on tumor cells [75]. The circulation time of the nanoparticles in the blood is determined by the rate of biological clearance from the blood, spleen, liver, and kidney (Figure 9.1) [10,76]. While biological clearance is a natural protection mechanism against invasion by foreign organisms, it is a significant obstacle for the application of nanoparticles in cancer therapy. In addition, the localization of drugs or nanoparticles in filtering organs such as the liver and the spleen can cause hypersensitivity reactions such as exaggerated immune responses, making the concept of targeted drug delivery to eliminate secondary toxicity pointless [77–80]. To evade rapid biological clearance, various approaches have been developed focusing on surface modification or controlling the physiochemical properties (size, shape, surface charge, and deformability) of nanoparticles, which play critical roles in their biological clearance [81–84].

9.3.1 Blood Clearance

9.3.1.1 Protein Adsorption and Opsonization. Plasma proteins immediately cover the surface of administered nanoparticles, which they recognize as foreign materials [85,86]. Mechanisms of plasma protein binding to nanoparticles are not yet fully understood due to the complexity of the binding events between the proteins and nanoparticles. Proteins associated with various nanoparticles have been determined by protein separation and identification techniques such as centrifugation, gel filtration, two-dimensional polyacrylamide gel electrophoresis (2D-PAGE), and atomic force microscopy [87–89]. The composition of proteins adsorbed to the nanoparticles primarily depends on the protein concentration, kinetics of protein binding, and surface characteristics of the nanoparticles (Table 9.3) [90–104]. Abundant plasma proteins such as albumin, immunoglobulin G (IgG), fibrinogen, and apolipoproteins are first adsorbed on most types of nanoparticles. Over time, these proteins are sequentially exchanged with other

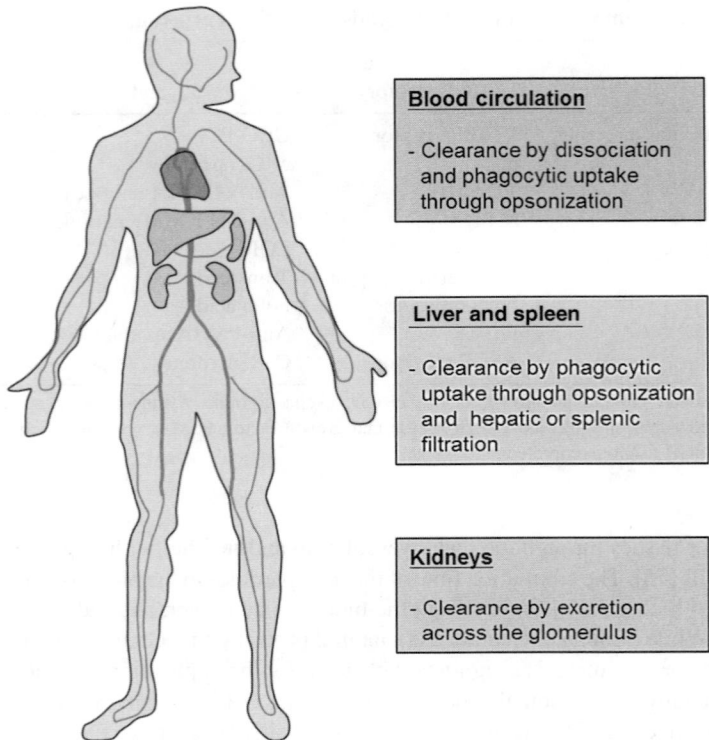

Blood circulation

- Clearance by dissociation and phagocytic uptake through opsonization

Liver and spleen

- Clearance by phagocytic uptake through opsonization and hepatic or splenic filtration

Kidneys

- Clearance by excretion across the glomerulus

FIGURE 9.1 Biological clearance of nanoparticles in blood circulation, liver, spleen, and kidneys. (*Source:* Modified from Reference [10].)

high affinity proteins. The exchange rates are relative to the chemical properties and stability of the protein [105]. For instance, fibrinogen binds immediately onto the surface of solid lipid nanoparticles and is rapidly displaced by apolipoproteins after 0.5 min incubation in blood [106]. Protein binding to iron oxide particles also changes with the plasma concentration and incubation time [107].

Physiochemical properties of nanoparticles that influence protein adsorption to their surface include surface charge, hydrophobicity, particle size, surface curvature, and surface area. Electrically neutral nanoparticles have distinctly less protein adsorption than charged nanoparticles [108–110], whereas protein adsorption increases in general for negatively charged nanoparticles [111]. The composition of adsorbed proteins depends on the functional groups and its charge. For example, proteins with an isoelectric point below 5.5 (e.g., albumin) have affinity for positively charged nanoparticles with amine groups ($-NH_2$), while those with isoelectric point above 5.5 (e.g., IgG) was shown to be predominantly adsorbed to negatively charged nanoparticles with carboxyl acid ($-COOH$) [112]. Macrophage uptake of positively charged nanoparticles was greatly increased by albumin adsorption [113]. Hydrophobicity of a nanoparticle surface is another key factor

TABLE 9.3   Nanoparticle Systems and Opsonins

System	Identified Oposnins	Reference
Polystyrene with poloxamer 184, 188, 407	Factor B, transferrin, albumin, fibrinogen, IgG, apolipoproteins	[92]
Liposomes	Albumin, fibrinogen, apolipoproteins, IgG, α1-antitrypsin, α2-macroglobulin, IgM	[93]
Solid lipid nanoparticles with Tween 80	Fibrinogen, IgG, IgM, apolipoproteins (including ApoE), transthyretin	[94]
Solid lipid nanoparticles with poloxamer 188	Fibrinogen, IgG, IgM, apolipoproteins (excluding ApoE), transthyretin, albumin	[94]
Poly(lactic acid) nanoparticles with PEG	Albumin, fibrinogen, apolipoproteins, IgG	[95]
Poly(hexadecyl cyanoacrylate) nanoparticles	Albumin, apolipoproteins, IgG, transferrin	[96]
Poly(ε-caprolactone) nanoparticles	IgG, apolipoproteins	[97]
Polycyanoacrylate nanoparticles	Albumin, IgG, IgM, fibrinogen, apolipoproteins	[98]
Iron oxide nanoparticles	Albumin, IgG, IgM, fibrinogen	[99]
Poly(D,L-lactic acid) nanoparticles	Albumin, IgG, fibrinogen, IgM, apolipoproteins, antithrombin III	[100]
Polystyrene with Rhodamine B	Albumin, IgG, fibrinogen, apolipoproteins, PLS:6	[101]
Poly(isobutylcyanoacrylate) with Dextran	Albumin, IgG, fibrinogen, apolipoproteins, serotransferrin, transthyretin	[102]
Poly(butyl cyanoacrylate) nanoparticles with polysorbate 80	Albumin, IgG, fibrinogen, IgM, apolipoproteins	[103]
Solid lipid nanoparticles with poloxamer or poloxamine coating	Albumin, fibrinogen, apolipoproteins	[104]

PEG, poly(ethylene glycol); IgG, immunoglobulin G; IgM, immunoglobulin M; PLS, unidentified protein.
Source: Modified from Reference [87].

in plasma protein adsorption [108]. Higher hydrophobicity of the nanoparticle surface resulted in greater and faster protein binding [114–116]. The effects of size, surface curvature, and surface area of the nanoparticles on protein adsorption have also been evaluated. A study using *N*-isopropylacrylamide (NIPAM) and *N-tert*-butylacrylamide (BAM) copolymer nanoparticles with sizes ranging from 70 to 700 nm found the same amount of protein adsorption [117]. In contrast, another study on poly methoxypolyethyleneglycol cyanoacrylate-*co-n*-hexadecyl cyano-acrylate (mPEG–PHDCA) copolymer nanoparticles with three different sizes (80, 170, and 240 nm) demonstrated that larger nanoparticles had greater amounts of protein binding. The amount of proteins adsorbed to nanoparticles with a particle size of 80, 170, and 240 nm was 6, 23, and 34% of serum protein, respectively [118]. These conflicting results may be ascribed to the difference in material systems. Alternatively, change in one property of nanoparticles may have been accompanied by inadvertent change in other properties, which can influence the protein adsorption profile.

9.3.1.2 Disintegration of Nanoparticles by Protein Adsorption. Protein adsorption can trigger the disintegration of nanoparticles, compromising their function as drug carriers. The instability of polymeric micelles by plasma protein binding has been investigated [119]. The disintegration of poly(ethylene glycol)–poly(caprolactone) (PEG–PCL) micelles in serum solution was shown using fluorescence self-quenching method [120]. *In vitro* and *in vivo* disintegration of poly(ethylene glycol)–poly(DL-lactic acid) (PEG–PDLLA) micelles by α- and β-globulins was also shown using the fluorescence resonance energy transfer (FRET) technique [121,122]. There are three possible mechanisms associated with the disintegration of micelles by plasma protein: protein adsorption, protein penetration, and drug extraction by proteins (Figure 9.2) [123]. Unstable nanoparticles may be quickly eliminated in systemic circulation by protein-mediated disintegration [121].

9.3.1.3 Opsonization-Mediated Nanoparticle Uptake by Immune Cells. Another main clearance pathway of nanoparticles is the uptake of nanoparticles by immune

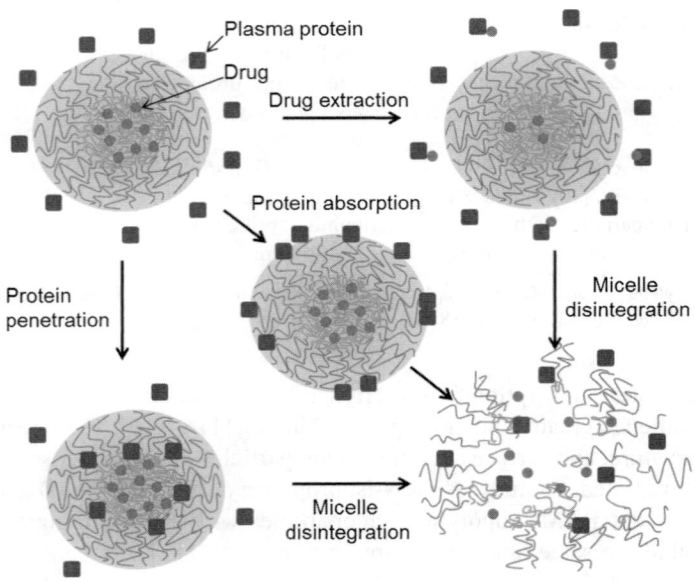

FIGURE 9.2 Disintegration of nanoparticles by plasma protein; protein adsorption, protein penetration, and drug extraction by proteins (*Source:* Modified from Reference [123].) (*See insert for color representation of the figure.*)

cells. Immune cells both in the blood stream (leukocytes, monocytes, platelets, and dendritic cells) and in organs (Kupffer cells in liver, macrophage and B cells in spleen, and dendritic cells in lymph nodes) engulf and remove the nanoparticles from the blood circulation. This process is generally mediated by binding of specific receptor proteins called opsonins to the surface of nanoparticles [124]. The bound opsonins, such as plasma proteins, immunoglobulins, and component factors in complement system, facilitate the nanoparticle uptake into immune cells, negatively affecting the biodistribution of nanoparticles and hence drug delivery to target tumor tissues [125,126]. The opsonins on the nanoparticle undergo conformational changes and are recognized by the immune cells via specialized receptors [108].

Macrophages are mainly responsible for the clearance of large molecules like nanoparticles. The receptor-mediated phagocytic pathways are classified into four categories (mannose receptor-, complement receptor-, immunoglobulin Fcγ receptor-, and scavenger receptor- mediated phagocytosis) according to the opsonins adsorbed onto the nanoparticles [124]. Through these mechanisms, opsonins such as IgG, apolipoproteins, and fibrinogen promote phagocytosis of nanoparticles by macrophages in the reticuloendothelial system (RES) [106,107,127]. These nanoparticles are subject to localization in RES organs such as the liver and spleen, and eventually eliminated by resident immune cells [113,128,129]. Adsorption of fetuin and fibronectin triggers the phagocytosis of polystyrene nanoparticles by Kupffer cells via the scavenger receptor [130,131]. Increasing the amount of C3 and IgG adsorbed on lecithin-coated polystyrene nanoparticles caused an increased uptake of Kupffer cells [132]. Preopsonization of poloxamine-coated polystyrene nanoparticles by incubation in serum solution accelerated the phagocytosis by macrophages [133,134]. Furthermore, during the clearance process, immune cells produce proinflammatory responses, which involve the secretion of cytokines such as tumor necrosis factor-α, interleuckin-12, and interferon-γ. These cytokines recruit more immune cells to take up nanoparticles. This can cause immunotoxicity and severe damages to the RES organs [77]. However, the effect of nanoparticle opsonization is not always destructive. Some studies have shown that opsonins can enhance tumor targeting and blood circulation time of nanoparticles. For example, dysopsonins, such as albumin, extended the retention time of nanoparticles in the blood [135,136]. In addition, various apolipoproteins enabled nanoparticles to transport across the blood–brain barrier (BBB) for brain tumor therapy [137].

9.3.2 Organ Filtrations

Some organs and tissues have naturally permeable vasculature in order to conduct their physiological functions. The inherent gaps between endothelial cells in vessels of those organs, called fenestration (Figure 9.3) [138], play a major role in determining the biological fate of nanoparticles.

While the fenestration of nanoparticles can be used for tumor targeting similar to the EPR effect, it can also cause rapid clearance and nontargeting effect of

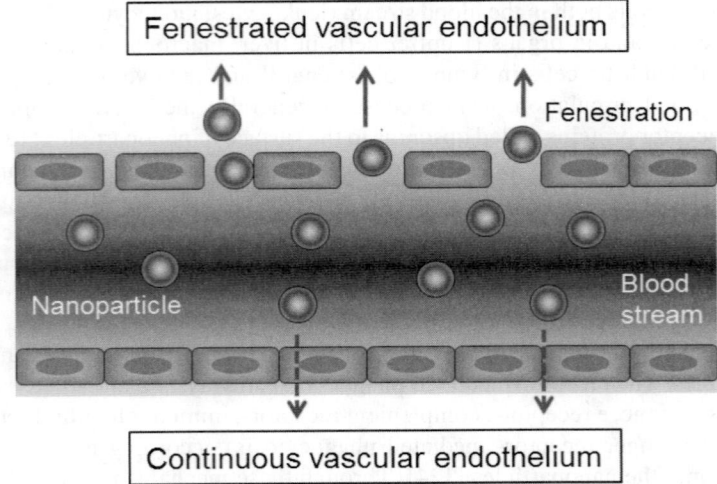

FIGURE 9.3 Schematic illustration of vascular extravasations of nanoparticles in either continuous or fenestrated tissues.

nanoparticles. The size and structure of a fenestration depend on the physiological functions of the organs and tissues (Table 9.4) [5,47,139–150]. Furthermore, the size of a fenestration can change under various pathological conditions [151]. In the spleen, nanoparticles should pass through splenic fenestrations (splenic sinuses) to remain in blood circulation, while, in the liver, the nanoparticles should avoid going through hepatic fenestrations (hepatic sinuses). Otherwise, nanoparticles may be captured by the hepatocyte tissues (hepatocytes cluster), where the nanoparticles are eliminated by Kupffer cells and hepatocytes. Therefore, to improve the blood circulation and tumor targeting of nanoparticles, an understanding of anatomical fenestrations is essential [138]. Of many organs and tissues that contain

TABLE 9.4   Sizes of Fenestration in Different Organs and Tissues

Organ or Tissue	Fenestration Size	Animal Model	Reference
Spleen	150 nm	Mice	[139,140]
Liver	100–150 nm	Rat	[141,142]
Kidney	4.5–5 nm	Rat	[143]
Lung	1–400 nm	Dog	[144]
Bone marrow	85–150 nm	Guinea-pig, rabbit, rat	[145,146]
Skeletal, cardiac, and smooth muscle	<6 nm	Mice	[147]
Skin, subcutaneous, and mucous membrane	<6 nm	Mice	[147]
Blood–brain barrier (BBB)	No fenestrations	—	[148,149]
Tumor	200–800 nm	Mice	[5,47]
Inflamed organs	80 nm–1.4 μm	Hamster	[150]

Source: Modified from Reference [138].

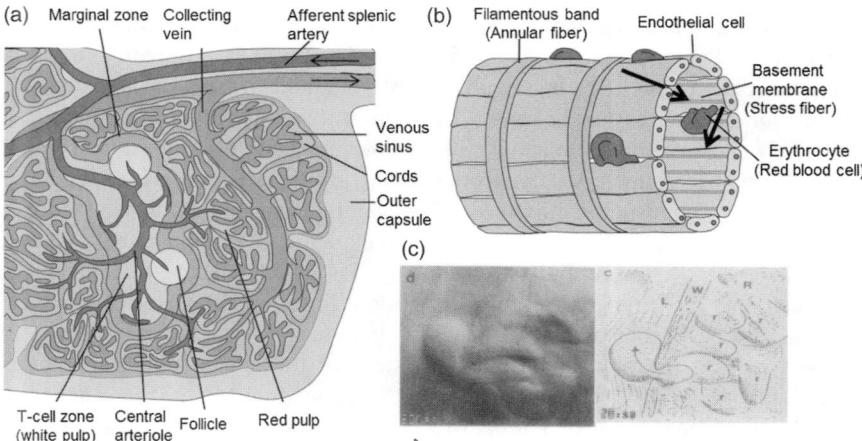

FIGURE 9.4 Structure of spleen. (a) Schema of the spleen, (b) venous sinuses in the red pulp of the spleen (*Source:* Modified from References [154 and 155]), (c) photomicrograph of an erythrocyte squeezing through sinus wall (*Source:* Reproduced with permission from Reference [155]).

fenestrations, we focus on the organs mainly responsible for the clearance of nanoparticles, such as spleen, liver, and kidneys.

9.3.2.1 Spleen. The main function of the spleen is to filter the blood [152,153]. The spleen possesses tree-shaped branching arterial vessels that are linked to a venous sinusoidal system (Figure 9.4a) [154,155]. The frame of the spleen is mainly composed of fibrous connective tissue called trabecula. The smaller arterial vessels are surrounded by the lymphoid tissues, which form the white pulp of the spleen. Some of the smallest arterial branches end in the marginal sinus located between the white pulp and the marginal zone, while other branches cross the marginal zone to form the venous system of the red pulp, where mechanical blood filtration occurs [154]. To efficiently filter antigens, microorganisms, and old red blood cells (erythrocytes) from the blood, the venous system of the red pulp has a unique microanatomical structure. Arterial vessels eventually reach into cords in the red pulp, where they are structured as an open blood system without an endothelial barrier [156]. A large number of splenic macrophages are resident in the cords to remove unfiltered organisms. After the blood is purified in the venous sinuses of the red pulp, it is collected in the efferent veins, and finally leaves the spleen through the collecting veins.

To transport through the venous sinuses, red blood cells must pass through the sinus wall. The sinus wall consists of a lining of endothelial cells with basement membrane (stress fiber) and filamentous bands (annular fiber). The rod-shaped endothelial cells are positioned in parallel to the longitudinal axis of the sinus and connect to each other by the stress fiber running along the cellular axis. The annular fiber composed of extracellular matrix components connects to the basement membrane along the interendothelial slits and forms the endothelial cells into tubular pattern (Figure 9.4b) [157,158]. The unique arrangement of the endothelial cells lining the sinus wall forces the blood to pass through the slits to travel from the cords into the sinus [155]. The slits seldom exceed 0.2–0.5 μm in width [157]. However, these narrow slits do not hinder the passage of healthy red blood cells, whose diameters range from 6 to 8 μm, because of their high degree of deformability, whereas unhealthy blood cells having rigid membranes get stuck in these slits and are ultimately eliminated by the splenic macrophages in the cords (Figure 9.4c) [155]. Large and stiff nanoparticles when injected into the blood stream can also be removed by the same splenic filtration process. Although the recognition of pathogens by macrophages occurs via specific opsonin receptors as discussed earlier, the physical filtration system of the spleen potentially helps splenic macrophages to recognize pathogens stuck in sinus slits [152]. Therefore, physiochemical properties of nanoparticles such as size, flexibility, shape, and surface properties influence passage across the sinus slits in the spleen. It has been reported that splenic clearance of polystyrene nanoparticles increased with their size [159,160]. Also, rigid chitosan nanoparticles with the highest degree of substitution (DS) of 5β-cholanic acid showed rapid accumulation in the mouse spleen via the filtration mechanism, compared to softer chitosan nanoparticles with low DS of 5β-cholanic acid [11]. These studies indicate that the physical properties of nanoparticles such as size and deformability play a pivotal role in splenic passage.

9.3.2.2 Liver. The liver has multiple functions including metabolic control, nutrient (e.g., glycogen) storage, plasma protein synthesis, hormone production, and the elimination of foreign or toxic substances (detoxification) [161]. Here, we focus on the detoxification process in the liver that directly relates to the clearance of nanoparticles. The liver receives blood from both the hepatic portal vein and the hepatic arteries. Around 75% of the liver's blood supply is obtained through the hepatic portal vein that drains the blood from the small intestine, stomach, spleen, and pancreas. The remaining 25% of the blood supply comes from the hepatic arteries. The hepatic portal vein and the hepatic arteries branch out in the liver and are combined at the entrance to the hepatic sinusoids [161]. The blood then passes through the sinusoids and collects in the central vein that drains into the hepatic vein (Figure 9.5a) [162].

The basic functional unit of the liver is the hepatic lobule (Figure 9.5b), which consists of a concentric arrangement of hepatocytes (hepatic cords) radiating outward from the central vein and those separated by the sinusoids. The portal triads include a bile duct, the terminal branches of the hepatic artery, and portal vein, which are regularly located at the vertices of the lobule. The sinusoids are lined by fenestrated endothelial cells and have no recognizable basement membrane. These unique fenestrations allow hepatocytes to have direct exposure to the blood in the space of Disse. The space of Disse generally contains extracellular matrix proteins (e.g., reticular fibers), fibroblasts (e.g., hepatic stellate cells), and dendritic cells. Hepatic macrophages (Kupffer cells) reside in the sinusoidal endothelium [162].

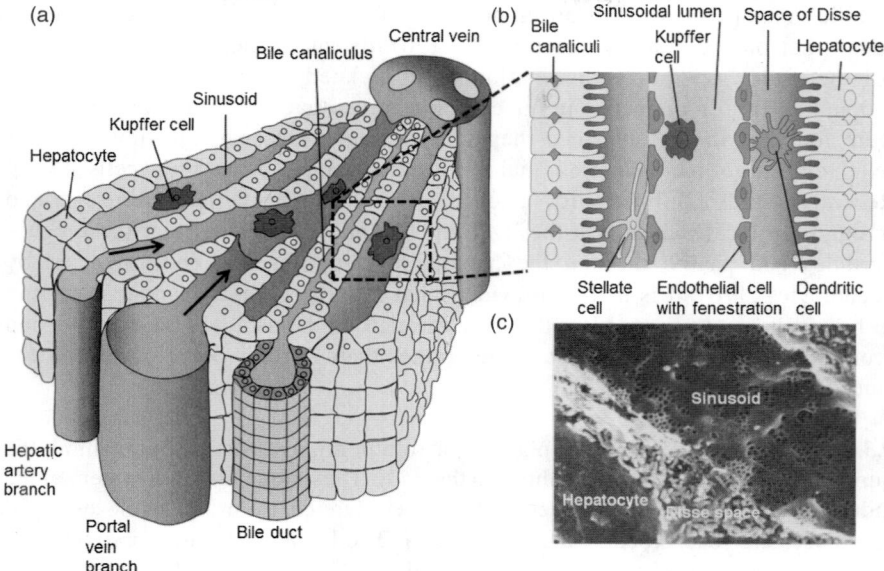

FIGURE 9.5 Structure of a liver lobule. (a) Schema of the liver lobule, (b) sinusoid in the liver lobule (*Source:* Modified from Reference [162]), (c) scanning electron micrograph (SEM) of the sinusoid (*Source:* Professor Robin Fraser, University of Otago, New Zealand; reproduced per the copyright release to the public domain).

During the detoxification process, Kupffer cells exclusively phagocytose large particles (e.g., 800 nm in diameter), whereas hepatocytes mainly endocytose small particles (e.g., 50 nm in diameter) transported through the endothelial fenestrations [163,164]. Particles taken up by hepatocytes are excreted through the bile by biliary excretion [165]. The size of the fenestration for rats is reported to be around 100 nm in mean diameter under physiological conditions, and the fenestrations are clustered in groups of 10–100 fenestration occupying 2–20% of the sinusoidal endothelial cell surfaces (Figure 9.5c) [142,166,167]. These fenestrations provide a sieving function of sinusoidal blood. Since the liver is a key organ that stores nutrients and removes toxic substances, the internal blood flow rate in the liver is quite slow. This enhances the absorption of plasma nutrients and foreign organisms by the hepatocytes inside the fenestration [161].

This mechanism also contributes to the clearance of small nanoparticles that can pass through the hepatic fenestrations. Most small sized nanoparticles (<100 nm in diameter) have shown substantial accumulation in the liver. Examples are shell cross-linked PEG–poly(*t*-butyl acrylate) nanoparticles (18–37 nm in diameter) [168], DNA–PEI complex with crystallized ligand shell (<50 nm in diameter) [169], PEGylated iron oxide (SPIO) (6.2 nm in average diameter) [170], cyclo-dextrin–camptothecin nanoparticles (30–40 nm in diameter) [171], core cross-linked *N*-isopropylacrylamide (NIPAm) nanoparticles (around 50 nm in diameter) [172], and quantum dots (3.4–5.5 nm in diameter) [173]. Hepatotoxicity testing of silica nanoparticles with different sizes (70, 300, and 1000 nm in diameter) found that only the small nanoparticles (70 nm) caused hepatic toxicity [174]. Another study on the biodistribution of PEGylated polystyrene nanoparticles (50 and 140 nm in diameter) in mice demonstrated that most small nanoparticles were found in the cytoplasm of hepatocytes after 24 h post injection, whereas relatively large nanoparticles mainly internalized into Kupffer cells [175]. Recently, a study exhibited silica nanoparticles (~20 nm in diameter) were excreted through the fecal matter (via the liver and the bile), indicating that these nanoparticles had passed through the hepatic fenestrations and endocytosized by hepatocytes [176]. It is therefore evident that relatively small nanoparticles tend to be mechanically trapped in the liver through the sinusoidal fenestrations and preferentially removed by hepatocytes rather than Kupffer cells located on the sinusoidal endothelium [177].

9.3.2.3 Kidneys. The kidneys play major role in homeostasis by maintaining the purity and constancy of internal fluids in the body. The unique filtration system of the kidneys allows wastes and excess ions to be excreted by urine, while essential substances are reabsorbed to the blood [161]. The kidneys cleanse approximately one-quarter of total blood supply in the body each minute. The blood enters the kidneys through the renal artery, which divides into several branches called interlobar arteries. Then, at the medulla-cortex border, the interlobar arteries branch into the arcuate arteries, which curve over the medullary pyramids. Smaller interlobar arteries branch off the arcuate arteries, which supply the cortex at the outer region of the kidneys. Once cleansed, the venous blood collects and leaves the kidneys through the renal vein [161]. Nephrons are the basic functional units of the kidneys responsible for eliminating toxins, wastes, and foreign materials from the body via urine. Each nephron consists of two main structures: a glomerulus and a renal tubule (Figure 9.6) [161]. The closed end of the renal tubule is cup-shaped and completely covers the glomerulus, which is called the Bowman's capsule. Entangled podocytes in the glomerulus form porous membranes called filtration slits. This slits allow small wastes to transport from the plasma to the renal tubule.

The renal filtration is a nonselective, passive, and mechanical process. Thus, glomerular filtration of an object is highly dependent on its size. The filtrate follows the renal tubule, reaches the bladder, and is eventually excreted as urine. Useful substances in the filtrates are reabsorbed by the tubule. The size of the renal filtration slits in rats is around 39 nm [179]. However, the actual slit size is about 4.5–5 nm due to the stratified nature of the glomerular capillary walls [143]. The filtration size threshold of PAMMA dendrimer for mice has been reported to be around 5.4 nm [180,181]. While quantum dots (QDs) with a diameter of 4.36–5.52 nm were rapidly cleared by renal clearance in mice [173,182], QDs >8 nm in diameter did not show the renal clearance but exhibited substantial accumulation in the liver. Glomerular filtration is also influenced by other properties such as charge, shape, and deformability of filtrate. For example, anionic glycolated Fab fragments showed significantly less renal filtration compared to cationic glycolated Fab fragments owing to the electrostatic repulsion between the anionic Fab fragments and the anionic

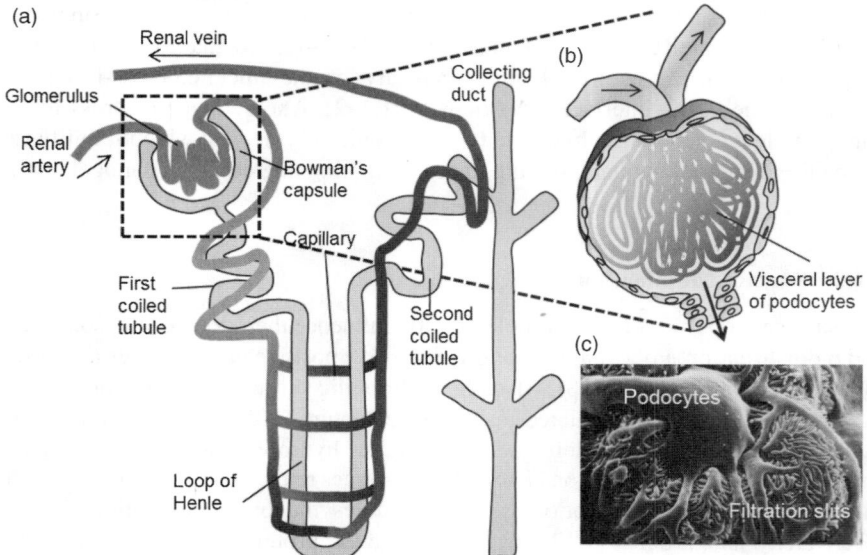

FIGURE 9.6   Structure of a nephron. (a) Schema of the nephron, (b) glomerulus in the nephron (*Source:* Modified from Reference [161]), (c) SEM of the podocytes in the glomerulus (*Source:* Reproduced with permission from Reference [178]).

membrane of the glomerular capillary walls [183]. Opsonization stimulated by the surface charge of nanoparticles increases the size, which precludes renal filtration [173]. On the other hand, flexible Ficoll and asymmetrically shaped dextran demonstrated higher renal filtration efficacy than globular proteins, even though their sizes (4.6 and 5.5 nm, respectively) are relatively larger than that of the proteins (3.75 nm) [184]. Although these studies used biological molecules, the results are applicable for nanoparticles in renal clearance.

9.4   STRATEGIES TO MINIMIZE THE BIOLOGICAL CLEARANCE OF NANOPARTICLES

Since the emergence of nanoparticles in the 1950s, various strategies to minimize their biological clearance have been developed. Some of them are highlighted in the following discussion.

9.4.1   Chemical Cross-Linking

As we have discussed earlier in Section 9.3.1.2, polymeric micelles are unstable under physiological conditions. To solve the problem, various covalent cross-linking methods have been introduced such as photo cross-linking [185] and disulfide cross-linking [186].

There are generally three cross-linkable locations for a triblock copolymer micelle: shell, interface, and core [187–190]. However, to maintain the surface properties and increase drug loading capacity of the micelle, cross-linking is typically done at the interfacial positions [191,192]. Another type of cross-linked nanoparticle is a nano-sized hydrogel (e.g., Pluronic polymer), of which stability can be readily increased by manipulating the cross-linking degree of the nanogel matrix [193,194].

9.4.2 Surface Modification

The surface properties of nanoparticles greatly influence their degree of opsonization and biodistribution. For example, neutral or weakly positive nanoparticles are shown to evade opsonization [81,108]. Also, the hydrophilic surface of nanoparticles helps them to escape opsonin-mediated macrophage capture [195,196]. PEG, a nonionic hydrophilic polymer, is widely used to increase hydrophilicity of nanoparticles. Brush configurations of PEG on nanoparticle surfaces reduce protein adsorption. The resistance to protein adsorption greatly increases the blood-circulation time of PEGlyated nanoparticles [9,12,197,198]. However, proteins like albumin, α-chymotrypsin, and hen-egg-white lysozyme can adsorb to the PEG-coated surfaces by forming hydrogen bonding interactions between their carboxyl groups and oxygen atoms in PEG [199–201]. Despite this protein adsorption, PEG surface coating is still the most widely accepted method to reduce opsonization onto nanoparticles. On the other hand, albumin-coated liposomes have demonstrated more prolonged circulation time in blood compared to PEG-coated liposomes, indicating that albumin coating can prevent the adsorption of other proteins more efficiently [202]. Therefore, various albumin-coated nanoparticles have been developed to extend their circulation time in blood [203,204].

9.4.3 Control of the Size, Shape, and Deformability of Nanoparticles

The size of nanoparticles is a key attribute that can be manipulated in order to reduce the biological clearance associated with anatomical fenestrations discussed earlier. Although the vascular heterogeneity across different animal models needs to be considered, the diameter range 100–200 nm has been considered an appropriate size in rodent models for both long-term circulation and passive accumulation in tumors [22]. Other properties of nanoparticles such as shape and deformability (flexibility) are also believed to play a role [83,84]. Recent advances in nanotechnology such as the template technology have enabled the development of particles with various sizes and shapes [205,206]. The template technology, which utilizes a template for making homogenous nanoparticles in size and shape, has been used to investigate the effect of particle shape on the phagocytosis by macrophages [207,208]. Macrophages contacting elliptical shaped particles along their major axis rapidly phagocytosed the particles within 6 min, while macrophages that initially contacted the same particles along their minor axis did not internalize the particles. This result suggests that macrophage uptake depends on the particle shape upon initial contact. Another example pertaining to the effect of nanoparticle shape on biological clearance is filamentous micelles (filomicelles) that have two-dimensional shape ($\sim$18 μm in length and $\sim$20–60 nm in diameter). These filamentous micelles exhibited a significantly prolonged blood circulation time (circulation half-lives: $\sim$5 days) and improved drug delivery [30,209]. This shape not only facilitates micelles passage through RES organs such as the spleen and liver but also minimizes their internalization into macrophages. In addition, red blood cells (RBCs) mimicking polymeric particles have recently been engineered using a poly(lactic-co-glycolic acid) (PLGA) RBC-shaped template [210]. Although the deformability of these particles was less than that of mouse red blood cells, it was still more flexible than spherical PLGA particles due to their special shape. These particles also demonstrated oxygen and drug carrying capabilities [210]. Another recent study on the deformability of particles showed that micro-sized N-isopropylacrylamide (NIPAm) hydrogel particles ($\sim$1 μm in diameter) were able to pass through membrane pores at least 10 times smaller in size [211].

It was also reported that the difference in deformability of nanoparticles affected how the nanoparticles were internalized into macrophages [212]. In this study, soft nanoparticles were preferentially internalized via macropinocytosis while rigid nanoparticle uptake was found to be mediated by clathrin-mediated routes. As for nanoparticles with intermediate flexibility, they were internalized via multiple mechanisms. Furthermore, soft chitosan nanoparticles have shown prolonged blood circulation and higher tumor targeting efficacy compared to rigid chitosan nanoparticles [11]. Chitosan nanoparticles with higher degree of substitution of 5β-cholanic acid have demonstrated lower deformability and, thus, higher accumulation in the liver and spleen even though they had a similar particle size (300–400 nm).

9.5 FUTURE PERSPECTIVES

We have discussed mechanisms by which nanoparticles are biologically cleared from the body and current strategies to enhance the blood-circulation and tumor targeting effect of nanoparticles. The surface properties of nanoparticles are indirectly related to their blood clearance through plasma protein adsorption and opsonization, while physical properties such as size, shape, and deformability directly impact the removal of nanoparticles via the filtering organs. To extend the retention time of nanoparticles in blood circulation, a number of strategies manipulating the physiochemical properties of nanoparticles have been developed. Nevertheless, the fraction of nanoparticles that actually reach the target tissues still remains very small ($\sim$5%), and the majority of administered nanoparticles are found in the RES organs [213].

To overcome this critical challenge, it is necessary to revisit nanoparticle-based delivery strategies and key factors in the design of nanoparticles. One of the most important prerequisites in the design of a nanoparticle system is to stably retain the loaded drug in blood circulation before accessing cancer cells. Disintegration of nanoparticles by plasma protein adsorption is the first clearance mechanism encountered by the systemically administered nanoparticles. A vast majority of nanoparticle studies focus on postextravasation events such as cellular uptake and intracellular trafficking, but what is the use if most drug has leached out of the system and blank nanoparticles finally reach the targets? Without solving the first problem, targeted drug delivery strategies will have difficulty in taking the next step.

In this regard, the absence of appropriate techniques to investigate whether the nanoparticles remained intact after systemic administration is the major obstacle in the development of new nanoparticle systems. Optical imaging of fluorescently labeled nanoparticles can provide information on pharmacokinetics and biodistribution of the nanoparticles; however, from this technique it is not obvious whether the fluorescent signals come from the intact nanoparticles loaded with drug or disintegrated pieces of nanoparticles. Fluorescence imaging of labeled nanoparticles does not provide sufficient evidence that the nanoparticle has been successfully delivered to the tumor. In an effort to overcome these limitations, our group has designed and synthesized disulfide cross-linked nanoparticles [214]. Using FRET imaging, we monitor the intactness of these nanoparticles in blood vessels and tissues spatiotemporally. The FRET imaging has allowed us to prove that our nanoparticles could stably retain anticancer drugs in bloodstream and increase its blood half-life and delivery to tumors.

REFERENCES

1. Ashley, C. E., Carnes, E. C., Phillips, G. K., Padilla, D., Durfee, P. N., Brown, P. A., et al. (2011). The targeted delivery of multicomponent cargos to cancer cells by nanoporous particle-supported lipid bilayers. *Nature Materials 10*, 389–397.

2. Cabral, H., Matsumoto, Y., Mizuno, K., Chen, Q., Murakami, M., Kimura, M., et al. (2011). Accumulation of sub-100 nm polymeric micelles in poorly permeable tumours depends on size. *Nature Nanotechnology 6*, 815–823.

3. von Maltzahn, G., Park, J. -H., Lin, K. Y., Singh, N., Schwöppe, C., Mesters, R., et al. (2011). Nanoparticles that communicate in vivo to amplify tumour targeting. *Nature Materials 10*, 545–552.

4. Matsumura, Y., Maeda, H. (1986). A new concept for macromolecular therapeutics in cancer chemotherapy: mechanism of tumoritropic accumulation of proteins and the antitumor agent smancs. *Cancer Research 46*, 6387–6392.

5. Torchilin, V. (2011). Tumor delivery of macromolecular drugs based on the EPR effect. *Advanced Drug Delivery Reviews 63*, 131–135.

6. Byrne, J. D., Betancourt, T., Brannon-Peppas, L. (2008). Active targeting schemes for nanoparticle systems in cancer therapeutics. *Advanced Drug Delivery Reviews 60*, 1615–1626.

7. Wang, J., Tian, S., Petros, R. A., Napier, M. E., DeSimone, J. M. (2010). The complex role of multivalency in nanoparticles targeting the transferrin receptor for cancer therapies. *Journal of the American Chemical Society 132*, 11306–11313.

8. Moghimi, S. M., Hunter, A. C., Murray, J. C. (2001). Long-circulating and target-specific nanoparticles: theory to practice. *Pharmacological Reviews 53*, 283–318.

9. Lee, S., Lee, S. -Y., Park, S., Ryu, J. H., Na, J. H., Koo, H., et al. (2012). In vivo NIRF imaging of tumor targetability of nanosized liposomes in tumor-bearing mice. *Macromolecular Bioscience 12*, 849–856.

10. Alexis, F., Pridgen, E., Molnar, L. K., Farokhzad, O. C. (2008). Factors affecting the clearance and biodistribution of polymeric nanoparticles. *Molecular Pharmaceutics 5*, 505–515.

11. Na, J. H., Lee, S. -Y., Lee, S., Koo, H., Min, K. H., Jeong, S. Y., et al. (2012). Effect of the stability and deformability of self-assembled glycol chitosan nanoparticles on tumor-targeting efficiency. *Journal of Controlled Release 163*, 2–9.

12. Perry, J., Reuter, K. G., Kai, M. P., Herlihy, K. P., Jones, S. W., Luft, J. C., et al. (2012). PEGylated PRINT nanoparticles: the impact of PEG density on protein binding, macrophage association, biodistribution, and pharmacokinetics. *Nano Letters 12*, 5304–5310.

13. Meng, H., Xue, M., Xia, T., Ji, Z., Tarn, D. Y., Zink, J. I., et al. (2011). Use of size and a copolymer design feature to improve the biodistribution and the enhanced permeability and retention effect of doxorubicin-loaded mesoporous silica nanoparticles in a murine xenograft tumor model. *ACS Nano 5*, 4131–4144.

14. Shroff, K., Kokkoli, E. (2012). PEGylated liposomal doxorubicin targeted to α5β1-expressing MDA-MB-231 breast cancer cells. *Langmuir 28*, 4729–4736.

15. Hu, C. -M. J., Zhang, L. (2012). Nanoparticle-based combination therapy toward overcoming drug resistance in cancer. *Biochemical Pharmacology 83*, 1104–1111.

16. Blanco, E., Hsiao, A., Ruiz-Esparza, G. U., Landry, M. G., Meric-Bernstam, F., Ferrari, M. (2011). Molecular-targeted nanotherapies in cancer: enabling treatment specificity. *Molecular Oncology 5*, 492–503.

17. Anderson, L. J. E., Hansen, E., Lukianova-Hleb, E. Y., Hafner, J. H., Lapotko, D. O. (2010). Optically guided controlled release from liposomes with tunable plasmonic nanobubbles. *Journal of Controlled Release 144*, 151–158.

18. Huynh, L., Neale, C., Pomès, R., Allen, C. (2012). Computational approaches to the rational design of nanoemulsions, polymeric micelles, and dendrimers for drug delivery. *Nanomedicine: nanotechnology, biology and medicine 8*, 20–36.

19. Zatzkewitz, H. (1954). Incorporation of physiologically-active substances into a colloidal blood plasma substitute. I. Incorporation of mescaline peptide into polyvinylpyrrolidone. *Hoppe-Seylers Journal of Physiological Chemistry 297*, 149–156.

20. Bangham, A. D., Horne, R. W. (1964). Negative staining of phospholipids and their structural modification by surface-active agents as observed in the electron microscope. *Journal of Molecular Biology 8*, 660–668.

21. Yallapu, M. M., Jaggi, M., Chauhan, S. C. (2011). Design and engineering of nanogels for cancer treatment. *Drug Discovery Today 16*, 457–463.

22. Petros, R. A., DeSimone, J. M. (2010). Strategies in the design of nanoparticles for therapeutic applications. *Nature Reviews Drug Discovery 9*, 615–627.

23. Gref, R., Minamitake, Y., Peracchia, M., Trubetskoy, V., Torchilin, V., Langer, R. (1994). Biodegradable long-circulating polymeric nanospheres. *Science 263*, 1600–1603.

24. Amoozgar, Z., Park, J., Lin, Q., Yeo, Y. (2012). Low molecular-weight chitosan as a pH-sensitive stealth coating for tumor-specific drug delivery. *Molecular Pharmaceutics 9*, 1262–1270.

25. Choi, K. Y., Yoon, H. Y., Kim, J. -H., Bae, S. M., Park, R. -W., Kang, Y. M., et al. (2011). Smart nanocarrier based on PEGylated hyaluronic acid for cancer therapy. *ACS Nano 5*, 8591–8599.

26. Discher, B. M., Won, Y. -Y., Ege, D. S., Lee, J. C. -M., Bates, F. S., Discher, D. E., et al. (1999). Polymersomes: tough vesicles made from diblock copolymers. *Science 284*, 1143–1146.

27. Ahmed, F., Pakunlu, R. I., Srinivas, G., Brannan, A., Bates, F., Klein, M. L., et al. (2006). Shrinkage of a rapidly growing tumor by drug-loaded polymersomes: pH-triggered release through copolymer degradation. *Molecular Pharmaceutics 3*, 340–350.

28. Smart, T. P., Ryan, A. J., Howse, J. R., Battaglia, G. (2009). Homopolymer induced aggregation of poly(ethylene oxide)n-b-poly(butylene oxide)m polymersomes. *Langmuir 26*, 7425–7430.

29. Shen, J., Zhan, C., Xie, C., Meng, Q., Gu, B., Li, C., et al. (2011). Poly(ethylene glycol)-block-poly(D,L-lactide acid) micelles anchored with angiopep-2 for brain-targeting delivery. *Journal of Drug Targeting 19*, 197–203.

30. Loverde, S. M., Klein, M. L., Discher, D. E. (2012). Nanoparticle shape improves delivery: rational coarse grain molecular dynamics (rCG-MD) of taxol in worm-like PEG-PCL micelles. *Advanced Materials 24*, 3823–3830.

31. Yoo, H. S., Park, T. G. (2001). Biodegradable polymeric micelles composed of doxorubicin conjugated PLGA–PEG block copolymer. *Journal of Controlled Release 70*, 63–70.

32. Batrakova, E. V., Li, S., Brynskikh, A. M., Sharma, A. K., Li, Y., Boska, M., et al. (2010). Effects of pluronic and doxorubicin on drug uptake, cellular metabolism, apoptosis and tumor inhibition in animal models of MDR cancers. *Journal of Controlled Release 143*, 290–301.

33. Kojima, C., Kono, K., Maruyama, K., Takagishi, T. (2000). Synthesis of polyamido-amine dendrimers having poly(ethylene glycol) grafts and their ability to encapsulate anticancer drugs. *Bioconjugate Chemistry 11*, 910–917.

34. Morgan, M. T., Carnahan, M. A., Immoos, C. E., Ribeiro, A. A., Finkelstein, S., Lee, S. J., et al. (2003). Dendritic molecular capsules for hydrophobic compounds. *Journal of the American Chemical Society 125*, 15485–15489.

35. King, H. D., Dubowchik, G. M., Mastalerz, H., Willner, D., Hofstead, S. J., Firestone, R. A., et al. (2002). Monoclonal antibody conjugates of doxorubicin prepared with branched peptide linkers: inhibition of aggregation by methoxytriethyleneglycol chains. *Journal of Medicinal Chemistry 45*, 4336–4343.

36. Serrano-Luna, J., Gutiérrez-Meza, M., Mejía-Zepeda, R., Galindo-Gómez, S., Tsutsumi, V., Shibayama, M. (2010). Effect of phosphatidylcholine–cholesterol liposomes on *Entamoeba histolytica* virulence. *Canadian Journal of Microbiology 56*, 987–995.

37. Yoshizawa, Y., Kono, Y., Ogawara, K. -i., Kimura, T., Higaki, K. (2011). PEG liposomalization of paclitaxel improved its in vivo disposition and anti-tumor efficacy. *International Journal of Pharmaceutics 412*, 132–141.

38. Giljohann, D. A., Seferos, D. S., Daniel, W. L., Massich, M. D., Patel, P. C., Mirkin, C. A. (2010). Gold nanoparticles for biology and medicine. *Angewandte Chemie International Edition 49*, 3280–3294.

39. Mahmoudi, M., Sant, S., Wang, B., Laurent, S., Sen, T. (2011). Superparamagnetic iron oxide nanoparticles (SPIONs): development, surface modification and applications in chemotherapy. *Advanced Drug Delivery Reviews 63*, 24–46.

40. Gradishar, W. J. (2006). Albumin-bound paclitaxel: a next-generation taxane. *Expert Opinion on Pharmacotherapy 7*, 1041–1053.

41. Ge, J., Neofytou, E., Lei, J., Beygui, R. E., Zare, R. N. (2012). Protein–polymer hybrid nanoparticles for drug delivery. *Small 8*, 3573–3578.

42. Kim, J. H., Noh, Y. -W., Heo, M. B., Cho, M. Y., Lim, Y. T. (2012). Multifunctional hybrid nanoconjugates for efficient in vivo delivery of immunomodulating oligonucleotides and enhanced antitumor immunity. *Angewandte Chemie International Edition 51*, 9670–9673.

43. Yang, H., Tyagi, P., Kadam, R. S., Holden, C. A., Kompella, U. B. (2012). Hybrid dendrimer hydrogel/PLGA nanoparticle platform sustains drug delivery for one week and antiglaucoma effects for four days following one-time topical administration. *ACS Nano 6*, 7595–7606.

44. Lee, S. J., Huh, M. S., Lee, S. Y., Min, S., Lee, S., Koo, H., et al. (2012). Tumor-homing poly-siRNA/glycol chitosan self-cross-linked nanoparticles for systemic siRNA delivery in cancer treatment. *Angewandte Chemie International Edition 51*, 7203–7207.

45. Murakami, M., Cabral, H., Matsumoto, Y., Wu, S., Kano, M. R., Yamori, T., et al. (2011). Improving drug potency and efficacy by nanocarrier-mediated subcellular targeting. *Science Translational Medicine 3*, 64ra2.

46. Ko, J. Y., Park, S., Lee, H., Koo, H., Kim, M. S., Choi, K., et al. (2010). pH-sensitive nanoflash for tumoral acidic pH imaging in live animals. *Small 6*, 2539–2544.

47. Fang, J., Nakamura, H., Maeda, H. (2011). The EPR effect: unique features of tumor blood vessels for drug delivery, factors involved, and limitations and augmentation of the effect. *Advanced Drug Delivery Reviews 63*, 136–151.

48. Maeda, H. (2010). Tumor-selective delivery of macromolecular drugs via the EPR effect: background and future prospects. *Bioconjugate Chemistry 21*, 797–802.

49. Maeda, H. (2001). SMANCS and polymer-conjugated macromolecular drugs: advantages in cancer chemotherapy. *Advanced Drug Delivery Reviews 46*, 169–185.

50. Simonsen, T. G., Gaustad, J. -V., Leinaas, M. N., Rofstad, E. K. (2012). High interstitial fluid pressure is associated with tumor-line specific vascular abnormalities in human melanoma xenografts. *PLoS ONE 7*, e40006.

51. Kratz, F., Warnecke, A. (2012). Finding the optimal balance: challenges of improving conventional cancer chemotherapy using suitable combinations with nano-sized drug delivery systems. *Journal of Controlled Release*. 23178950.

52. Heldin, C. -H., Rubin, K., Pietras, K., Ostman, A. (2004). High interstitial fluid pressure—an obstacle in cancer therapy. *Nature Reviews Cancer 4*, 806–813.

53. Yu, M., Tannock, Ian F. (2012). Targeting tumor architecture to favor drug penetration: a new weapon to combat chemoresistance in pancreatic cancer? *Cancer Cell 21*, 327–329.

54. Lee, H., Hoang, B., Fonge, H., Reilly, R., Allen, C. (2010). In vivo distribution of polymeric nanoparticles at the whole-body, tumor, and cellular levels. *Pharmaceutical Research 27*, 2343–2355.

55. Nagamitsu, A., Greish, K., Maeda, H. (2009). Elevating blood pressure as a strategy to increase tumor-targeted delivery of macromolecular drug SMANCS: cases of advanced solid tumors. *Japanese Journal of Clinical Oncology 39*, 756–766.

56. Seki, T., Fang, J., Maeda, H. (2009). Enhanced delivery of macromolecular antitumor drugs to tumors by nitroglycerin application. *Cancer Science 100*, 2426–2430.

57. Maeda, H. (2012). Vascular permeability in cancer and infection as related to macro-molecular drug delivery, with emphasis on the EPR effect for tumor-selective drug targeting. *Proceedings of the Japan Academy, Series B: Physical and Biological Sciences 88*, 53–71.

58. Gullotti, E., Yeo, Y. (2009). Extracellularly activated nanocarriers: a new paradigm of tumor targeted drug delivery. *Molecular Pharmaceutics 6*, 1041–1051.

59. Yu, M. K., Park, J., Jon, S. (2012). Targeting strategies for multifunctional nanoparticles in cancer imaging and therapy. *Theranostics 2*, 3–44.

60. Farokhzad, O. C., Langer, R. (2009). Impact of nanotechnology on drug delivery. *ACS Nano 3*, 16–20.

61. Ruoslahti, E., Bhatia, S. N., Sailor, M. J. (2010). Targeting of drugs and nanoparticles to tumors. *The Journal of Cell Biology 188*, 759–768.

62. Miele, E., Spinelli, G. P., Miele, E., Fabrizio, E. D., Ferretti, E., Tomao, S., et al. (2012). Nanoparticle-based delivery of small interfering RNA: challenges for cancer therapy. *International Journal of Nanomedicine 7*. 3637–3657.

63. Deshayes, S., Maurizot, V., Clochard, M. -C., Baudin, C., Berthelot, T., Esnouf, S., et al. (2011). "Click" conjugation of peptide on the surface of polymeric nanoparticles for targeting tumor angiogenesis. *Pharmaceutical Research 28*, 1631–1642.

64. Danhier, F., Pourcelle, V., Marchand-Brynaert, J., Jérôme, C., Feron, O., Préat, V., (2012). Targeting of tumor endothelium by RGD-grafted PLGA-nanoparticles, *Methods in Enzymology 508*, 157–175.

65. Tsourkas, A., Shinde-Patil, V. R., Kelly, K. A., Patel, P., Wolley, A., Allport, J. R., et al. (2005). In vivo imaging of activated endothelium using an anti-VCAM-1 magnetooptical probe. *Bioconjugate Chemistry 16*, 576–581.

66. Sano, K., Temma, T., Azuma, T., Nakai, R., Narazaki, M., Kuge, Y., et al. (2011). A pre-targeting strategy for MR imaging of functional molecules using dendritic Gd-based contrast agents. *Molecular Imaging and Biology 13*, 1196–1203.

67. Li, X., Qiu, L., Zhu, P., Tao, X., Imanaka, T., Zhao, J., et al. (2012). Epidermal growth factor–ferritin H-chain protein nanoparticles for tumor active targeting. *Small 8*, 2505–2514.

68. Li, Y., He, H., Jia, X., Lu, W. -L., Lou, J., Wei, Y. (2012). A dual-targeting nanocarrier based on poly(amidoamine) dendrimers conjugated with transferrin and tamoxifen for treating brain gliomas. *Biomaterials 33*, 3899–3908.

69. Zhang, H., Cai, Z., Sun, Y., Yu, F., Chen, Y., Sun, B. (2012). Folate-conjugated β-cyclodextrin from click chemistry strategy and for tumor-targeted drug delivery. *Journal of Biomedical Materials Research Part A 100*, 2441–2419.

70. Javier, D. J., Nitin, N., Levy, M., Ellington, A., Richards-Kortum, R. (2008). Aptamer-targeted gold nanoparticles as molecular-specific contrast agents for reflectance imaging. *Bioconjugate Chemistry 19*, 1309–1312.

71. Wang, Y. -C., Liu, X. -Q., Sun, T. -M., Xiong, M. -H., Wang, J. (2008). Functionalized micelles from block copolymer of polyphosphoester and poly(ε-caprolactone) for receptor-mediated drug delivery. *Journal of Controlled Release 128*, 32–40.

72. Canton, I., Battaglia, G. (2012). Endocytosis at the nanoscale. *Chemical Society Reviews 41*, 2718–2739.

73. Bartlett, D. W., Su, H., Hildebrandt, I. J., Weber, W. A., Davis, M. E. (2007). Impact of tumor-specific targeting on the biodistribution and efficacy of siRNA nanoparticles

measured by multimodality in vivo imaging. *Proceedings of the National Academy of Sciences of the United States of America 104*, 15549–15554.

74. Kirpotin, D. B., Drummond, D. C., Shao, Y., Shalaby, M. R., Hong, K., Nielsen, U. B., et al. (2006). Antibody targeting of long-circulating lipidic nanoparticles does not increase tumor localization but does increase internalization in animal models. *Cancer Research 66*, 6732–6740.

75. Torchilin, V. P., Passive and active drug targeting: drug delivery to tumors as an example drug delivery, In *Handbook of Experimental Pharmacology*, ed. Schäfer-Korting, M., Springer, Berlin, 2010, pp. 3–53.

76. Huang, R. B., Mocherla, S., Heslinga, M. J., Charoenphol, P., Eniola-Adefeso, O. (2010). Dynamic and cellular interactions of nanoparticles in vascular-targeted drug delivery (review). *Molecular Membrane Biology 27*, 312–327.

77. Desai, N. (2012). Challenges in development of nanoparticle-based therapeutics. *The AAPS Journal 14*, 282–295.

78. Vega-Villa, K. R., Takemoto, J. K., Yáñez, J. A., Remsberg, C. M., Forrest, M. L., Davies, N. M. (2008). Clinical toxicities of nanocarrier systems. *Advanced Drug Delivery Reviews 60*, 929–938.

79. Jong, W. H. D., Borm, P. J. (2008). Drug delivery and nanoparticles: applications and hazards. *International Journal of Nanomedicine 3*, 133–149.

80. Maynard, A. D., Warheit, D. B., Philbert, M. A. (2011). The new toxicology of sophisticated materials: nanotoxicology and beyond. *Toxicological Sciences 120*, S109–S129.

81. Hirn, S., Semmler-Behnke, M., Schleh, C., Wenk, A., Lipka, J., Schäffler, M., et al. (2011). Particle size-dependent and surface charge-dependent biodistribution of gold nanoparticles after intravenous administration. *European Journal of Pharmaceutics and Biopharmaceutics 77*, 407–416.

82. Huang, K., Ma, H., Liu, J., Huo, S., Kumar, A., Wei, T., et al. (2012). Size-dependent localization and penetration of ultrasmall gold nanoparticles in cancer cells, multi-cellular spheroids, and tumors in vivo. *ACS Nano 6*, 4483–4493.

83. Longmire, M. R., Ogawa, M., Choyke, P. L., Kobayashi, H. (2011). Biologically optimized nanosized molecules and particles: more than just size. *Bioconjugate Chemistry 22*, 993–1000.

84. Venkataraman, S., Hedrick, J. L., Ong, Z. Y., Yang, C., Ee, P. L. R., Hammond, P. T., et al. (2011). The effects of polymeric nanostructure shape on drug delivery. *Advanced Drug Delivery Reviews 63*, 1228–1246.

85. Cedervall, T., Lynch, I., Lindman, S., Berggård, T., Thulin, E., Nilsson, H., et al. (2007). Understanding the nanoparticle–protein corona using methods to quantify exchange rates and affinities of proteins for nanoparticles. *Proceedings of the National Academy of Sciences of the United States of America 104*, 2050–2055.

86. Walkey, C. D., Chan, W. C. W. (2012). Understanding and controlling the interaction of nanomaterials with proteins in a physiological environment. *Chemical Society Reviews 41*, 2780–2799.

87. Aggarwal, P., Hall, J. B., McLeland, C. B., Dobrovolskaia, M. A., McNeil, S. E. (2009). Nanoparticle interaction with plasma proteins as it relates to particle biodistribution, biocompatibility and therapeutic efficacy. *Advanced Drug Delivery Reviews 61*, 428–437.

88. Lartigue, L., Wilhelm, C., Servais, J., Factor, C., Dencausse, A., Bacri, J. -C., et al. (2012). Nanomagnetic sensing of blood plasma protein interactions with iron oxide nanoparticles: impact on macrophage uptake. *ACS Nano 6*, 2665–2678.

89. Schaefer, J., Schulze, C., Marxer, E. E. J., Schaefer, U. F., Wohlleben, W., Bakowsky, U., et al. (2012). Atomic force microscopy and analytical ultracentrifugation for probing nanomaterial protein interactions. *ACS Nano 6*, 4603–4614.

90. Lundqvist, M., Stigler, J., Elia, G., Lynch, I., Cedervall, T., Dawson, K. A. (2008). Nanoparticle size and surface properties determine the protein corona with possible implications for biological impacts. *Proceedings of the National Academy of Sciences of the United States of America 105*, 14265–14270.

91. Hulander, M., Lundgren, A., Berglin, M., Ohrlander, M., Lausmaa, J., Elwing, H. (2011). Immune complement activation is attenuated by surface nanotopography. *International Journal of Nanomedicine 6*, 2653–2666.

92. Blunk, T., Hochstrasser, D. F., Sanchez, J. C., Muller, B. W., Muller, R. H. (1993). Colloidal carriers for intravenous drug targeting: plasma protein adsorption patterns on surface-modified latex particles evaluated by two-dimensional polyacrylamide gel electrophoresis. *Electrophoresis 14*, 1382–1387.

93. Diederichs, J. E. (1996). Plasma protein adsorption patterns on liposomes: establishment of analytical procedure. *Electrophoresis 17*, 607–611.

94. Goppert, T. M., Muller, R. H. (2003). Plasma protein adsorption of Tween 80- and poloxamer 188-stabilized solid lipid nanoparticles. *Journal of Drug Targeting 11*, 225–231.

95. Gref, R., Lück, M., Quellec, P., Marchand, M., Dellacherie, E., Harnisch, S., et al. (2000). 'Stealth' corona-core nanoparticles surface modified by polyethylene glycol (PEG): influences of the corona (PEG chain length and surface density) and of the core composition on phagocytic uptake and plasma protein adsorption. *Colloids and Surfaces B: Biointerfaces 18*, 301–313.

96. Kim, H. R., Andrieux, K., Gil, S., Taverna, M., Chacun, H., Desmaële, D., et al. (2007). Translocation of poly(ethylene glycol-co-hexadecyl)cyanoacrylate nanoparticles into rat brain endothelial cells: role of apolipoproteins in receptor-mediated endocytosis. *Biomacromolecules 8*, 793–799.

97. Lemarchand, C., Gref, R., Passirani, C., Garcion, E., Petri, B., Müller, R., et al. (2006). Influence of polysaccharide coating on the interactions of nanoparticles with biological systems. *Biomaterials 27*, 108–118.

98. Peracchia, M. T., Harnisch, S., Pinto-Alphandary, H., Gulik, A., Dedieu, J. C., Desmaële, D., et al. (1999). Visualization of in vitro protein-rejecting properties of PEGylated stealth® polycyanoacrylate nanoparticles. *Biomaterials 20*, 1269–1275.

99. Thode, K., Lück, M., Semmler, W., Müller, R. H., Kresse, M. (1997). Determination of plasma protein adsorption on magnetic iron oxides: sample preparation. *Pharmaceutical Research 14*, 905–910.

100. AlléAaemann, E., Gravel, P., Leroux, J. -C., Balant, L., Gurny, R. (1997). Kinetics of blood component adsorption on poly(D,L-lactic acid) nanoparticles: Evidence of complement C3 component involvement. *Journal of Biomedical Materials Research 37*, 229–234.

101. Müller, R. H., Rühl, D., Lück, M., Paulke, B. R. (1997). Influence of fluorescent labelling of polystyrene particles on phagocytic uptake, surface hydrophobicity, and plasma protein adsorption. *Pharmaceutical Research 14*, 18–24.

102. Labarre, D., Vauthier, C., Chauvierre, C., Petri, B., Müller, R., Chehimi, M. M. (2005). Interactions of blood proteins with poly(isobutylcyanoacrylate) nanoparticles decorated with a polysaccharidic brush. *Biomaterials 26*, 5075–5084.

103. Göppert, T. M., Müller, R. H. (2005). Polysorbate-stabilized solid lipid nanoparticles as colloidal carriers for intravenous targeting of drugs to the brain: Comparison of plasma protein adsorption patterns. *Journal of Drug Targeting 13*, 179–187.

104. Göppert, T. M., Müller, R. H. (2005). Protein adsorption patterns on poloxamer- and poloxamine-stabilized solid lipid nanoparticles (SLN). *European Journal of Pharmaceutics and Biopharmaceutics 60*, 361–372.

105. Noh, H., Vogler, E. A. (2007). Volumetric interpretation of protein adsorption: Competition from mixtures and the Vroman effect. *Biomaterials 28*, 405–422.

106. Göppert, T. M., Müller, R. H. (2005). Adsorption kinetics of plasma proteins on solid lipid nanoparticles for drug targeting. *International Journal of Pharmaceutics 302*, 172–186.

107. Jansch, M., Stumpf, P., Graf, C., Rühl, E., Müller, R. H. (2012). Adsorption kinetics of plasma proteins on ultrasmall superparamagnetic iron oxide (USPIO) nanoparticles. *International Journal of Pharmaceutics 428*, 125–133.

108. OwensIII, D. E., Peppas, N. A. (2006). Opsonization, biodistribution, and pharmacokinetics of polymeric nanoparticles. *International Journal of Pharmaceutics 307*, 93–102.

109. Roser, M., Fischer, D., Kissel, T. (1998). Surface-modified biodegradable albumin nano- and microspheres. II: effect of surface charges on in vitro phagocytosis and biodistribution in rats. *European Journal of Pharmaceutics and Biopharmaceutics 46*, 255–263.

110. Salvador-Morales, C., Zhang, L., Langer, R., Farokhzad, O. C. (2009). Immunocompatibility properties of lipid–polymer hybrid nanoparticles with heterogeneous surface functional groups. *Biomaterials 30*, 2231–2240.

111. Gessner, A., Lieske, A., Paulke, B. R., Müller, R. H. (2002). Influence of surface charge density on protein adsorption on polymeric nanoparticles: analysis by two-dimensional electrophoresis. *European Journal of Pharmaceutics and Biopharmaceutics 54*, 165–170.

112. Gessner, A., Lieske, A., Paulke, B. -R., Müller, R. H. (2003). Functional groups on polystyrene model nanoparticles: Influence on protein adsorption. *Journal of Biomedical Materials Research Part A 65*, 319–326.

113. Xiao, K., Li, Y., Luo, J., Lee, J. S., Xiao, W., Gonik, A. M., et al. (2011). The effect of surface charge on in vivo biodistribution of PEG-oligocholic acid based micellar nanoparticles. *Biomaterials 32*, 3435–3446.

114. Yadav, K. S., Jacob, S., Sachdeva, G., Chuttani, K., Mishra, A. K., Sawant, K. K. (2011). Long circulating PEGylated PLGA nanoparticles of cytarabine for targeting leukemia. *Journal of Microencapsulation 28*, 729–742.

115. Vij, N., Min, T., Marasigan, R., Belcher, C., Mazur, S., Ding, H., et al. (2010). Development of PEGylated PLGA nanoparticle for controlled and sustained drug delivery in cystic fibrosis. *Journal of Nanobiotechnology 8*, 22.

116. Ratzinger, G., Länger, U., Neutsch, L., Pittner, F., Wirth, M., Gabor, F. (2009). Surface modification of PLGA particles: the interplay between stabilizer, ligand size, and hydrophobic interactions. *Langmuir 26*, 1855–1859.

117. Cedervall, T., Lynch, I., Foy, M., Berggård, T., Donnelly, S. C., Cagney, G., et al. (2007). Detailed identification of plasma proteins adsorbed on copolymer nanoparticles. *Angewandte Chemie International Edition 46*, 5754–5756.

118. Fang, C., Shi, B., Pei, Y. -Y., Hong, M. -H., Wu, J., Chen, H. -Z. (2006). In vivo tumor targeting of tumor necrosis factor-α-loaded stealth nanoparticles: effect of MePEG molecular weight and particle size. *European Journal of Pharmaceutical Sciences 27*, 27–36.

119. Yokoyama, M. (2010). Polymeric micelles as a new drug carrier system and their required considerations for clinical trials. *Expert Opinion on Drug Delivery 7*, 145–158.

120. Savić, R., Azzam, T., Eisenberg, A., Maysinger, D. (2006). Assessment of the integrity of poly(caprolactone)-b-poly(ethylene oxide) micelles under biological conditions: a fluorogenic-based approach. *Langmuir 22*, 3570–3578.

121. Chen, H., Kim, S., He, W., Wang, H., Low, P. S., Park, K., et al. (2008). Fast release of lipophilic agents from circulating PEG-PDLLA micelles revealed by in vivo Förster resonance energy transfer imaging. *Langmuir 24*, 5213–5217.

122. Chen, H., Kim, S., Li, L., Wang, S., Park, K., Cheng, J. -X. (2008). Release of hydrophobic molecules from polymer micelles into cell membranes revealed by Förster resonance energy transfer imaging. *Proceedings of the National Academy of Sciences of the United States of America 105*, 6596–6601.

123. Kim, S., Shi, Y., Kim, J. Y., Park, K., Cheng, J. -X. (2010). Overcoming the barriers in micellar drug delivery: loading efficiency, in vivo stability, and micelle–cell interaction. *Expert Opinion on Drug Delivery 7*, 49–62.

124. Dobrovolskaia, M. A., McNeil, S. E. (2007). Immunological properties of engineered nanomaterials. *Nature Nanotechnology 2*, 469–478.

125. Moghimi, S. M., Patel, H. M. (1998). Serum-mediated recognition of liposomes by phagocytic cells of the reticuloendothelial system—the concept of tissue specificity. *Advanced Drug Delivery Reviews 32*, 45–60.

126. Frank, M. M., Fries, L. F. (1991). The role of complement in inflammation and phagocytosis. *Immunology Today 12*, 322–326.

127. Camner, P., Lundborg, M., Låstbom, L., Gerde, P., Gross, N., Jarstrand, C. (2002). Experimental and calculated parameters on particle phagocytosis by alveolar macrophages. *Journal of Applied Physiology 92*, 2608–2616.

128. Kim, K., Kim, J. H., Park, H., Kim, Y. -S., Park, K., Nam, H., et al. (2010). Tumor-homing multifunctional nanoparticles for cancer theragnosis: Simultaneous diagnosis, drug delivery, and therapeutic monitoring. *Journal of Controlled Release 146*, 219–227.

129. Tong, L., He, W., Zhang, Y., Zheng, W., Cheng, J. -X. (2009). Visualizing systemic clearance and cellular level biodistribution of gold nanorods by intrinsic two-photon luminescence. *Langmuir 25*, 12454–12459.

130. Nagayama, S., Ogawara, K. -i., Minato, K., Fukuoka, Y., Takakura, Y., Hashida, M., et al. (2007). Fetuin mediates hepatic uptake of negatively charged nanoparticles via scavenger receptor. *International Journal of Pharmaceutics 329*, 192–198.

131. Dutta, D., Sundaram, S. K., Teeguarden, J. G., Riley, B. J., Fifield, L. S., Jacobs, J. M., et al. (2007). Adsorbed proteins influence the biological activity and molecular targeting of nanomaterials. *Toxicological Sciences 100*, 303–315.

132. Nagayama, S., Ogawara, K. -i., Fukuoka, Y., Higaki, K., Kimura, T. (2007). Time-dependent changes in opsonin amount associated on nanoparticles alter their hepatic uptake characteristics. *International Journal of Pharmaceutics 342*, 215–221.

133. Moghimi, S. M., Hedeman, H., Muir, I., Illum, L., Davis, S. (1933). An investigation of the filtration capacity and the fate of large filtered sterically-stabilized microspheres in rat spleen. *Biochimica et Biophysica Acta 1157*, 233–240.

134. Moghimi, S. M. (1995). Mechanisms of splenic clearance of blood cells and particles: towards development of new splenotropic agents. *Advanced Drug Delivery Reviews 17*, 103–115.

135. Estevanato, L., Cintra, D., Baldini, N., Portilho, F., Barosa, L., Martins, O., et al. (2011). Preliminary biocompatibility investigation of magnetic albumin nanosphere designed as a potential versatile drug delivery system. *International Journal of Nanomedicine 6*, 1709–1717.

136. Piao, L., Li, H., Teng, L., Yung, B. C., Sugimoto, Y., Brueggemeier, R. W., et al. (2013). Human serum albumin-coated lipid nanoparticles for delivery of siRNA to breast cancer. *Nanomedicine: Nanotechnology, Biology and Medicine 9*, 122–129.

137. van Rooy, I., Cakir-Tascioglu, S., Hennink, W., Storm, G., Schiffelers, R., Mastrobattista, E. (2011). In vivo methods to study uptake of nanoparticles into the brain. *Pharmaceutical Research 28*, 456–471.

138. Gaumet, M., Vargas, A., Gurny, R., Delie, F. (2008). Nanoparticles for drug delivery: the need for precision in reporting particle size parameters. *European Journal of Pharmaceutics and Biopharmaceutics 69*, 1–9.

139. Takakura, Y., Mahato, R. I., Hashida, M. (1998). Extravasation of macromolecules. *Advanced Drug Delivery Reviews 34*, 93–108.

140. Wadenvik, H., Kutti, J. (1988). The spleen and pooling of blood cells. *European Journal of Haematology 41*, 1–5.

141. Cogger, V. C., McNerney, G. P., Nyunt, T., DeLeve, L. D., McCourt, P., Smedsrød, B., et al. (2010). Three-dimensional structured illumination microscopy of liver sinusoidal endothelial cell fenestrations. *Journal of Structural Biology 171*, 382–388.

142. Fahimi, H. D., *The Liver: Biology and Pathobiology*, Raven Press, New York, 1982.

143. Ohlson, M., Sörensson, J., Haraldsson, B. (2001). A gel-membrane model of glomerular charge and size selectivity in series. *American Journal of Physiology—Renal Physiology 280*, F396–F405.

144. Conhaim, R. L., Eaton, A., Staub, N. C., Heath, T. D. (1988). Equivalent pore estimate for the alveolar-airway barrier in isolated dog lung. *Journal of Applied Physiology 64*, 1134–1142.

145. Nakaoka, R., Tabata, Y., Yamaoka, T., Ikada, Y. (1997). Prolongation of the serum half-life period of superoxide dismutase by poly(ethylene glycol) modification. *Journal of Controlled Release 46*, 253–261.

146. Moghimi, S. M. (1995). Exploiting bone marrow microvascular structure for drug delivery and future therapies. *Advanced Drug Delivery Reviews 17*, 61–73.

147. Seymour, L. W. (1992). Passive tumor targeting of soluble macromolecules and drug conjugates. *Critical Reviews in Therapeutic Drug Carrier Systems 9*, 135–187.

148. Cucullo, L., McAllister, M. S., Kight, K., Krizanac-Bengez, L., Marroni, M., Mayberg, M. R., et al. (2002). A new dynamic in vitro model for the multidimensional study of

astrocyte–endothelial cell interactions at the blood–brain barrier. *Brain Research 951*, 243–254.

149. Kas, H. S. (2004). Drug delivery to brain by microparticulate systems. *Advances in Experimental Medicine and Biology 553*, 221–230.

150. Arfors, K. E., Rutili, G., Svensiö, E. (1979). Microvascular transport of macromolecules in normal and inflammatory conditions. *Acta Physiologica Scandinavica, Supplementum 463*, 93–103.

151. Hirano, A., Kawanami, T., Llena, J. F. (1994). Electron microscopy of the blood-brain barrier in disease. *Microscopy Research and Technique 27*, 543–556.

152. Kraal, G. (1992). Cells in the marginal zone of the spleen. *International Review of Cytology 132*, 31–74.

153. Steiniger, B., Barth, P. (2000). Microanatomy and function of the spleen. *Advances in Anatomy, Embryology and Cell Biology 151*, 1–101.

154. Mebius, R. E., Kraal, G. (2005). Structure and function of the spleen. *Nature Reviews Immunology 5*, 606–616.

155. MacDonald, I. C., Ragan, D. M., Schmidt, E. E., Groom, A. C. (1987). Kinetics of red blood cell passage through interendothelial slits into venous sinuses in rat spleen, analyzed by in vivo microscopy. *Microvascular Research 33*, 118–134.

156. Groom, A. C., Schmidt, E. E., MacDonald, I. C. (1991). Microcirculatory pathways and blood flow in spleen: new insights from washout kinetics, corrosion casts, and quantitative intravital videomicroscopy. *Scanning Microscopy 5*, 159–173.

157. Chen, L. -T., Weiss, L. (1973). The role of the sinus wall in the passage of erythrocytes through the spleen. *Blood 41*, 529–537.

158. Drenckhahn, D., Wagner, J. (1986). Stress fibers in the splenic sinus endothelium in situ: molecular structure, relationship to the extracellular matrix, and contractility. *The Journal of Cell Biology 102*, 1738–1747.

159. Moghimi, S. M., Porter, C. J. H., Muir, I. S., Illum, L., Davis, S. S. (1991). Non-phagocytic uptake of intravenously injected microspheres in rat spleen: influence of particle size and hydrophilic coating. *Biochemical and Biophysical Research Communications 177*, 861–866.

160. Demoy, M., Andreux, J. -P., Weingarten, C., Gouritin, B., Guilloux, V., Couvreur, P. (1999). Spleen capture of nanoparticles: influence of animal species and surface characteristics. *Pharmaceutical Research 16*, 37–41.

161. Marieb, E. N., *Essentials of Human Anatomy and Physiology*, Benjamin Cummings, San Francisco, 2002.

162. Adams, D. H., Eksteen, B. (2006). Aberrant homing of mucosal T cells and extra-intestinal manifestations of inflammatory bowel disease. *Nature Reviews Immunology 6*, 244–251.

163. Dan, C., Wake, K. (1985). Modes of endocytosis of latex particles in sinusoidal endothelial and Kupffer cells of normal and perfused rat liver. *Experimental Cell Research 158*, 75–85.

164. Wisse, E., *Ultrastructure and function of Kupffer cells and other sinusoidal cells in the liver. In: Kupffer cells and other liver sinusoidal cells*, Elsevier, Amsterdam, 1977.

165. Kuntz, E., Kuntz, H. -D. *Hepatology: Principles and Practice: History, Morphology, Biochemistry, Diagnostics, Clinic, Therapy*, Springer, Berlin, 2006.

166. Vidal-Vanaclocha, F., Barberá-Guillem, E. (1985). Fenestration patterns in endothelial cells of rat liver sinusoids. *Journal of Ultrastructure Research 90*, 115–123.

167. Fraser, R., Dobbs, B. R., Rogers, G. W. (1995). Lipoproteins and the liver sieve—the role of the fenestrated sinusoidal endothelium in lipoprotein metabolism, atherosclerosis, and cirrhosis. *Hepatology 21*, 863–874.

168. Sun, X., Rossin, R., Turner, J. L., Becker, M. L., Joralemon, M. J., Welch, M. J., et al. (2005). An assessment of the effects of shell cross-linked nanoparticle size, core composition, and surface PEGylation on in vivo biodistribution. *Biomacromolecules 6*, 2541–2554.

169. Kren, B. T., Unger, G. M., Sjeklocha, L., Trossen, A. A., Korman, V., Diethelm-Okita, B. M., et al. (2009). Nanocapsule-delivered Sleeping Beauty mediates therapeutic Factor VIII expression in liver sinusoidal endothelial cells of hemophilia A mice. *The Journal of Clinical Investigation 119*, 2086–2099.

170. Glaus, C., Rossin, R., Welch, M. J., Bao, G. (2010). In vivo evaluation of 64Cu-labeled magnetic nanoparticles as a dual-modality PET/MR imaging agent. *Bioconjugate Chemistry 21*, 715–722.

171. Schluep, T., Hwang, J., Hildebrandt, I. J., Czernin, J., Choi, C. H. J., Alabi, C. A., et al. (2009). Pharmacokinetics and tumor dynamics of the nanoparticle IT-101 from PET imaging and tumor histological measurements. *Proceedings of the National Academy of Sciences of the United States of America 106*, 11394–11399.

172. Hoshino, Y., Koide, H., Furuya, K., Haberaecker, W. W., Lee, S. -H., Kodama, T., et al. (2012). The rational design of a synthetic polymer nanoparticle that neutralizes a toxic peptide in vivo. *Proceedings of the National Academy of Sciences of the United States of America 109*, 33–38.

173. Choi, H. S., Liu, W., Liu, F., Nasr, K., Misra, P., Bawendi, M. G., et al. (2010). Design considerations for tumour-targeted nanoparticles. *Nature Nanotechnology 5*, 42–47.

174. Nishimori, H., Kondoh, M., Isoda, K., Tsunoda, S. -i., Tsutsumi, Y., Yagi, K. (2009). Silica nanoparticles as hepatotoxicants. *European Journal of Pharmaceutics and Biopharmaceutics 72*, 496–501.

175. Popielarski, S. R., Hu-Lieskovan, S., French, S. W., Triche, T. J., Davis, M. E. (2005). A nanoparticle-based model delivery system to guide the rational design of gene delivery to the liver. 2. In vitro and in vivo uptake results. *Bioconjugate Chemistry 16*, 1071–1080.

176. Kumar, R., Roy, I., Ohulchanskky, T. Y., Vathy, L. A., Bergey, E. J., Sajjad, M., et al. (2010). In vivo biodistribution and clearance studies using multimodal organically modified silica nanoparticles. *ACS Nano 4*, 699–708.

177. Longmire, M., Choyke, P. L., Kobayashi, H. (2008). Clearance properties of nano-sized particles and molecules as imaging agents: considerations and caveats. *Nanomedicine 3*, 703–717.

178. Smoyer, W. E., Mundel, P. (1998). Regulation of podocyte structure during the development of nephrotic syndrome. *Journal of Molecular Medicine 76*, 172–183.

179. Deen, W. M., Lazzara, M. J., Myers, B. D. (2001). Structural determinants of glomerular permeability. *American Journal of Physiology—Renal Physiology 281*, F579–F596.

180. Kobayashi, H., Brechbiel, M. W. (2004). Dendrimer-based nanosized MRI contrast agents. *Current Pharmaceutical Biotechnology 5*, 539–549.

181. Kobayashi, H., Brechbiel, M. W. (2005). Nano-sized MRI contrast agents with dendrimer cores. *Advanced Drug Delivery Reviews 57*, 2271–2286.

182. Choi, H. S., Ashitate, Y., Lee, J. H., Kim, S. H., Matsui, A., Insin, N., et al. (2010). Rapid translocation of nanoparticles from the lung airspaces to the body. *Nature Biotechnology 28,* 1300–1303.

183. Kobayashi, H., Le, N., Kim, I. -s., Kim, M. -K., Pie, J. -E., Drumm, D., et al. (1999). The pharmacokinetic characteristics of glycolated humanized anti-Tac Fabs are determined by their isoelectric points. *Cancer Research 59,* 422–430.

184. Venturoli, D., Rippe, B. (2005). Ficoll and dextran vs. globular proteins as probes for testing glomerular permselectivity: effects of molecular size, shape, charge, and deformability. *American Journal of Physiology—Renal Physiology 288,* F605–F613.

185. van Nostrum, C. F. (2011). Covalently cross-linked amphiphilic block copolymer micelles. *Soft Matter 7,* 3246–3259.

186. Cheng, R., Feng, F., Meng, F., Deng, C., Feijen, J., Zhong, Z. (2011). Glutathione-responsive nano-vehicles as a promising platform for targeted intracellular drug and gene delivery. *Journal of Controlled Release 152,* 2–12.

187. Sun, Y., Yan, X., Yuan, T., Liang, J., Fan, Y., Gu, Z., et al. (2010). Disassemblable micelles based on reduction-degradable amphiphilic graft copolymers for intracellular delivery of doxorubicin. *Biomaterials 31,* 7124–7131.

188. Sun, P., Zhou, D., Gan, Z. (2011). Novel reduction-sensitive micelles for triggered intracellular drug release. *Journal of Controlled Release 152,* Supplement 1, e85–e87.

189. Kim, J. O., Sahay, G., Kabanov, A. V., Bronich, T. K. (2010). Polymeric micelles with ionic cores containing biodegradable cross-links for delivery of chemotherapeutic agents. *Biomacromolecules 11,* 919–926.

190. O'Reilly, R. K., Hawker, C. J., Wooley, K. L. (2006). Cross-linked block copolymer micelles: functional nanostructures of great potential and versatility. *Chemical Society Reviews 35,* 1068–1083.

191. Li, Y., Xiao, K., Luo, J., Xiao, W., Lee, J. S., Gonik, A. M., et al. (2011). Well-defined, reversible disulfide cross-linked micelles for on-demand paclitaxel delivery. *Biomaterials 32,* 6633–6645.

192. Koo, A. N., Min, K. H., Lee, H. J., Lee, S. -U., Kim, K., Chan Kwon, I., et al. (2012). Tumor accumulation and antitumor efficacy of docetaxel-loaded core-shell-corona micelles with shell-specific redox-responsive cross-links. *Biomaterials 33,* 1489–1499.

193. Choi, W. I., Yoon, K. C., Im, S. K., Kim, Y. H., Yuk, S. H., Tae, G. (2010). Remarkably enhanced stability and function of core/shell nanoparticles composed of a lecithin core and a pluronic shell layer by photo-crosslinking the shell layer: In vitro and in vivo study. *Acta Biomaterialia 6,* 2666–2673.

194. Choi, W. I., Lee, J. H., Kim, J. -Y., Kim, J. -C., Kim, Y. H., Tae, G. (2012). Efficient skin permeation of soluble proteins via flexible and functional nano-carrier. *Journal of Controlled Release 157,* 272–278.

195. Moghimi, S. M., Szebeni, J. (2003). Stealth liposomes and long circulating nano-particles: critical issues in pharmacokinetics, opsonization and protein-binding properties. *Progress in Lipid Research 42,* 463–478.

196. Chen, J., Yan, G. -j., Hu, R. -r., Gu, Q. -w., Chen, M. -l., Gu, W., et al. (2012). Improved pharmacokinetics and reduced toxicity of brucine after encapsulation into stealth liposomes: role of phosphatidylcholine. *International Journal of Nanomedicine 7,* 3567–3577.

197. Rojnik, M., Kocbek, P., Moret, F., Compagnin, C., Celotti, L., Bovis, M. J., et al. (2012). In vitro and in vivo characterization of temoporfin-loaded PEGylated PLGA nanoparticles for use in photodynamic therapy. *Nanomedicine 7,* 663–677.

198. Jiang, W., Lionberger, R., Yu, L. X. (2011). In vitro and in vivo characterizations of PEGylated liposomal doxorubicin. *Bioanalysis 3*, 333–344.

199. Ragi, C., Sedaghat-Herati, M. R., Ouameur, A. A., Tajmir-Riahi, H. A. (2005). The effects of poly(ethylene glycol) on the solution structure of human serum albumin. *Biopolymers 78*, 231–236.

200. Topchieva, I. N., Sorokina, E. M., Efremova, N. V., Ksenofontov, A. L., Kurganov, B. I. (1999). Noncovalent adducts of poly(ethylene glycols) with proteins. *Bioconjugate Chemistry 11*, 22–29.

201. Furness, E. L., Ross, A., Davis, T. P., King, G. C. (1998). A hydrophobic interaction site for lysozyme binding to polyethylene glycol and model contact lens polymers. *Biomaterials 19*, 1361–1369.

202. Yokoe, J. -i., Sakuragi, S., Yamamoto, K., Teragaki, T., Ogawara, K. -i., Higaki, K., et al. (2008). Albumin-conjugated PEG liposome enhances tumor distribution of liposomal doxorubicin in rats. *International Journal of Pharmaceutics 353*, 28–34.

203. Xie, J., Wang, J., Niu, G., Huang, J., Chen, K., Li, X., et al. (2010). Human serum albumin coated iron oxide nanoparticles for efficient cell labeling. *Chemical Communications 46*, 433–435.

204. Wu, Y., Chakrabortty, S., Gropeanu, R. A., Wilhelmi, J., Xu, Y., Er, K. S., et al. (2010). pH-responsive quantum dots via an albumin polymer surface coating. *Journal of the American Chemical Society 132*, 5012–5014.

205. Acharya, G., Shin, C. S., McDermott, M., Mishra, H., Park, H., Kwon, I. C., et al. (2010). The hydrogel template method for fabrication of homogeneous nano/microparticles. *Journal of Controlled Release 141*, 314–319.

206. Merkel, T. J., Chen, K., Jones, S. W., Pandya, A. A., Tian, S., Napier, M. E., et al. (2012). The effect of particle size on the biodistribution of low-modulus hydrogel PRINT particles. *Journal of Controlled Release 162*, 37–44.

207. Champion, J. A., Mitragotri, S. (2006). Role of target geometry in phagocytosis. *Proceedings of the National Academy of Sciences of the United States of America 103*, 4930–4934.

208. Sharma, G., Valenta, D. T., Altman, Y., Harvey, S., Xie, H., Mitragotri, S., et al. (2010). Polymer particle shape independently influences binding and internalization by macrophages. *Journal of Controlled Release 147*, 408–412.

209. Geng, Y., Dalhaimer, P., Cai, S., Tsai, R., Tewari, M., Minko, T., et al. (2007). Shape effects of filaments versus spherical particles in flow and drug delivery. *Nature Nanotechnology 2*, 249–255.

210. Doshi, N., Zahr, A. S., Bhaskar, S., Lahann, J., Mitragotri, S. (2009). Red blood cell-mimicking synthetic biomaterial particles. *Proceedings of the National Academy of Sciences of the United States of America 106*, 21495–21499.

211. Hendrickson, G. R., Lyon, L. A. (2010). Microgel translocation through pores under confinement. *Angewandte Chemie International Edition 49*, 2193–2197.

212. Banquy, X., Suarez, F., Argaw, A., Rabanel, J. -M., Grutter, P., Bouchard, J.-F., et al. (2009). Effect of mechanical properties of hydrogel nanoparticles on macrophage cell uptake. *Soft Matter 5*, 3984–3991.

213. Bae, Y. H., Park, K. (2011). Targeted drug delivery to tumors: myths, reality and possibility. *Journal of Controlled Release 153*, 198–205.

214. Lee, S.-Y., Kim, S., Tyler, J. Y., Park, K., Cheng, J.-X. (2013). Blood-stable, tumor-adaptable disulfide bonded mPEG-(Cys)4-PDLLA micelles for chemotherapy. *Biomaterials 34*, 552–561.

10

DRUG DELIVERY STRATEGIES FOR COMBATING MULTIPLE DRUG RESISTANCE

JOSEPH W. NICHOLS

Department of Bioengineering, University of Utah, Salt Lake City, UT, USA

YOU HAN BAE

Department of Pharmaceutics, University of Utah, Salt Lake City, UT, USA

10.1 INTRODUCTION

It has been 40 years since the United States Congress passed the National Cancer Act, which directed a great deal of effort into research combating cancer. Since that time, only modest gains have been made in reducing cancer mortality rates in the United States and elsewhere, and the most significant gains have come as a result of prevention campaigns and declining smoking rates [1,2]. Table 10.1 shows that death rates for many cancers have actually increased since 1975 due to rising incidence rates. A major challenge facing researchers has been multiple drug resistance (MDR) in solid tumors [3–6]. Multiple drug resistance is defined as the simultaneous resistance to a wide range of drugs that are unrelated in structure or mechanism of action, and is encountered in up to 80% of nonresectable and 40% of all cancer cases [7,8]. The presence of MDR markers in a patient is strongly associated with poor clinical outcomes [9]. The mechanisms of MDR are extremely varied and can include both chemical and physical barriers to drug action. An MDR tumor generally displays some combination of several MDR phenotypes. The expression of these phenotypes may also vary from cell to cell within a tumor [10].

Nanoparticulate Drug Delivery Systems: Strategies, Technologies, and Applications, First Edition.
Edited by Yoon Yeo.

TABLE 10.1   Cancer Incidence, Survival, and Death Rates in the United States for the Years from the Mid-1970s to Early 2000s

	Incidence Rates[a]		Mortality Rates[b]		Survival Rates[a]	
	1975	2003	1975[c]	2000–2004	1975 (%)	2003 (%)
All sites	400.4	472.9	162.2	192.7	48.9	66.7
Brain, nervous system	5.9	6.6	3.8	4.4	23.0	33.7
Breast	105.1	126.5	14.6	14.5	75.5	89.9
Colon and rectum	59.5	50.6	21.7	19.4	48.7	65.6
Leukemia	12.8	13.3	6.6	7.5	33.4	55.6
Lung and bronchus	52.3	64.5	36.8	54.7	11.5	16.0
Non-Hodgkin lymphoma	11.1	20.6	4.7	7.6	46.0	70.2
Ovary	16.3	13.6	4.7	5.0	33.7	44.1
Pancreas	11.8	11.6	8.6	10.6	3.0	4.8
Prostate	94.0	169.0	8.4	10.5	66.4	99.4

Incidence and mortality rates reported as per 100,000 population per year. Survival rates are the percent of patients alive 5 years from diagnosis.
[a]Reference [1].
[b]Reference [2].
[c]As reported in Reference [11].

Many innovative strategies are in development to overcome individual resistance mechanisms. Most modern drug delivery technologies use nanoparticles as the main delivery platform because of their versatility in both form and function. New treatments are frequently successful in overcoming individual drug delivery barriers (such as resistance mechanisms) and show promise in preclinical studies, but successful application of these technologies in a clinical setting has been rare. Transferring new technologies from laboratory to clinic may require a broader perspective in formulating a drug delivery strategy. Drug carrier characteristics should be optimized to deliver the highest concentration possible from drug administration to drug action, rather than narrowly focus on singular characteristics.

This chapter outlines major mechanisms of drug resistance in tumors, emphasizing the contribution of each mechanism to the whole classification called MDR. The chapter then samples innovative strategies to circumvent or overcome these resistance mechanisms.

10.2   PATHWAYS TO DRUG RESISTANCE

An effective cancer treatment must deliver sufficient quantities of drug to its site of action for a sufficiently long period of time to allow drug action to be effective. Otherwise, the treatment will fail to annihilate the cancer regardless of the drug's predicted effectiveness. Achieving a lethal drug dosage at the intracellular site of action for all cancerous cells in a tumor is the perennial challenge in applied drug delivery research. The difficulty of drug delivery arises from the relatively large

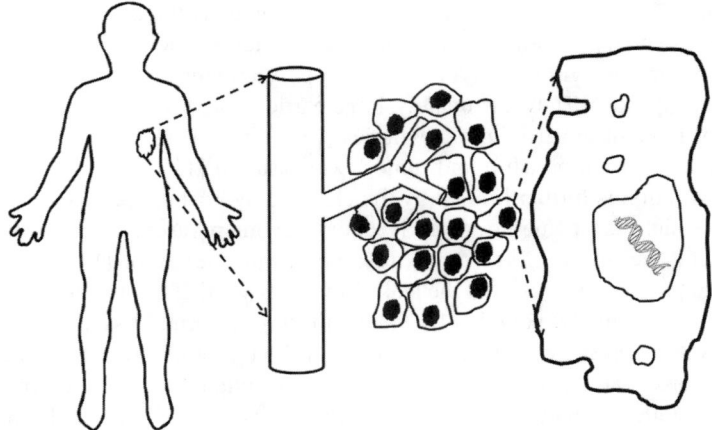

FIGURE 10.1   Before reaching an intracellular site of action, a drug must first pass through several stages of drug distribution. The first is systemic circulation, which distributes the drug carriers throughout the entire circulatory system. In this stage, carriers are subject to clearance through renal filtration or the MPS. A small number of particles will avoid clearance and extravasate out of the disorganized neovasculature into the interstitial space of the tumor. Drug carrier concentration in the tumor may be high relative to other tissues due to the EPR effect but most of the administered drug will not encounter the tumor. Once in the tumor the drug carriers must pass through the tumor interstitial space and extracellular matrix to encounter a cell. The final phase of distribution involves crossing the lipid bilayer membrane to enter the cell then diffusing to the site of action. Drugs targeted to the nucleus must also enter the nuclear envelope. (*Source:* Adapted from Reference [10] and modified.)

distances and diverse environments through which an individual drug particle must traverse before reaching its target. The total path of a drug can be summarized in Figure 10.1, showing the various environments a particle must travel through before an intracellular target may be reached.

10.2.1   Resistance of the Tumor Microenvironment

A great deal of drug resistance is conferred to tumors by characteristics of the tumor microenvironment which prevent a drug from ever reaching the subcellular site of action. Most cancer therapeutics are administered via intravenous injection and within minutes are distributed throughout the entire circulatory system. The first major barrier to drug delivery is to move the drug out of the circulatory system and into the tumor before the drug can be cleared by renal filtration [12] and the monocyte phagocyte system (MPS) [13]. Long circulating particles have the opportunity to collect in the tumor by means of the enhanced permeability and retention (EPR) effect, which is the tendency of drugs to preferentially accumulate in a solid tumor due to the tumor's leaky vasculature and poor lymphatic drainage [14,15]. The EPR effect is widely used in modern drug therapeutics, but its usefulness can be mitigated by several countervailing trends [16].

To take advantage of the EPR effect, a drug delivery system must be long circulating, which generally means using large particles which are able to avoid renal filtration modified with a polymer such as poly(ethylene glycol) (PEG) to slow phagocytic uptake [17]. Unfortunately, large particles are less able to extravasate into the tumor interstitium from the capillaries [18]. Thus, there is a trade off between extended circulation half-life and tumor extravasation [19].

Extravasation is further limited by the elevated hydrostatic pressure of the tumor interior, which can sometimes nearly equal the microvascular pressure, severely diminishing the driving force for extravasation into the tumor [17]. This elevated interstitial pressure is a result of the leaky blood vessels that freely allow proteins and other solutes normally excluded by the capillaries of normal tissues. This causes a higher osmotic pressure in interstitial space of the tumor and results in edema [20]. The additional water cannot be cleared because the lymph vessels are often destroyed by a high static pressure resulting from rapidly dividing cells in the limited tumor space [21]. Volume expansion cannot equalize the pressure because of the dense highly cross-linked extracellular matrix (ECM) of the tumor [22].

The pressure at the tumor periphery is lower than that in the center of the tumor creating a driving force for convective flow from the tumor center to the outside [23]. The vasculature is not uniformly distributed in the tumor but is mostly concentrated at the tumor periphery. This is because the solid stress of the tumor destroys the original blood vessels [24], which must be replaced by new vasculature through angiogenesis. These new vessels form from existing vessels outside the tumor and are disorganized, rarely penetrate to the tumor center, and leave large spaces between capillaries [25–27]. This distance becomes a large diffusive barrier and results in chronically hypoxic and undernourished regions in the tumor, which can be seen in Figure 10.2 [28], which is aggravated by the convective flow away from the center of the tumor. This presents a critical challenge for researchers to consider because the conditions of the hypoxic regions select for the most dangerous cell phenotypes such as aggressive growth and metastatic behavior, but these are the very cells least likely to be exposed to lethal drug concentrations [29].

Diffusion is made more difficult by the dense collagen network of the tumor, which may slow or even stop the movement of large particles entirely [30,31]. In this capacity, the collagen matrix can be considered a gate or filter, physically excluding some particles based on size. Those particles that are not excluded must travel a highly tortuous path navigating both the collagen fibers and the extremely dense cell mass. This tends to exacerbate the distance barrier to particle diffusion.

10.2.2   Cellular Resistance

Once a drug reaches its site of action, it must enter the cell and reach the site of action before it can be effective. Drug-resistant cells have several mechanisms that can prevent the drug from entering the cell and distributing to the target site. One of the principal and best understood mechanisms of MDR is the active efflux of small molecule drugs by membrane-bound, nonspecific, ATP-driven molecular pumps, of which P-glycoprotein (Pgp) is the most common [32–35]. Pgp pumps are normally

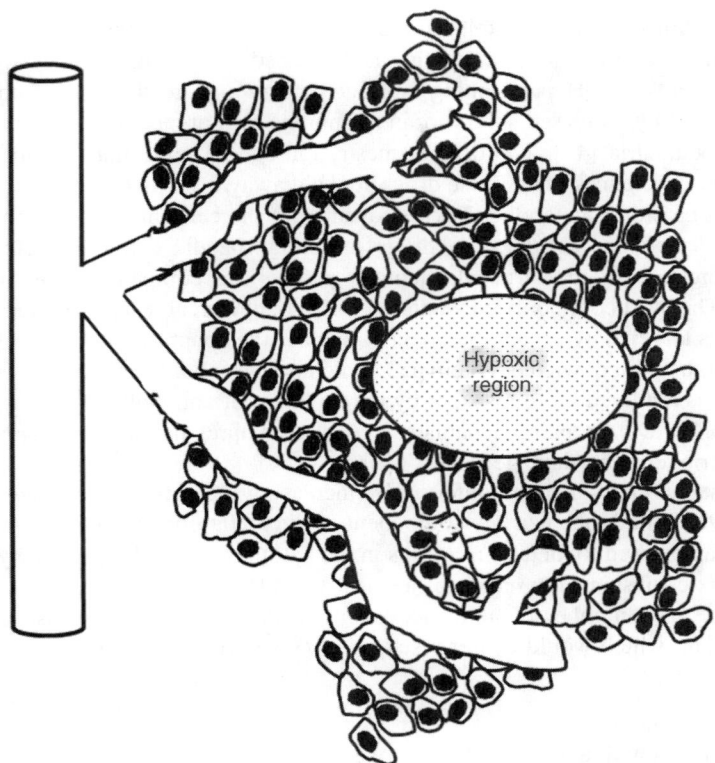

FIGURE 10.2 Angiogenesis originates from existing blood vessels outside the tumor. New blood vessels tend to be highly irregular and disorganized; they also permeate primarily at the tumor periphery rather than penetrating directly into the tumor interior. This leaves some cells distant from the nearest capillaries creating hypoxic regions in the interior regions of the tumor. Also note that the pressure gradient runs toward the outside of the tumor, so a slight convective flow is expected to run away from the tumor core.

found in the liver, pancreas, kidney, colon, testes, and jejunum of healthy tissues and play an important role in protecting and detoxifying cells commonly exposed to toxic environments [36,37]. It is part of a large family of proteins with similar functions encoded by the MDR1 gene known as the ATP-binding cassette (ABC) family [38]. Pgp can transport a large variety of unrelated compounds from the membrane to the cell exterior and is capable of maintaining large concentration gradients across the cell membrane [39,40].

Another mechanism for drug efflux not related to the ABC pumps is the lung resistance-related protein (LRP) most commonly found in lung carcinomas [41]. LRP is not a membrane pump, but rather a major vault protein. Vaults are thought to be associated with the nuclear pore complex and regulate entry into the nucleus. Drugs can be sequestered into a vault at the nuclear pore complex and then transported into the cytoplasm where it is subsequently exocytosed [42].

Many drugs may also be sequestered into a number of intracellular compartments especially the mitochondria, Golgi, and endosomes. Drugs collect into these compartments by pH partitioning and active pumping in the case of endosomal vesicles and by increased solubility in membrane-rich organelles such as the Golgi and mitochondria [43]. The drug sequestration can be quite marked due to large solubility differences [44]. These drugs are held away from their sites of action until they can be either metabolized or removed from the cell entirely via exocytosis.

Cells can gain resistance to drugs targeting specific signaling pathways by modifying part of the targeted chemical pathway to compensate for the drug's action. The most commonly encountered forms of chemical resistance modulate key pathways targeted by cancer therapeutics such as the apoptosis pathway. Almost all cancers must shut down the protein p53, which is a key choke point in apoptosis [45]. The cell is rendered more resistant by upregulation of antiapoptosis factors such as bcl2 [46,47]. DNA repair mechanisms may also be upregulated to weaken apoptotic signals triggered by damaged DNA [48].

Resistance may also be conferred by increasing the cellular metabolism of the drug [49]. Glutathione is a key component of the glutathione detoxification system, which can be used to metabolize drugs in the cytoplasm [50]. It operates by binding to the toxic substance by means of a sulfur bond [51]. This mechanism can confer resistance to a number of anticancer compounds [52–55]. It can also act as an antioxidant, which would confer resistance toward drugs designed to generate free radicals [51].

10.2.3 Perspective on Drug Resistance

This review of MDR mechanisms is not meant to be comprehensive and does not provide depth on any of the mechanisms described. The intent of this brief description of MDR mechanisms is to provide perspective on the breadth of the challenge in designing a drug delivery system to overcome the diverse mechanisms of resistance. Any approach to combat MDR that only takes a single mechanism into account is unlikely to succeed in a clinical setting. Oftentimes, a delivery strategy targeted to overcome one MDR mechanism will leave it especially vulnerable to another. For example, using large particles as drug carriers will help increase circulation time and accumulation in the tumor; however, once in the tumor the diffusion of large particles is severely limited and is unlikely to reach the majority of tumor cells. Large particles become helpful again as a way to circumvent Pgp pumps because they can cross the cell membrane via endocytosis rather than diffusion. All drug delivery systems must compromise between these countervailing effects.

Different cells within the same tumor will have different phenotypes, express Pgp at varying levels, have different apoptotic thresholds, or metabolize toxins at different rates [56]. In the face of treatment, natural selection favors resistant cell types upon tumor recurrence [57]. Any plausible technology to combat MDR should account for the diversity of mechanisms present in a single tumor and seek to combat the cancer as holistically as possible. Otherwise, each successive treatment will increase the risk of a more resistant recurrent tumor.

10.3   COMBATING MDR

The diversity of MDR mechanisms requires that any drug delivery system designed to combat MDR be extremely versatile. Nanoparticles are among the most versatile tools available to researcher for combating cancer [58]. Platforms for nanotherapeutics can take many forms as shown in Figure 10.3. The different classes of these nanoparticles include dendrimers [59], linear polymeric carriers [60], inorganic nanoparticles [61], micelles [62], polymersomes [63], and liposomes [64]. Attaching drugs to or encapsulating them in nanocarriers helps solubilize and stabilize most drug formulations [65]. The large particle size of drug delivery systems excludes clearance by renal filtration, which increases circulation time, allowing for drug

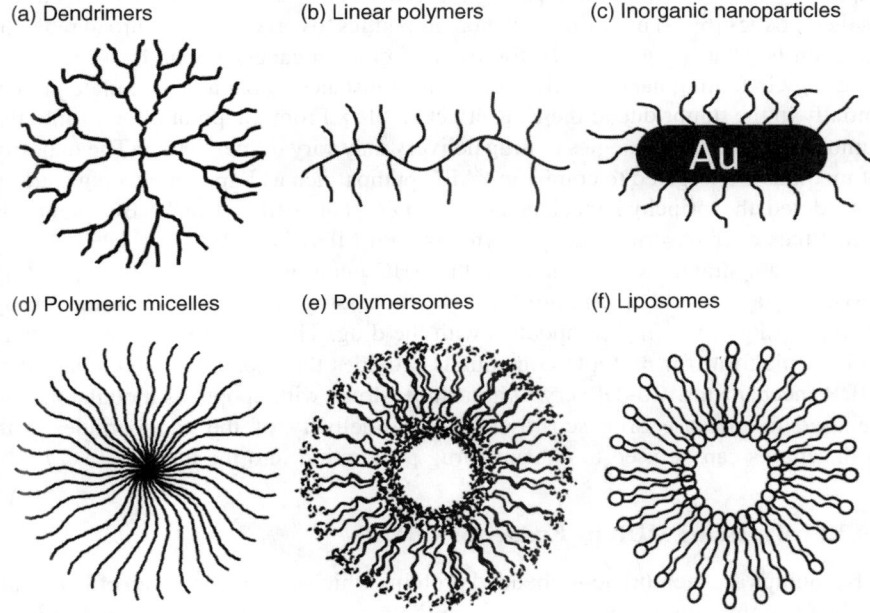

(a) Dendrimers   (b) Linear polymers   (c) Inorganic nanoparticles

(d) Polymeric micelles   (e) Polymersomes   (f) Liposomes

FIGURE 10.3   Various physical forms of nanoparticles used in cancer therapy. (a) Dendrimers are highly branched polymers that form into a tree or bush-like structure. (b) Linear macromolecular carriers are composed of a degradable polymer backbone to which drugs or targeting moieties can be attached. (c) Inorganic nanoparticles can be made from various metals or other compounds such as SiO; the surface of these particles is often modified with PEG, targeting moieties or drugs. (d) Micelles are often composed of block copolymers bearing a hydrophilic and hydrophobic region. The hydrophobic region of these polymers interacts in aqueous environments forming a stable core into which drugs can be loaded. The hydrophilic region of the polymer can be modified in other ways. (e) Polymersomes are generally made from triblock copolymers that assemble to form a vesicular structure with a hydrophilic core. Compounds may be loaded into both the hydrophilic and hydrophobic regions. (f) Liposomes are similar in structure to polymersomes but are made using phospholipids similar to those in natural lipid bilayer membranes.

accumulation through the EPR effect. The circulation time of nano-sized drug carriers can be further enhanced by modifying the surface to inhibit MPS uptake. The most common and effective surface modification for increasing circulation time is grafting PEG on the particle surface. PEG is a highly hydrophilic polymer that extends out and sterically prevents proteins from adsorbing the particle surface and MPS uptake [66]. Other modifications can also be added to a nanoparticle to confer the desired properties. Targeting moieties are a common modification designed to increase tumor localization and cellular uptake [67]. Drug can also be attached to the surface of the particle or be stored in the internal hydrophobic or hydrophilic compartments. This versatility makes co-delivery of multiple compounds with different characteristics possible [68]. In addition to drug compounds, dyes or other imaging agents can also be included with the drug carrier. This provides the capability to image the tumor during the course of treatment, a capability known as theranostics [69]. The ability of nanotherapeutics to carry on such a broad range of function is what brings it to the forefront of modern cancer research.

Long circulating nanoparticles are, in most instances, able to accumulate preferentially in the tumor due to the EPR effect [70,16]. From the point of entry into the tumor, however, the strategies of drug delivery may vary tremendously. The majority of methods are targeted to combating ABC pumps such as Pgp, which are generally considered the principal mechanism of MDR [11]. Efflux from ABC pumps is sometimes even described as synonymous with MDR [71,72].

The main strategies for combating the ABC pump mechanism of multiple drug resistance are to either circumvent it by bypassing membrane diffusion or by delivering Pgp disabling compounds with the drug. The capacity of a nanoparticle to be a platform for multiple compounds provides the capability to combat other MDR mechanisms. Co-delivery of anticancer drugs with apoptosis modulators can be used to increase drug sensitivity, and co-delivery of the nanoparticles with collagenases can be used to enhance drug penetration into the tumor.

10.3.1 Bypassing MDR by Endocytosis

ABC pumps are membrane-embedded proteins which capture and then efflux small molecule compounds while dissolved in the membrane. The process of diffusing across the membrane exposes small molecular weight drugs to efflux [73]. Encapsulating the drug in a large particle allows bulk uptake through endocytic pathways, thus bypassing MDR pumps such as Pgp. Bringing the intact drug carrier into the cytoplasm through endocytosis and phagocytosis enables the drug to be released directly into the cytoplasm resulting in a higher sustained cytoplasmic concentration [74].

A major challenge to achieving sufficiently cytotoxic-free drug concentrations is for the cell to uptake a sufficient quantity of drug carriers. The rate of endocytosis is dependent on a number of factors including size, shape, and particle–membrane interactions. Optimal particle uptake appears to occur for particles in the 30–50 nm diameter range [75,76]. The aspect ratio of the particles also plays a role in active uptake by the cell. Studies done with silica nanoparticles have shown that slightly elongated particles (with an aspect ratio of 2.1–2.5) show the highest cellular uptake rates [77,78].

Active targeting improves cellular uptake by conjugating ligands to the nanoparticle, which will interact with receptors on the cell membrane. These targeting moieties can increase the frequency and strength of membrane interactions, giving greater likelihood of internalization. Targeting moieties are often antibodies against proteins that are generally overexpressed on cancer cells. A broad range of antibodies can be used depending on the cancer type [79,80]. Among the most common targets for active targeting are folate receptors [81] and Her2 receptor [82], which are overexpressed in many cancers. Targeting to specific protein receptors suffers from the differential expression of these receptors within a single tumor. An example of this phenomenon is the drug Trastuzumab (Herceptin®) which targets the Her2 antigen which is overexpressed in some 20–30% breast cancer patients. The use of this treatment governed by the immunohistochemistry (IHC) scoring system which rates the overall Her2 expression in a tumor. The highest score under this system is assigned to tumors in which only 10% of tumor cells stain strongly for Her2. Among patients with the highest levels of Her2 expression, only half appear to benefit from Herceptin targeted treatment [83]. Other targeted cancer treatments also must labor under this burden of differential protein expression in cancer cells.

Nonspecific targeting can also be used to enhance cellular uptake. The cell penetrating peptide TAT is associated with HIV infectivity and allows it to infect a wide variety of cells [84]. The effectiveness of TAT at entering cells has led to its use as a means of encouraging endocytosis in drug delivery [85]. TAT works in an enormous variety of cells, but this lack of specificity risks increasing the nonspecific toxicity of the drug [86]. Specificity can be increased (and negative side effects reduced) by shielding the TAT until the drug carrier reaches the tumor site. Deshielding can be tied to environmental stimuli unique to the tumor, such as acidity or hypoxia [87].

The effectiveness of a drug carrier is in part dependent on the mechanism by which it is internalized. Nanoparticles can enter the cell in a variety of ways including diffusion, clathrin-mediated endocytosis, caveolae-mediated endocytosis, and pinocytosis [88]. The means by which the particle enters the cell will affect the drug release and intracellular distribution of the drug. Particles that diffuse across the membrane will emerge near the membrane on the intracellular side. Drug release then is likely to occur close to the membrane where the drug molecules can then reenter the membrane by diffusion and be effluxed [89]. Most drug carriers are actively brought into the cell through either endocytosis or phagocytosis and then enter the endocytic pathway [90]. The lysosome is the end stage of the endocytic pathway and presents a severe problem for drug release. The harsh, acidic, and degradative environment in the lysosome is designed to degrade any phagocytosed particles before they are allowed entry into the cytosol to prevent harm to the cell [91].

Avoiding the lysosome is critical for effective delivery from endocytosed particles. Caveolae-mediated endocytosis operates with a unique pathway that may not involve the lysosome [92,93], suggesting a strategy for circumventing drug destruction by targeting the caveolae-mediated pathway [94]. Targeting a single endocytic pathway can be challenging; pinocytosis is a common uptake mechanism and is nonspecific. Unfortunately, the data was unclear as to whether the lysosomal phase is actually avoided [95,96]. If the lysosome is the end result of all endocytic pathways, it is necessary to design the drug delivery system to exit the endosome before they can merge with the lysosome. pH-sensitive micelles provide one strategy for endosomal escape [97,98]. Micelles can be tuned to break up at the endosomal pH releasing the contents into the vesicle. The sudden release of solute drug and micelle polymer in the endosome causes a spike in osmotic pressure driving water to rush into the vesicle. The vesicle may then swell to the breaking point, emptying the contents of the vesicle directly into the cytoplasm.

10.3.2   Co-Delivery of Drugs with Other Compounds

One of the major strengths of using nanoparticles as a platform for drug delivery is the ability to deliver finely tuned doses of multiple drugs into the local environment. Many nanoparticle drug formulations have been designed and studied to target MDR by co-delivering chemotherapeutic agents with chemosensitizers, oligonucleotides, proteins, and so on. Co-delivery of drugs with other compounds can allow synergistic cooperation, maximizing the effect of the delivered drug.

Drugs can be incorporated in nanoparticles in a variety of ways, which can enable various strategies for combinational drug therapy [99]. Drugs can be covalently linked to the nanoparticle or linear polymer chains, often by a degradable linker which is tuned to release the drug at the desired time [100]. Often, the drug is stored in hydrophilic and/or hydrophobic compartments of the nanoparticle. Micelles often store drugs in their hydrophobic core, and liposomes and polymersomes can store dissolved hydrophobic drug in the membrane and can additionally store hydrophilic drugs in the aqueous center [68]. Drug can also be physically entrapped in a polymer mesh such as in a dendrimer or hydrogel [101,102].

Co-delivery of drug with a Pgp modulator is a common strategy to combat ABC pump-mediated MDR [103]. This strategy has been employed for more than 30 years since Pgp was identified as a major MDR mechanism. Several generations of ABC pump modulators have been developed to deal with the weaknesses of the first modulators [104]. Use of these modulators has effectively increased the intracellular concentration and drug sensitivity of the cells [105]. On the other hand, ABC pumps are also important features of cells in healthy tissues, which are routinely exposed to toxins such as the liver, kidneys, and intestines [36]. Therefore, using Pgp modulators leads to nonspecific toxicity issues that limit the clinical relevance of the method [106]. Third generation modulators, currently being developed, have been shown to more effectively sensitize cells while maintaining tolerable toxicity [107,108]. Drugs delivered in nanoparticles collect in the liver, kidney, and spleen where ABC pumps play an important role in detoxifying healthy tissues. This makes ABC pump modulation a less than ideal tactic unless true tumor targeting can be achieved.

One important strategy to mitigate emergence of MDR is to deliver drugs of very different structures and functions simultaneously [68]. The ability to incorporate drugs in a nanoparticle platform by several mechanisms makes it possible to deliver disparate drugs. This strategy is used in agriculture, where applying multiple pesticides to a field simultaneously slows evolution by creating divergent pressures and mitigating the advantage of individuals with resistant traits [109]. Furthermore, well-chosen drug combinations can take advantage of synergistic effects among drugs to increase sensitivity of the target cells allowing the total amount of drug to be minimized [110].

The sensitivity of cells can be greatly increased by co-delivering apoptosis modulators with the drug. Apoptosis modulators can take the form of antisense oligonucleotides or other chemical compounds such as ceramide [111]. The anti-sense oligonucleotides generally work by knocking down expression of proteins, such as bcl2, which inhibit the apoptosis pathway. The resulting drug combination can be as much as 100-fold more potent than using drug alone [112]. Of course the apoptosis pathway is common to all cells, so carefully targeted drug release is again a challenge as lowering the apoptotic threshold in normal cells could lead to tissue damage.

Nanoparticles can also be used to modify the tumor microenvironment and improve drug distribution through the tissue. The most commonly explored means of improving drug distribution at the tissue and microenvironmental levels is to deliver collagenases with the nanoparticles, which can degrade the collagen matrix of the tumor tissue [113,31]. Degrading the collagen matrix has a twofold benefit. The most obvious is that the dense network of protein fibers can be a significant barrier to particle diffusion, especially the relatively large particles used as drug delivery systems. Drug bound nanoparticles fail to move more than a few microns from the point of extravasation into the tumor. The collagen matrix also contributes to the elevated interstitial pressure observed in most tumors by preventing volume expansion and providing resistance to bulk fluid flow [30]. The interstitial pressure in the tumor can approach that of the microvasculature, thus moving the convective driving force for extravasation of the capillary into the tumor and reducing overall tumor drug accumulation [24,114,115]. Degradation of the tumor protein matrix temporarily lowers the oncotic pressure of the tumor and restores the driving force for extravasation [115]. Collagen matrix degradation has also been shown to improve diffusion within the tumor tissue [31], thus improving overall distribution. One concern with using this approach is that a host of cells may be released and metastasize. Further study is needed to evaluate this risk.

10.3.3 Stimuli-Based Methods

Nanoparticles may be designed to respond to a variety of stimuli. Micelles sensitive to surrounding pH were discussed as a means to deliver drug directly to the cytoplasm of the cell in a previous section [98]. Nanoparticles may also be designed to respond to external stimuli such as radiation, ultrasound, and temperature, which can then be used in the clinic to localize drug release, sensitize cells, or directly kill cells [116–118]. Cancer treatments relying on external stimuli have some tremendous advantages over more passive treatment methods. The main advantage of using external stimuli as a trigger for treatment is the level of control it offers the physician. The therapeutic effect occurs precisely where and when the physician determines is appropriate.

Photothermal ablation is a form of external-stimuli-based nanotherapy and has an additional advantage of not requiring drug sensitive tumor cells. Treatment by photothermal ablation is accomplished by administering nanoparticles to the patient, then focusing particular wavelengths of light on a tumor. The light is then absorbed by the nanoparticles, causing heating of the surrounding tissue [119]. Ablative therapy can result in local temperatures within the tissue approaching the boiling point of water [120]. This gives ablative therapy the advantage of not requiring cellular entry, and the specificity of the treatment can be consciously controlled by the practitioner. Particle distribution can be a significant challenge just as for drug-based therapies; however, heat diffusion may be less hindered than particle diffusion depending on the tumor type and composition [121]. The effectiveness of the therapy depends in part on the ability to predict and control the heat distribution, which generally requires high-level mathematical modeling of the tumor heat profile during treatment [122,123]. Achieving sufficient tissue penetration to treat deep tumors is also a distinct challenge for ablation therapy, but a broad range of stimuli have been shown to accomplish therapeutic heat ranges with different particles providing a large range of treatment options [116,123,124]. In situations where necrosis is undesirable, more gentle heating (in the 45°C range) can be used to sensitize cells to traditional drug therapies [125]. Cell sensitization in hyperthermic conditions results from the inhibition of intracellular repair mechanisms due to heat shock [126].

There exist other forms of external-stimuli-based therapies designed to improve therapeutic efficacy. This includes using ferromagnetic particles, which can actively move deeper into a tumor in response to a magnetic field [127], and ultrasound-sensitive drug packets, which can be released specifically at a specific site with focused ultrasound [117]. Many of these therapies have theranostic potential, meaning that the particles used can be simultaneously used as imaging contrast agents to visualize the tumor even as the treatment is occurring [128].

Unfortunately, these methods are only able to treat "visible" tumors, or tumors that are large enough and can collect enough particles to stand out against the background noise for imaging. Smaller tumors and metastatic cells must be treated by more traditional drug-based therapies. Also, the inorganic nanoparticles, which are frequently used for stimuli-based therapies, are not readily cleared by the body and may accumulate permanently in the liver and spleen posing serious long-term toxicity issues [129].

10.4   FUTURE PERSPECTIVES

MDR is a grave development in the lifetime of a tumor. MDR tumors are generally composed of more resilient and potentially more metastatic cells, which lead to a poor prognosis for a patient. The diversity of mechanisms contributing to MDR creates a difficult problem in designing effective strategies to treat resistant cancers. Research has revealed an impressive arsenal of technologies capable of combating individual resistance mechanisms. Unfortunately, the multimodal nature of resistance can severely limit the effectiveness of narrowly designed drug delivery systems. This in part explains the limited success of modern therapies in clinical settings. The next step in drug delivery to resistant tumors will require a more holistic view of distribution and resistance, accounting for each stage of delivery from circulatory distribution to efficacy at the site of action. This approach may require trade-offs to cope with the exacting demands of tumor delivery and distribution, cellular uptake, and drug efficacy. Designing drug delivery systems from a comprehensive perspective requires a more profound understanding of the interplay between the MDR mechanisms and drug delivery.

Acknowledgments

This work is supported by NIH CA 101850.

REFERENCES

1. Howlader, N. et al. *SEER Cancer Statistics Review, 1975–2009*. National Cancer Institute, Bethesda, MD, 2010.
2. American Cancer Society. *Cancer Facts & Figures 2012*. American Cancer Society, Atlanta. 2012, p. 18.
3. Ling, V. (1997). Multidrug resistance: molecular mechanisms and clinical relevance. *Cancer Chemotherapy and Pharmacology 40*, 3–8.
4. Raderer, M., Scheithaur, W. (1993). Clinical trials of agents that reverse multidrug resistance, a literature review. *Cancer 72*, 3553–3563.
5. Wind, N. S., Holen, I. (2011). Multidrug resistance in breast cancer: from *in vitro* models to clinical studies. *International Journal of Breast Cancer 2011*, 1–12.
6. Szakács, G., Paterson, J. K., Ludwig, J. A., Booth-Genthe, C., Gottesman, M. M. (2006). Targeting multidrug resistance in cancer. *Nature Reviews Drug Discovery 5*, 219–234.
7. Gottesman, M. (1993). How cancer cells evade chemotherapy: sixteenth Richard and Hinda Rosenthal Foundation Award Lecture. *Cancer Research 53*, 747–754.
8. Gottesman, M. M., Fojo, T., Bates, S. E. (2002). Multidrug resistance in cancer: role of ATP–dependent transporters. *Nature Reviews Cancer 2*, 48–58.
9. Sotiriou, C. (2003). Breast cancer classification and prognosis based on gene expression profiles from a population-based study. *Proceedings of the National Academy of Sciences 100*, 10393–10398.
10. Mikhail, A. S., Allen, C. (2009). Block copolymer micelles for delivery of cancer therapy: transport at the whole body, tissue and cellular levels. *Journal of Controlled Release 138*, 214–223.
11. Bailar, J. C. (1987). Rethinking the war on cancer. *Issues in Science and Technology 4*, 16–21.

12. Choi, H. S. et al. (2007). Renal clearance of quantum dots. *Nature Biotechnology 25*, 1165–1170.

13. Hume, D. (2006). The mononuclear phagocyte system. *Current Opinion in Immunology 18*, 49–53.

14. Matsumura, Y., Maeda, H. (1986). A new concept for macromolecular therapeutics in cancer chemotherapy: mechanism of tumoritropic accumulation of proteins and the antitumor agent smancs. *Cancer Research 46*, 6387–6392.

15. Maeda, H. (2001). SMANCS and polymer-conjugated macromolecular drugs: advantages in cancer chemotherapy. *Advanced Drug Delivery Reviews 46*, 169–185.

16. Boucher, Y., Salehi, H., Witwer, B., Harsh, G. R., Jain, R. K. (1997). Interstitial fluid pressure in intracranial tumours in patients and in rodents. *British Journal of Cancer 75*, 829–836.

17. Gabizon, A., Shmeeda, H., Horowitz, A. T., Zalipsky, S. (2004). Tumor cell targeting of liposome-entrapped drugs with phospholipid-anchored folic acid–PEG conjugates. *Advanced Drug Delivery Reviews 56*, 1177–1192.

18. Woodle, M. C. (1995). Sterically stabilized liposome therapeutics. *Advanced Drug Delivery Reviews 16*, 249–265.

19. Dreher, M. R. et al. (2006). Tumor vascular permeability, accumulation, and penetration of macromolecular drug carriers. *Journal of the National Cancer Institute 98*, 335–344.

20. Sylven, B., Bois, I. (1960). Protein content and enzymatic assays of interstitial fluid from some normal tissues and transplanted mouse tumors. *Cancer Research 20*, 831–836.

21. Roose, T., Netti, P. A., Munn, L. L., Boucher, Y., Jain, R. K. (2003). Solid stress generated by spheroid growth estimated using a linear poroelasticity model. *Microvascular Research 66*, 204–212.

22. Venn, M., Maroudas, A. (1977). Chemical composition and swelling of normal and osteoarthrotic femoral head cartilage. I. Chemical composition. *Annals of the Rheumatic Diseases 36*, 121–129.

23. Boucher, Y., Baxter, L. T., Jain, R. K. (1990). Interstitial pressure gradients in tissue-isolated and subcutaneous tumors: implications for therapy. *Cancer Research 50*, 4478–4484.

24. Boucher, Y., Jain, R. K. (1992). Microvascular pressure is the principal driving force for interstitial hypertension in solid tumors: implications for vascular collapse. *Cancer Research 52*, 5110–5114.

25. Fukumura, D., Duda, D. G., Munn, L. L., Jain, R. K. (2010). Tumor microvasculature and microenvironment: novel insights through intravital imaging in pre-clinical models. *Microcirculation (New York, NY, 1994) 17*, 206–225.

26. Holash, J., Wiegand, S. J., Yancopoulos, G. D. (1999). New model of tumor angiogenesis: dynamic balance between vessel regression and growth mediated by angiopoietins and VEGF. *Oncogene 18*, 5356–5362.

27. Schofield, J. W., Gaffney, E. A., Gatenby, R. A., Maini, P. K. (2011). Tumour angiogenesis: the gap between theory and experiments. *Journal of Theoretical Biology 274*, 97–102.

28. Primeau, A. J., Rendon, A., Hedley, D., Lilge, L., Tannock, I. F. (2005). The distribution of the anticancer drug Doxorubicin in relation to blood vessels in solid tumors. *Clinical Cancer Research 11*, 8782–8788.

29. Ameri, K. et al. (2010). Circulating tumour cells demonstrate an altered response to hypoxia and an aggressive phenotype. *British Journal of Cancer 102*, 561–569.

30. Netti, P. A., Berk, D. A., Swartz, M. A., Grodzinsky, A. J., Jain, R. K. (2000). Role of extracellular matrix assembly in interstitial transport in solid tumors. *Cancer Research 60*, 2497–2503.

31. Ramanujan, S. et al. (2002). Diffusion and convection in collagen gels: implications for transport in the tumor interstitium. *Biophysical Journal 83*, 1650–1660.

32. Sharom, F. J. (1997). The P-glycoprotein efflux pump: how does it transport drugs? *The Journal of Membrane Biology 160*, 161–175.

33. Eckford, P. D. W., Sharom, F. J. (2009). ABC efflux pump-based resistance to chemotherapy drugs. *Chemical Reviews 109*, 2989–3011.

34. Endicott, J. A., Ling, V. (1989). The biochemistry of P-glycoprotein-mediated multidrug resistance. *Annual Review of Biochemistry 58*, 137–171.

35. Nielsen, D., Skovsgaard, T. (1992). P-glycoprotein as multidrug transporter: a critical review of current multidrug resistant cell lines. *Biochimica et Biophysica Acta (BBA)— Molecular Basis of Disease 1139*, 169–183.

36. Thiebaut, F. et al. (1987). Cellular localization of the multidrug-resistance gene product P-glycoprotein in normal human tissues. *Proceedings of the National Academy of Sciences of the United States of America 84*, 7735–7738.

37. Tanigawara, Y. (2000). Role of P-glycoprotein in drug disposition. *Therapeutic Drug Monitoring 22*, 137–140.

38. Dean, M., Hamon, Y., Chimini, G. (2002). The human ATP-binding cassette (ABC) transporter superfamily. *The Human ATP-Binding Cassette (ABC) Transporter Super-family 42*, 1007–1017.

39. Seelig, A. (1998). A general pattern for substrate recognition by P-glycoprotein. *European Journal of Biochemistry/FEBS 251*, 252–261.

40. Mayer, L. D., Bally, M. B., Hope, M. J., Cullis, P. R. (1985). Uptake of antineoplastic agents into large unilamellar vesicles in response to a membrane potential. *Biochimica et Biophysica Acta 816*, 294–302.

41. Izquierdo, M. A. et al. (1996). Broad distribution of the multidrug resistance-related vault lung resistance protein in normal human tissues and tumors. *The American Journal of Pathology 148*, 877–887.

42. Chugani, D. C., Rome, L. H., Kedersha, N. L. (1993). Evidence that vault ribonucleoprotein particles localize to the nuclear pore complex. *Journal of Cell Science 106*(Pt 1), 23–29.

43. Duvvuri, M., Krise, J. P. (2005). Intracellular drug sequestration events associated with the emergence of multidrug resistance: a mechanistic review. *Frontiers in Bioscience 10*, 1499–1509.

44. Hurwitz, S. J., Terashima, M., Mizunuma, N., Slapak, C. A. (1997). Vesicular anthracy-cline accumulation in doxorubicin-selected U-937 cells: participation of lysosomes. *Blood 89*, 3745–3754.

45. Kerr, J. F., Winterford, C. M., Harmon, B. V. (1994). Apoptosis. Its significance in cancer and cancer therapy. *Cancer 73*, 2013–2026.

46. Dole, M. G. et al. (1995). Bcl-xL is expressed in neuroblastoma cells and modulates Bcl-xL is expressed in neuroblastoma cells and modulates chemotherapy-induced apoptosis. *Cancer Research 55*, 2576–2582.

47. Teixeira, C., Reed, J., Pratt, M. (1995). Estrogen promotes chemotherapeutic drug resistance by a mechanism involving Bcl-2 proto-oncogene expression in human breast cancer cells. *Cancer Research 55*, 3902–3907.

48. Masuda, H. et al. (1988). Increased DNA repair as a mechanism of acquired resistance to cis-diamminedichloroplatinum (II) in human ovarian cancer cell lines. *Cancer Research 48*, 5713–5716.

49. Tew, K. D. (1994). Glutathione-associated enzymes in anticancer drug resistance. *Cancer Research 54*, 4313–4320.

50. Raijmakers, M. T. M., Bruggeman, S. W. M., Steegers, E. A. P., Peters, W. H. M. (2002). Distribution of components of the glutathione detoxification system across the human placenta after uncomplicated vaginal deliveries. *Placenta 23*, 490–496.

51. Townsend, D. M., Tew, K. D. (2003). The role of glutathione-S-transferase in anti-cancer drug resistance. *Oncogene 22*, 7369–7375.

52. Lewis, A. D. et al. (1988). Amplification and increased expression of alpha class glutathione S-transferase-encoding genes associated with resistance to nitrogen mustards. *Proceedings of the National Academy of Sciences of the United States of America 85*, 8511–8515.

53. Black, S. M. et al. (1990). Expression of human glutathione S-transferases in *Saccharomyces cerevisiae* confers resistance to the anticancer drugs adriamycin and chlorambucil. *The Biochemical Journal 268*, 309–315.

54. Clapper, M., Hoffman, S., Tew, K. (1990). Sensitization of colon tumor xenografts to L-phenylalanine mustard using ethacrynic acid. *Journal of Cell Pharmacology 1*, 71–78.

55. Efferth, T., Volm, M. (2005). Glutathione-related enzymes contribute to resistance of tumor cells and low toxicity in normal organs to artesunate. *In Vivo (Athens, Greece) 19*, 225–232.

56. Heppner, G., Dexter, D., DeNucci, T., Calabresi, P. (1978). Heterogeneity in drug sensitivity among tumor cell subpopulations of a single mammary tumor. *Cancer Research 38*, 3758–3763.

57. Richardson, M. E., Siemann, D. W. (1997). Tumor cell heterogeneity: impact on mechanisms of therapeutic drug resistance. *International Journal of Radiation Oncology, Biology, Physics 39*, 789–795.

58. Gindy, M. E., Prud'homme, R. K. (2009). Multifunctional nanoparticles for imaging, delivery and targeting in cancer therapy. *Expert Opinion on Drug Delivery 6*, 865–878.

59. Gillies, E. R., Fréchet, J. M. J. (2005). Dendrimers and dendritic polymers in drug delivery. *Drug Discovery Today 10*, 35–43.

60. Minko, T., Kopecková, P., Kopecek, J. (2000). Efficacy of the chemotherapeutic action of HPMA copolymer-bound doxorubicin in a solid tumor model of ovarian carcinoma. *International Journal of Cancer. Journal International du Cancer 86*, 108–117.

61. Ghosh, P., Han, G., De, M., Kim, C. K., Rotello, V. M. (2008). Gold nanoparticles in delivery applications. *Advanced Drug Delivery Reviews 60*, 1307–1315.

62. Nishiyama, N., Kataoka, K. (2006). Current state, achievements, and future prospects of polymeric micelles as nanocarriers for drug and gene delivery. *Pharmacology & Therapeutics 112*, 630–648.

63. Christian, D. A. et al. (2009). Polymersome carriers: from self-assembly to siRNA and protein therapeutics. *European Journal of Pharmaceutics and Biopharmaceutics:*

Official Journal of Arbeitsgemeinschaft für Pharmazeutische Verfahrenstechnik e.V 71, 463–474.

64. Gabizon, A. A. (1995). Liposome circulation time and tumor targeting: implications for cancer chemotherapy. *Advanced Drug Delivery Reviews 16*, 285–294.

65. Sutton, D., Nasongkla, N., Blanco, E., Gao, J. (2007). Functionalized micellar systems for cancer targeted drug delivery. *Pharmaceutical Research 24*, 1029–1046.

66. Li, S.-D., Huang, L. (2008). Pharmacokinetics and biodistribution of nanoparticles. *Molecular Pharmaceutics 5*, 496–504.

67. Farokhzad, O. C. et al. (2006). Targeted nanoparticle-aptamer bioconjugates for cancer chemotherapy *in vivo*. *Proceedings of the National Academy of Sciences of the United States of America 103*, 6315–6320.

68. Zhang, L. et al. (2007). Co-delivery of hydrophobic and hydrophilic drugs from nanoparticle-aptamer bioconjugates. *ChemMedChem 2*, 1268–1271.

69. Kim, K. et al. (2010). Tumor-homing multifunctional nanoparticles for cancer theragnosis: simultaneous diagnosis, drug delivery, and therapeutic monitoring. *Journal of Controlled Release 146*, 219–227.

70. Noguchi, Y. et al. (1998). Early phase tumor accumulation of macromolecules: a great difference in clearance rate between tumor and normal tissues. *Cancer Science 89*, 307–314.

71. Ozben, T. (2006). Mechanisms and strategies to overcome multiple drug resistance in cancer. *FEBS Letters 580*, 2903–2909.

72. Choi, C.-H. (2005). ABC transporters as multidrug resistance mechanisms and the development of chemosensitizers for their reversal. *Cancer Cell International 5*, 30.

73. Romsicki, Y., Sharom, F. J. (1999). The membrane lipid environment modulates drug interactions with the P-glycoprotein multidrug transporter. *Biochemistry 38*, 6887–6896.

74. Wong, H. L. et al. (2006). A mechanistic study of enhanced doxorubicin uptake and retention in multidrug resistant breast cancer cells using a polymer-lipid hybrid nanoparticle system. *The Journal of Pharmacology and Experimental Therapeutics 317*, 1372–1381.

75. Zhang, S., Li, J., Lykotrafitis, G., Bao, G., Suresh, S. (2009). Size-dependent endocytosis of nanoparticles. *Advanced Materials (Deerfield Beach, Florida) 21*, 419–424.

76. Jiang, W., Kim, B. Y. S., Rutka, J. T., Chan, W. C. W. (2008). Nanoparticle-mediated cellular response is size-dependent. *Nature Nanotechnology 3*, 145–150.

77. Chithrani, B. D., Ghazani, A. A., Chan, W. C. W. (2006). Determining the size and shape dependence of gold nanoparticle uptake into mammalian cells. *Nano Letters 6*, 662–668.

78. Meng, H. et al. (2011). Aspect ratio determines the quantity of mesoporous silica nanoparticle uptake by a small GTPase-dependent macropinocytosis mechanism. *ACS Nano 5*, 4434–4447.

79. Cheever, M. A. et al. (2009). The prioritization of cancer antigens: a national cancer institute pilot project for the acceleration of translational research. *Clinical Cancer Research 15*, 5323–5337.

80. Schrama, D., Reisfeld, R. A., Becker, J. C. (2006). Antibody targeted drugs as cancer therapeutics. *Nature Reviews Drug Discovery 5*, 147–159.

81. Quintana, A. et al. (2002). Design and function of a dendrimer-based therapeutic nanodevice targeted to tumor cells through the folate receptor. *Pharmaceutical Research 19*, 1310–1316.

82. Liu, X., Wang, Y., Nakamura, K., Kubo, A., Hnatowich, D. J. (2008). Cell studies of a three-component antisense MORF/tat/Herceptin nanoparticle designed for improved tumor delivery. *Cancer Gene Therapy 15*, 126–132.

83. Vogel, C. L. et al. (2002). Efficacy and safety of trastuzumab as a single agent in first-line treatment of HER2-overexpressing metastatic breast cancer. *Journal of Clinical Oncology 20*, 719–726.

84. Vogel, J., Hinrichs, S. H., Reynolds, R. K., Luciw, P. A., Jay, G. (1988). The HIV tat gene induces dermal lesions resembling Kaposi's sarcoma in transgenic mice. *Nature 335*, 606–611.

85. Sethuraman, V. A., Bae, Y. H. (2007). TAT peptide-based micelle system for potential active targeting of anti-cancer agents to acidic solid tumors. *Journal of Controlled Release 118*, 216–224.

86. Mann, D. A., Frankel, A. D. (1991). Endocytosis and targeting of exogenous HIV-1 Tat protein. *The EMBO Journal 10*, 1733–1739.

87. van Vlerken, L. E., Vyas, T. K., Amiji, M. M. (2007). Poly(ethylene glycol)-modified nanocarriers for tumor-targeted and intracellular delivery. *Pharmaceutical Research 24*, 1405–1414.

88. Thurn, K. T. et al. (2011). Endocytosis of titanium dioxide nanoparticles in prostate cancer PC-3M cells. *Nanomedicine: Nanotechnology, Biology, and Medicine 7*, 123–130.

89. Colin de Verdière, A. et al. (1994). Uptake of doxorubicin from loaded nanoparticles in multidrug-resistant leukemic murine cells. *Cancer Chemotherapy and Pharmacology 33*, 504–508.

90. Khalil, I. A., Kogure, K., Akita, H., Harashima, H. (2006). Uptake pathways and subsequent intracellular trafficking in nonviral gene delivery. *Pharmacological Reviews 58*, 32–45.

91. Trouet, A., Deprez-de Campeneere, D., De Duve, C. (1972). Chemotherapy through lysosomes with a DNA-daunorubicin complex. *Nature: New Biology 239*, 110–112.

92. Ferrari, A. et al. (2003). Caveolae-mediated internalization of extracellular HIV-1 tat fusion proteins visualized in real time. *Molecular Therapy 8*, 284–294.

93. Le, P. U., Nabi, I. R. (2003). Distinct caveolae-mediated endocytic pathways target the Golgi apparatus and the endoplasmic reticulum. *Journal of Cell Science 116*, 1059–1071.

94. Gabrielson, N. P., Pack, D. W. (2009). Efficient polyethylenimine-mediated gene delivery proceeds via a caveolar pathway in HeLa cells. *Journal of Controlled Release 136*, 54–61.

95. Kiss, A. L., Botos, E. (2009). Endocytosis via caveolae: alternative pathway with distinct cellular compartments to avoid lysosomal degradation? *Journal of Cellular and Molecular Medicine 13*, 1228–1237.

96. Tran, D., Carpentier, J. L., Sawano, F., Gorden, P., Orci, L. (1987). Ligands internalized through coated or noncoated invaginations follow a common intracellular pathway. *Proceedings of the National Academy of Sciences of the United States of America 84*, 7957–7961.

97. Oishi, M., Kataoka, K., Nagasaki, Y. (2006). pH-responsive three-layered PEGylated polyplex micelle based on a lactosylated ABC triblock copolymer as a targetable and endosome-disruptive nonviral gene vector. *Bioconjugate Chemistry 17*, 677–688.

98. Mohajer, G., Lee, E. S., Bae, Y. H. (2007). Enhanced intercellular retention activity of novel pH-sensitive polymeric micelles in wild and multidrug resistant MCF-7 cells. *Pharmaceutical Research 24*, 1618–1627.

99. Cho, K., Wang, X., Nie, S., Chen, Z. G., Shin, D. M. (2008). Therapeutic nanoparticles for drug delivery in cancer. *Clinical Cancer Research 14*, 1310–1316.

100. Minko, T., Kopecková, P., Kopecek, J. (1999). Comparison of the anticancer effect of free and HPMA copolymer-bound adriamycin in human ovarian carcinoma cells. *Pharmaceutical Research 16*, 986–996.

101. Na, K., Park, K. H., Kim, S. W., Bae, Y. H. (2000). Self-assembled hydrogel nanoparticles from curdlan derivatives: characterization, anti-cancer drug release and interaction with a hepatoma cell line (HepG2). *Journal of Controlled Release 69*, 225–236.

102. D'Emanuele, A., Attwood, D. (2005). Dendrimer-drug interactions. *Advanced Drug Delivery Reviews 57*, 2147–2162.

103. Patil, Y., Sadhukha, T., Ma, L., Panyam, J. (2009). Nanoparticle-mediated simultaneous and targeted delivery of paclitaxel and tariquidar overcomes tumor drug resistance. *Journal of Controlled Release 136*, 21–29.

104. Modok, S., Mellor, H. R., Callaghan, R. (2006). Modulation of multidrug resistance efflux pump activity to overcome chemoresistance in cancer. *Current Opinion in Pharmacology 6*, 350–354.

105. Wu, J. et al. (2007). Reversal of multidrug resistance by transferrin-conjugated liposomes co-encapsulating doxorubicin and verapamil. *Journal of Pharmacy & Pharmaceutical Sciences: A Publication of the Canadian Society for Pharmaceutical Sciences, Société Canadienne des Sciences Pharmaceutiques 10*, 350–357.

106. Krishna, R., Mayer, L. D. (2000). Multidrug resistance (MDR) in cancer. Mechanisms, reversal using modulators of MDR and the role of MDR modulators in influencing the pharmacokinetics of anticancer drugs. *European Journal of Pharmaceutical Sciences: Official Journal of the European Federation for Pharmaceutical Sciences 11*, 265–283.

107. Pusztai, L. et al. (2005). Phase II study of tariquidar, a selective P-glycoprotein inhibitor, in patients with chemotherapy-resistant, advanced breast carcinoma. *Cancer 104*, 682–691.

108. Newman, M. J. et al. (2000). Discovery and characterization of OC144-093, a novel inhibitor of P-glycoprotein-mediated multidrug resistance. *Cancer Research 60*, 2964–72.

109. Tabashnik, B. E. (1989). Managing resistance with multiple pesticide tactics: theory, evidence, and recommendations. *Journal of Economic Entomology 82*, 1263–1269.

110. Martello, L. A. et al. (2000). Taxol and discodermolide represent a synergistic drug combination in human carcinoma cell lines. *Clinical Cancer Research 6*, 1978–1987.

111. van Vlerken, L. E., Duan, Z., Seiden, M. V., Amiji, M. M. (2007). Modulation of intracellular ceramide using polymeric nanoparticles to overcome multidrug resistance in cancer. *Cancer Research 67*, 4843–4850.

112. Pakunlu, R. I. et al. (2006). *In vitro* and *in vivo* intracellular liposomal delivery of antisense oligonucleotides and anticancer drug. *Journal of Controlled Release 114*, 153–162.

113. Goodman, T. T., Olive, P. L., Pun, S. H. (2007). Increased nanoparticle penetration in collagenase-treated multicellular spheroids. *International Journal of Nanomedicine 2*, 265–274.

114. Netti, P. A., Baxter, L. T., Boucher, Y., Skalak, R., Jain, R. K. (1995). Time-dependent behavior of interstitial fluid pressure in solid tumors: implications for drug delivery. *Cancer Research 55*, 5451–5458.

115. Eikenes, L., Bruland, Ø. S., Brekken, C., Davies, C. D. L. (2004). Collagenase increases the transcapillary pressure gradient and improves the uptake and distribution of monoclonal antibodies in human osteosarcoma xenografts. *Cancer Research 64*, 4768–4773.

116. O'Neal, D. P., Hirsch, L. R., Halas, N. J., Payne, J. D., West, J. L. (2004). Photo-thermal tumor ablation in mice using near infrared-absorbing nanoparticles. *Cancer Letters 209*, 171–176.

117. Dromi, S. et al. (2007). Pulsed-high intensity focused ultrasound and low temperature-sensitive liposomes for enhanced targeted drug delivery and antitumor effect. *Clinical Cancer Research 13*, 2722–2727.

118. Kong, G., Braun, R. D., Dewhirst, M. W. (2000). Hyperthermia enables tumor-specific nanoparticle delivery: effect of particle size. *Cancer Research 60*, 4440–4445.

119. Zhang, J. Z. (2010). Biomedical applications of shape-controlled plasmonic nano-structures: a case study of hollow gold nanospheres for photothermal ablation therapy of cancer. *The Journal of Physical Chemistry Letters 1*, 686–695.

120. Hirsch, L. R. et al. (2003). Nanoshell-mediated near-infrared thermal therapy of tumors under magnetic resonance guidance. *Proceedings of the National Academy of Sciences of the United States of America 100*, 13549–13554.

121. Liu, Z. et al. (2005). Radiofrequency tumor ablation: insight into improved efficacy using computer modeling. *American Journal of Roentgenology 184*, 1347–1352.

122. Ahmed, M., Liu, Z., Humphries, S., Goldberg, S. N. (2008). Computer modeling of the combined effects of perfusion, electrical conductivity, and thermal conductivity on tissue heating patterns in radiofrequency tumor ablation. *International Journal of Hyperthermia 24*, 577–588.

123. Curley, S. A., Izzo, F., Ellis, L. M., Nicolas Vauthey, J., Vallone, P. (2000). Radiofrequency ablation of hepatocellular cancer in 110 patients with cirrhosis. *Annals of Surgery 232*, 381–391.

124. Hill, C. R., ter Haar, G. R. (1995). High intensity focused ultrasound--potential for cancer treatment. *The British Journal of Radiology 68*, 1296–1303.

125. Visaria, R. K. et al. (2006). Enhancement of tumor thermal therapy using gold nanoparticle-assisted tumor necrosis factor-alpha delivery. *Molecular Cancer Therapeutics 5*, 1014–1020.

126. Kampinga, H. H. (2006). Cell biological effects of hyperthermia alone or combined with radiation or drugs: a short introduction to newcomers in the field. *International Journal of Hyperthermia 22*, 191–196.

127. Sun, C., Lee, J. S. H., Zhang, M. (2008). Magnetic nanoparticles in MR imaging and drug delivery. *Advanced Drug Delivery Reviews 60*, 1252–1265.

128. McCarthy, J. R., Weissleder, R. (2008). Multifunctional magnetic nanoparticles for targeted imaging and therapy. *Advanced Drug Delivery Reviews 60*, 1241–1251.

129. Sadauskas, E. et al. (2009). Protracted elimination of gold nanoparticles from mouse liver. *Nanomedicine: Nanotechnology, Biology, and Medicine 5*, 162–169.

11

INTRACELLULAR TRAFFICKING OF NANOPARTICLES: IMPLICATIONS FOR THERAPEUTIC EFFICACY OF THE ENCAPSULATED DRUG

LIN NIU

Department of Pharmaceutics, College of Pharmacy, University of Minnesota, Minneapolis, MN, USA

JAYANTH PANYAM

Department of Pharmaceutics, College of Pharmacy, University of Minnesota, Minneapolis, MN, USA; Masonic Cancer Center, University of Minnesota, Minneapolis, MN, USA

11.1 INTRODUCTION

Many drug molecules have intracellular sites of action and, therefore, require entry into the cell to exert their therapeutic effects. These include both macromolecules such as nucleic acids and small molecules such as taxanes. However, the cell membrane presents a formidable barrier to the entry of many drugs, either because the molecules are too large and/or polar to diffuse across the membrane (nucleic acids, proteins) or because they are actively pumped out of the cell by membrane-bound efflux transporters (paclitaxel, doxorubicin). Delivery systems that facilitate the intracellular accumulation of these drug molecules could enhance their therapeutic efficacy. Additionally, it has become obvious that for certain drugs, the localization into specific intracellular organelles is critical for their biological performance. For instance, to achieve effective gene expression, plasmid DNA

Nanoparticulate Drug Delivery Systems: Strategies, Technologies, and Applications, First Edition.
Edited by Yoon Yeo.
© 2013 John Wiley & Sons, Inc. Published 2013 by John Wiley & Sons, Inc.

needs to be delivered not only into the cell but also into the nucleus. In the past decade, studies from several laboratories have shown that encapsulation of a drug in polymeric nanoparticles can improve its intracellular delivery. This chapter examines the intracellular trafficking of polymeric nanoparticles and the impact of modulating this trafficking on the therapeutic performance of the encapsulated drug.

11.2 PATHWAYS OF CELLULAR ENTRY

In order to understand the interaction of nanoparticles with cells, it is imperative to first introduce the lipid bilayer. The cell membrane defines the boundary between cellular contents and the external environment, separating intracellular biochemical activities from the extracellular surroundings. This seemingly excluding barrier is also the major interface where cells exchange information and materials with the surroundings. Phagocytosis and endocytosis are processes that either originate at or involve the plasma membrane to fulfill this mission. Similarly, at the subcellular level, it is again the compartmentalization by lipid bilayer membranes that enable various specialized biochemical processes proceeding simultaneously within the cells without interfering with one another. However, the necessity of exchanging materials and information still holds true at the organelle level. Intracellular membrane systems are involved in sorting and transporting materials within the cell in an orderly manner. Thus, plasma membrane and endomembrane system as a whole constitute dynamic continuums that relay material and information within the cell and with its surroundings. The interplay between nanoparticles and cells with regard to cellular uptake and intracellular trafficking is also intimately related with this dynamic cellular membrane system.

11.2.1 Endocytosis: Many Different Versions

Cells acquire large biomolecules and foreign substances that otherwise would be impermeable to the cell membrane through a process termed endocytosis. This term encompasses many different active (energy-dependent) membrane-based uptake processes. The major pathways involved in nanoparticles uptake are illustrated in Figure 11.1 [1]. Receptor-mediated endocytosis is a process by which cells internalize receptor-bound ligands. Clathrin-mediated endocytosis is so far the best-characterized pathway by which cells internalize specific molecules and modulate the expression level of membrane-bound receptors [2]. Morphologically, a salient feature of this endocytic mode is the formation of clathrin-coated pits and vesicles [3]. Recently, it was shown that this form of endocytic activity is initiated through the nucleation of FCHo proteins and various scaffold proteins such as eps15 and intersectin that promote the formation of cell membrane curvature, followed by recruitment of adaptor protein 2 (AP2) and clathrin triskelia [4]. Once the assembly of the clathrin lattice on the membrane is achieved, the scission of the invaginated clathrin-coated vesicle from the cell membrane is triggered. This process is facilitated by the mechanochemical enzyme dynamin, which prefers to localize

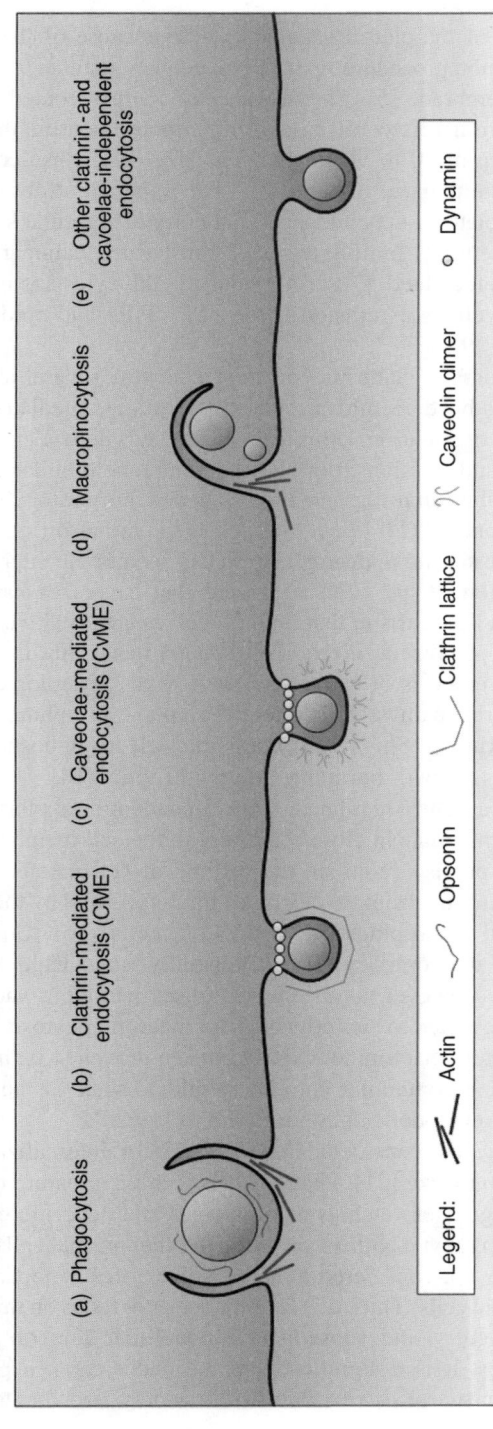

FIGURE 11.1 Principal nanocarrier internalization pathways in mammalian cells. (*Source:* Reproduced from Reference [1] with permission from Springer Science and Business Media.)

to the constricted neck of the clathrin-coated vesicle because of the curvature, to complete the vesicle fission procedure in a GTP-dependent manner, pinching off the vesicle from plasma membrane [5]. The diameter of clathrin-coated pits in mouse epithelial cells ranges from 120 to 150 nm; while in mouse neuron, the diameter of clathrin-coated pits is from 70 to 90 nm [6]. The size of clathrin-coated pits can increase to accommodate the size of cargo [7]. The upper limit observed so far is ~200 nm in the case of clathrin-dependent cellular entry of vesicular stomatitis virus [8], although the definite upper limit in terms of transporting nanoparticles through this pathway is not yet elucidated. Clathrin-mediated endocytosis is involved in the internalization and recycling of transferrin receptor (TfR) and epidermal growth factor receptor (EGFR) [9].

Distinct from the classical clathrin-dependent pathway, several alternative clathrin-independent routes have been identified [10]. Caveolae-mediated endocytosis has recently gained a lot of attention. Despite this pathway's dependence on actin and dynamin for assembly and scission from plasma membrane (similar to the clathrin pathway), caveolae exhibits an invaginated flask-shaped morphology with caveolin-1 as the defining component [11]. In addition, caveolae are also rich in cholesterol and sphingolipid and are similar to detergent-resistant membrane microdomain lipid rafts in lipid composition. It has been suggested that caveolae share a common endocytic pathway with lipid rafts in that both processes are sensitive to cholesterol depletion [11]. The size of caveolae is relatively smaller than clathrin-coated vesicle. The reported diameter range is 50–80 nm under normal physiological conditions [12,13]. Caveolae and raft pathways facilitate the uptake of sphingolipid binding toxins (e.g., cholera toxin and Shiga toxin), albumin, glycosylphosphatidylinositol (GPI)-anchored proteins, growth hormone, and SV40 virus [14].

Macropinocytosis is a tightly regulated, actin-dependent fluid phase process that allows cells to capture materials in close proximity to the cell membrane [15]. This so-called cell drinking process relies on the ruffling of cell membrane, engulfing solute molecules and nutrients from extracellular fluid, followed by the formation of a large vacuole termed macropinosome within the cells [16]. Unlike receptor-mediated endocytosis, macropinocytosis is generally considered a nonspecific sampling process [17]. However, a variety of viruses including vaccinia, adeno, and picorna have been reported to take advantage of macropinocytosis to gain access to cytosol [18]. In addition, macropinocytosis has been proven to be involved in the cellular entry of arginine containing cationic peptides, such as the TAT peptide derived from human immunodeficiency virus (HIV) [19,20].

Phagocytosis is typically considered a procedure for internalization of large particles that are 0.5 μm in size [21]. Phagocytosis is often a feature of distinct cell types. Professional phagocytes, such as macrophages and dendritic cells, typically possess much higher phagocytic ability compared to other normal cell types [22]. As a result, phagocytosis is not considered a major pathway for nanoparticle internalization in non-phagocytic cells. Thus, this chapter does not focus on this topic for the discussion of cellular entry and subsequent intracellular fate of nanoparticles. However, in most cases, it is a significant uptake and clearance pathway when nanoparticles are administered *in vivo* due to the widely documented role of the

TABLE 11.1 Agents Used to Inhibit Specific Endocytic Pathways

Dynamin Dependent	Clathrin	Caveolin	Macropinocytosis	Microtubule Dependent
Dynasore	Chlorpromazine	Filipin	Cytochalasin	Nocodazole
	Amantadine	Genistein	5-(N-ethyl-N-isopropyl)	
	K^+ depletion		amiloride (EIPA)	

reticuloendothelial system (RES) in nonselective elimination of nanoparticles [23]. Nanoparticle aggregates that are too large for endocytosis are likely cleared by professional phagocytes via phagocytosis. Moreover, phagocytosis could serve as a significant pathway for indirect nanoparticle acquisition. When successful intracellular nanoparticle delivery results in cell death, with an anticancer payload for example, the resulting apoptotic bodies and cellular debris will be cleared by professional phagocytes [24]. It is conceivable that a significant amount of nanoparticle components are still present within the apoptotic bodies. Thus, phagocytosis should not be overlooked for its contribution to the disposition of nanoparticles.

Often, several pathways contribute to the cellular uptake of nanoparticles. For instance, by using selective inhibitors to dissect endocytic pathways involved in the uptake of hydrophobically modified glycol chitosan nanoparticles by HeLa cells, caveolae-mediated pathway was found to account for 40% of uptake, clathrin-mediated pathway contributed to 20% of uptake, and another 30% was attributable to macropinocytosis [25]. A list of commonly used agents that selectively inhibit specific endocytic pathways is provided in Table 11.1.

11.2.2 Factors Affecting Nanoparticle Uptake into Cells

11.2.2.1 Size and Shape. It is generally accepted that size is the primary determinant of efficiency for the cellular entry of nanoparticles. It was reported that 50 nm gold nanoparticles show maximum uptake efficiency in HeLa cells compared to smaller or larger particles in the tested size range from 14 to 100 nm [26]. This trend was maintained in the presence or absence of transferrin, even though transferrin adsorption did facilitate uptake across the entire size range. Dos Santos et al. studied the effect of wider range of particle sizes (40, 100, 200, 500, 1000, 2000 nm) on the uptake efficiency of carboxyl-modified polystyrene nanoparticles in several cell lines (HeLa, A549 epithelial cells, 1321N1 astrocytes, HCMEC D3 endothelial cells, and RAW 264.7 macrophages) [27]. Regardless of the particle size, macrophages were the most efficient at internalizing the particles. Internalization of nanoparticles was highly size-dependent in all the cell types studied, with uptake efficiency decreasing with increasing particle size.

The size dependence of nanoparticle uptake and the presence of an optimal size for the highest uptake efficiency have been attributed to the smallest wrapping time to generate a suitable membrane curvature around nanoparticle/nanoparticle aggregates [28]. This size is predicted to be about 30 nm, which is in agreement with several experimental observations with both viral and nonviral nanoparticles [29,30].

On the basis of the wrapping time theory, it is also proposed that uptake of nanoparticles with optimal size could occur independent of clathrin- and caveolin-mediated endocytosis. A similar experimental study and mechanistic modeling have been conducted for single-walled carbon nanotubes (SWNTs). The model asserts that a size of 25 nm confers the optimal radius for membrane capture. For nanomaterials with a dimension smaller than 25 nm, clusters could form on the plasma membrane surface to generate sufficient enthalpic contribution via nanoparticle–cell membrane interaction to overcome the membrane elastic energy and entropic barriers related to vesicle formation. It was found that SWNTs demonstrate 1000 times higher endocytosis rate constant $(10^{-3}$ $min^{-1})$ than spherical gold nanoparticles with the same radius $(10^{-6} min^{-1})$ [31]. Moreover, a length dependence of SWNT uptake has been reported with a threshold of $\sim$189 nm, indicating nanotubes longer than this threshold could not be taken up by cells [32].

Conflicting data have been reported about the impact of shape on cellular uptake. The uptake of rod-shaped gold nanoparticles was reported as being less efficient compared with their spherical counterparts. Cells internalized 5 and 3.75 times more 74 nm and 14 nm spherical gold nanoparticles than 74 × 14 nm rod-shaped gold nanoparticles, respectively [26]. In addition, higher aspect ratio (1:5) nanoparticles were taken up poorly in comparison with lower aspect ratio (1:3) nanoparticles. In contrast, a study using cationic, cross-linked poly(ethylene glycol) hydrogel nanoparticles manufactured through PRINT technique has generated opposite results. HeLa cells internalized the high-aspect-ratio cylindrical particles (diameter, $d = 150$ nm, height, $h = 450$ nm) $\sim$4 times greater than the low-aspect-ratio symmetric cylindrical particles ($d = 200$ nm, $h = 200$ nm) [33]. Similarly, greater internalization of larger aspect ratio was observed for mesoporous silica nanoparticles in A375 melanoma cell lines [34]. On the other hand, Meng et al. showed that the internalization of rod-shaped silica nanoparticles in HeLa and A549 cell lines was much greater than that of spherical particles, but rod-shaped silica nanoparticles with aspect ratio of 2.1–2.5 demonstrated the highest uptake efficiency than either longer or shorter length rods. This was attributed to stronger activation by nanoparticles with this particular aspect ratio of GTP-binding protein Rac1, which is activated during macropinocytosis, and the Rho subfamily member protein CDC42 involved in regulating actin polymerization-dependent macropinocytosis processes [35].

11.2.2.2 Surface Properties. The surface characteristic of nanoparticles not only affects cellular uptake but also the intracellular fate following the cellular entry. Characteristics affecting nanoparticles uptake include surface charge, chemical composition, presence of targeting ligand, and the number of ligand molecules.

It is generally thought that neutral charge on the surface minimizes the interaction of nanoparticles with cells, partly because of weaker protein adsorption on the surface of nanoparticles in biological milieu [36,37]. PEGylation (incorporation of poly(ethylene glycol), PEG, chains on the surface) is the most well-established surface modification technique to provide "stealth" capability and improve the pharmacokinetic profile. The neutral and hydrophilic PEG coating lowers nonspecific protein adsorption, opsonization, and nanoparticle aggregation so that receptor-mediated endocytosis by cells of RES is dampened [37]. On the other hand, charged nanoparticles typically exhibit stronger ability to interact with cells. This is especially true for cationic nanoparticles that are highly charged, although higher cytotoxicity could also be expected [38]. Using an *in vitro* polarized epithelial Madin–Darby canine kidney (MDCK) cell model, it was found that both cationic and anionic nanoparticles are taken up by the epithelial cells through clathrin-dependent pathway and macropinocytosis, followed by significant transcytosis and accumulation at the basolateral membrane. Only a fraction of anionic but not cationic nanoparticles transited through the degradative lysosomal pathway [39]. In the context of *in vivo* distribution, since nanoparticles with high surface charge and larger size are cleared by macrophages more efficiently, 150 nm nanoparticles with slight negative surface charge has been shown to be the most effective in achieving prolonged circulation, a prerequisite for enhanced tumor accumulation [40].

A novel mechanism for how positive surface charge affects nanoparticle–cell interaction was recently proposed. Binding of positively charged nanoparticles to the negatively charged plasma membrane could modulate the membrane potential, induce membrane depolarization, and influx of Ca^{2+} [41]. To take advantage of the fact that positively charged surfaces lead to improved cellular entry, Poon et al. constructed a layer-by-layer nanoparticle system, consisting of quantum dot core, coated with iminobiotin-modified poly-L-lysine inner layer and biotin-functionalized PEG outer layer, bridged by neutravidin. Because of the pH-sensitive nature of iminobiotin–neutravidin interaction, it was expected that the outer PEG layer that conferred long circulating property would be shed in the acidic tumor microenvironment. The subsequent exposure of positively charged poly-L-lysine coating allowed for improved cellular uptake of nanoparticles. The benefit of this strategy was shown *in vivo* in MDA–MB-435 breast cancer model [42].

Another study suggested that it might be possible to modulate the mechanism of cellular entry by tuning the surface properties of nanoparticles. It was reported that 6 nm nanoparticles, decorated with a shell of alternating hydrophobic and anionic hydrophilic ligands in ribbon-like domains, display membrane penetration without bilayer disruption even at 4°C and altered intracellular localization compared to the "isomer" nanoparticle counterpart with identical surface functionalization but completely disordered surface pattern arrangement [43].

11.2.3 Targeting Ligand-Mediated Cellular Entry of Nanoparticles

In order to maximize efficacy and minimize off-target side effects, targeting proteins and receptors that are abundantly expressed on malignant cells have been routinely investigated. The mechanism of entry and intracellular trafficking of nanoparticles with the surface-bound ligand could, to some degree, mirror that of the specific target receptors [44]. Although overexpression of several different receptors has been reported, this chapter discusses transferrin and folic acid receptors as prototypical clathrin- and caveolae-based receptors, respectively.

11.2.3.1 Transferrin Receptor (TfR). Uptake of transferrin by TfR via clathrin-coated pit is one of the most well-characterized endocytic processes [45,46]. TfR is overexpressed in a wide variety of cancers [47]. Naturally, numerous groups have chosen TfR-mediated endocytosis as both a targeting and uptake enhancing strategy. Even though the presence of TfR targeting moiety on drug delivery carrier may not necessarily improve the overall accumulation in solid tumor, TfR targeting strategy has shown to significantly improve the amount of nanoparticles localized within cancer cells [48]. The study that provided the first evidence of successful RNA interference in human utilized TfR-targeted nanoparticle as a delivery platform for systemic administration of siRNA [49].

By using a 3-glysidoxypropyltrimethoxysilane (3-GPTMS) linker with epoxy terminal, the surface of mesoporous silica nanoparticles was decorated with Tf in another study. This allowed for enhanced accumulation of silica nanoparticles in TfR transfected HFF cell line compared to the wild type HFF cell line, indicating TfR-mediated endocytic uptake process in TfR overexpressed cells [50]. Cylindrical PEG-based PRINT nanoparticles conjugated with either human Tf or anti-TfR monoclonal antibody was internalized more readily in a panel of six common cancer cell lines relative to nanoparticles labeled with either bovine Tf or isotype control antibody, suggesting the potential for exploiting TfR-mediated endocytosis in improving intracellular delivery in TfR positive tumors [51]. TfR-targeted lipid nanoparticles encapsulated with antisense oligonucleotides against the antiapoptotic factor Bcl-2 were found to be internalized more effectively and possessed much better gene knockdown efficiency in leukemia cell lines than nontargeted lipid nanoparticle and free oligonucleotides [52]. Highest Bcl-2 downregulation and apoptosis-inducing efficacy was found in K562 cell line, which expressed the highest TfR level. The Bcl-2 downregulation and apoptosis induction were blocked by excessive free Tf. Combination of TfR targeting with pH responsiveness is particularly well suited to capitalize on the acidification of endosome subsequent to TfR-mediated endocytosis. Methacrylic acid copolymer complexed with poly(amido amine) (PAMAM) dendrimers and siRNA formed a core-shell polyelectrolyte micelle, which was decorated with anti-TfR Fab (antibody fragment of antigen binding) [53]. Targeted micelles showed stronger uptake by PC-3 prostate cancer cells than nontargeted control, and excess free anti-TfR antibody abolished this difference. Once the anionic portion of methacrylic acid copolymer was neutralized during acidification, polycationic dendrimer core containing siRNA was exposed, followed by endosomal escape and mRNA transcript silencing.

11.2.3.2 Folate Receptor. Unlike other transmembrane receptor proteins, folate receptor is a glycosylphosphatidylinositol (GPI)-anchored membrane protein, potentially associated with lipid raft [54]. Internalization of folate receptor follows a clathrin-independent mechanism. Folate receptor is sorted into recycling endosomal compartment, where GPI-anchored proteins are enriched and then trafficked to perinuclear sites before eventually returning folate receptor to the cell membrane. The synthesis of a disulfide-linked folate conjugate as a reduction-responsive fluorescence resonance energy transfer (FRET) probe enabled the elucidation of folate receptor life cycle, potential location of reduction reaction, and distinct sorting destination of cleaved yield products [55]. The disulfide disruption of folate-(BODIPY)-SS-rhodamine conjugate (folate-FRET) proceeded with a half-life of 6 h at recycling endosomal vesicle with reductive environment, which was independent of redox machinery in lysosome and Golgi apparatus. Furthermore, folate receptors (indicated by folate BODIPY fluorescent dye) were recycled back to plasma membrane, while released rhodamine was sorted into a distinct vesicle free of membrane-bound folate receptors. Since folate receptor is recycled to the plasma membrane, the timely release of the carrier-bound drug during the intracellular phase is crucial to attaining optimal intracellular drug delivery for a folate receptor-targeted system. Accordingly, folate receptor targeting drug delivery platforms with built-in reduction responsive drug unloading trigger have been formulated. In one of such systems, disulfide bond between hexaethylene glycol outer layer and cucurbit[6]uril (CB[6]) inner core was disrupted in the reducing environment of endosomes following folate receptor-mediated endocytic uptake, releasing doxorubicin entrapped within this self-assembled amphiphilic platform [56].

11.2.3.3 Other Receptors. It was shown that a delivery system based on beta-cyclodextrin-containing polycations possessed strong transfection efficiency against hepatoma cell line because of asialoglycoprotein receptor-mediated uptake [57]. The protein adsorption on the surface of nanoparticles also plays a major role in nanoparticle uptake. At the initial stages of nanoparticle development, unmodified nanoparticles were prone to opsonization by RES due to the protein adsorption and subsequent uptake by Kupffer cells in the liver. On the other hand, it was reported that polysorbate 80-coated poly(butyl cyanoacrylate) nanoparticles show promising ability to cross blood–brain barrier. Follow up studies elucidated the critical role of apolipoprotein adsorption that triggers receptor-mediated endocytosis by endothelial cells lining the blood–brain barrier [58].

11.3 INTRACELLULAR FATE OF INTERNALIZED NANOPARTICLES

Even though it is impossible to generalize a unifying theme of intracellular fate for distinct nanoparticle systems in distinct cell types, a set of potential itinerary repetitively observed for nanoparticles could be enumerated here. Upon cellular entry, a nanoparticle starts the intracellular journey either at cytosol directly via direct membrane fusogenic entry or from within membrane-confined vesicles such as early endosomes following endocytosis and macropinocytosis. As the early endosome undergoes maturation, a fraction of cargo within the compartment could be transported and sorted either to recycling pathway or toward multivesicular body and lysosome, the most widely documented intracellular sequestration destination for nanoparticles. Meanwhile, a typical trajectory of cargo transport mediated by motor molecules along the microtubule highway towards perinuclear region could also be observed. Throughout the aforementioned endosome maturation processes, nanoparticles with specific surface properties could escape endolysosomes and enter

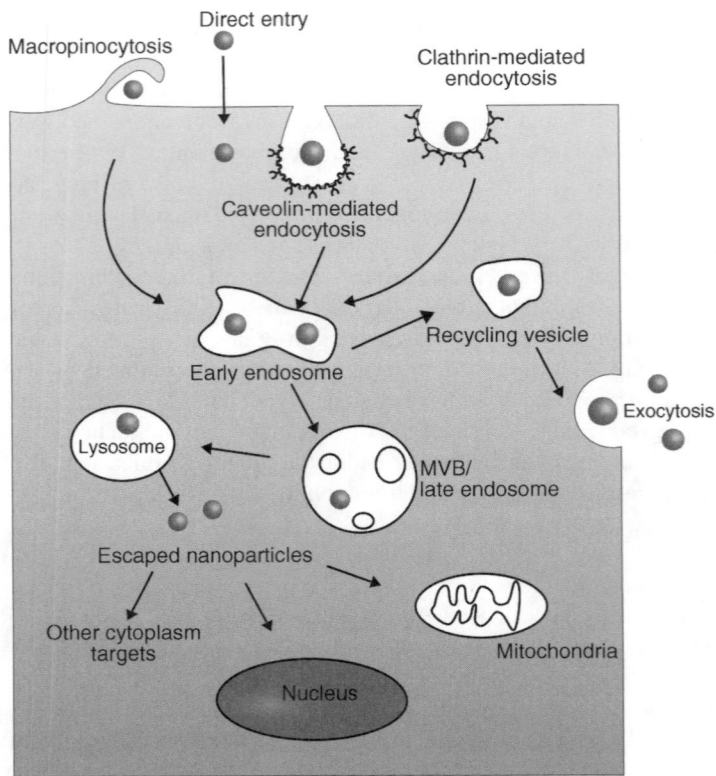

FIGURE 11.2   Potential intracellular trafficking pathways for nanoparticles. MVB: multi-vesicular body.

cytosol at this stage, leading to either intracellular localization with or without further distribution to other subcellular organelles. Some common pathways for nanoparticle intracellular trafficking are depicted in Figure 11.2.

The intracellular fate is intimately linked with the route of entry, which is dependent on the surface properties of nanoparticles. Targeting moiety often has major influence on both entry and intracellular trafficking since endocytosis route could dictate the subsequent intracellular sorting processes. Oba et al. have demonstrated that cyclic arginine-glycine-aspartic acid (RGD) targeting peptide not only selectively boosts PEG–poly-lysine–DNA polyplex cellular entry into cells expressing $\alpha_v\beta_3$ and $\alpha_v\beta_5$ integrin receptors but also ensures preferential perinuclear localization so that transfection efficiency against HeLa cells is significantly augmented. It was shown that this was not due to enhanced uptake but due to favorable intracellular trafficking following caveolae-mediated endocytosis, which directs more plasmid DNA cargo through caveosome pathway rather than degradative lysosome pathway as suggested by Figure 11.3 [59,60].

An elegant study on the uptake and intracellular fate of gold nanoparticles in HeLa cells demonstrated the correlation between surface chemistry and cellular fate

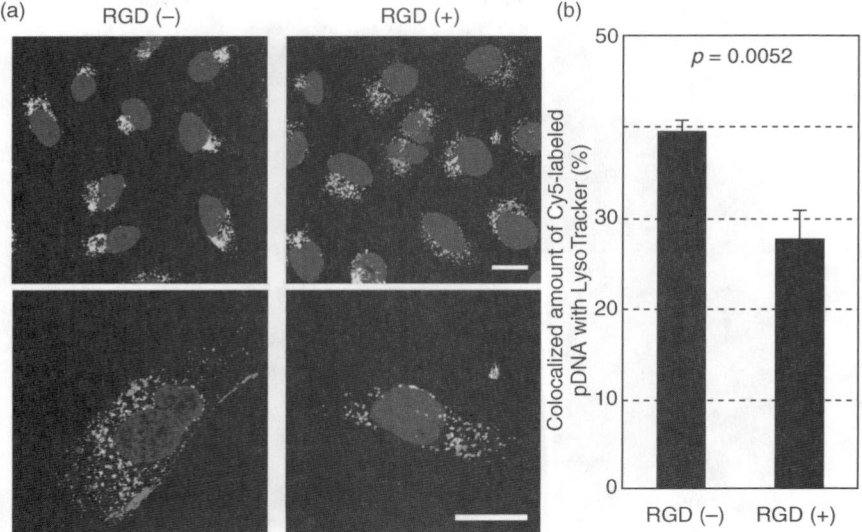

FIGURE 11.3 Distribution of RGD (+) and RGD (−) polyplex micelles in late endosomes and lysosomes. Polyplex micelles loading Cy5-labeled pDNA (red) were incubated with HeLa cells for 1 h. After replacement with fresh medium, the cells were reincubated for 11 h. The cell nuclei were stained with Hoechst 33342 (blue), and the acidic late endosomes and lysosomes were stained with LysoTracker Green (green). (a) CLSM images of the cells transfected with RGD (−) micelles (left) and RGD (+) micelles (right). The scale bars represent 20 μm. (b) Quantification of Cy5-labeled pDNA colocalized with LysoTracker Green in the inner cytoplasm. Error bars in the graph represent SEM ($n = 10$). (*Source:* Reprinted with permission from Reference [60]. Copyright (2008) American Chemical Society.) (*See insert for color representation of the figure.*)

of gold nanoparticles. Bare gold nanoparticles were taken up by endocytosis and were confined exclusively in endosomes, while PEGylation prevented cellular uptake completely. However, the inclusion of these PEGylated gold nanoparticles within liposomes or functionalization with cell penetrating peptide resulted in cytosolic distribution, circumventing the endosomal fate of unmodified counterpart as demonstrated in Figure 11.4 [61]. The cytosolic disposition was attributed to direct entry of nanoparticles across the cell membrane and/or disruptive surface property that damaged endosomal membranes [61].

11.3.1 Organelle Targeting

Specific organelle targeting is an important determinant of nucleic acid delivery efficiency of nonviral systems. RNA interference occurs in the cytosol while DNA transfection can only be achieved following transport of plasmid DNA into the nucleus. For successful siRNA delivery into cytosol, nanocarriers that enter the cell through endocytosis must exhibit endosomal escape [62,63]. For successful gene

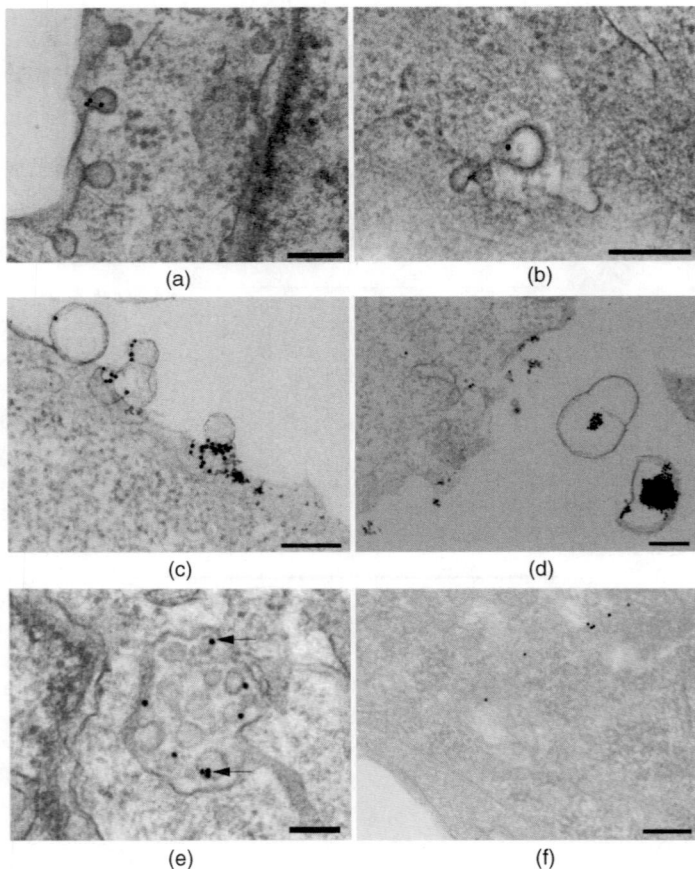

(a)   (b)

(c)   (d)

(e)   (f)

FIGURE 11.4 Uptake of PEG-modified gold nanoparticles in the presence of liposomes. Observed uptake mechanisms without apparent direct involvement of the liposomes include (a) caveolae and (b) clathrin-mediated endocytosis. Particle delivery by liposomes and subsequent uptake are shown in (c) and (d). Perhaps reflecting the wealth of different uptake mechanisms available under these conditions, nanoparticles (free) are found in the endosome (e) or free in the cytosol (f). Scale bars are 200 nm. (*Source:* Reprinted with permission from Reference [61]. Copyright (2008) American Chemical Society.)

delivery, nanocarriers must achieve nuclear entry after endosomal escape so that the genetic materials are delivered directly into the nucleus. To this end, nuclear localization signal (NLS) peptide has been utilized to enable nuclear targeting. Two different targeting ligands may be conjugated to the same nanoparticle surface to achieve sequential cell and organelle targeting. For example, conjugation with RGD and NLS peptides imparted both tumor targeting and nucleus targeting property to 30 nm gold nanoparticles [64].

Mitochondria serve as the powerhouse of the cell, producing ATP through oxidative phosphorylation. In addition, mitochondria contain a distinct set of DNA inherited from maternal source. Abnormal functionality in mitochondria has been linked with a diverse range of neurodegenerative disorders including Alzheimer's. Thus, a mitochondria-targeted delivery system is desirable for mitochondrial gene therapy, metabolic diseases, and proapoptotic strategy for cancer treatment. Lipid nanoparticles loaded with proapoptotic drug SV30, in spite of lacking a specific mitochondria targeting functionality, exhibited a preferential accumulation at sites adjacent to mitochondria, as suggested by colocalization of MitoTracker (a mitochondrion labeling fluorescent dye) and Nile red-labeled lipid nanoparticles [65]. On the other hand, liposomes functionalized with mitochondriotropic triphenylphosphonium cations promoted preferential localization of nanocarrier at mitochondria and improved the therapeutic efficacy of ceramide both *in vitro* and *in vivo* [66]. A multifunctional iron oxide "nanoworm" functionalized with both glioblastoma targeting peptide (CGKRK) and mitochondria targeting peptide ($_D[KLAKLAK]_2$) has shown promising efficacy in treating glioblastoma bearing mice by inducing proapoptotic processes. Neither tumor targeting peptide nor mitochondria targeting peptide alone could confer this beneficial activity, highlighting that the sequential, cooperative action of both peptides was indispensable to achieving intracellular delivery followed by modulation of intracellular trafficking and localization [67].

11.3.2 Exocytosis of Nanoparticles

The story of nanoparticle–cell interaction would be incomplete without a discussion of the elimination mechanisms of nanoparticles, which has important bearing on nanoparticles' retention, drug release profile, and toxicity. In addition to the possible disassembly and degradation of nanoparticles in the lysosomes, exocytosis is another clearance mechanism for internalized nanoparticles. Cargo that has been sorted to secretory/recycling vesicles is destined to the fate of exocytosis. The relatively scarce data on exocytosis of nanoparticles prevent generalization of rules governing nanoparticle excretion. Nevertheless, according to the studies on nanoparticle internalization and sorting, it is conceivable that for a specific nanoparticle of interest (with unique size, shape, and surface property) whether or not exocytosis occurs at all, to what extent, and how fast likely depends on the cell line under study and could be highly context dependent.

Even though the data with regard to exocytosis are not as abundant as internalization, exocytosis of poly(D,L-lactide-*co*-glycolide) nanoparticles has been studied *in vitro* using human arterial vascular smooth muscle cells (VSMCs). These studies showed that intracellular concentration of nanoparticles was maintained as long as the cells were in contact with excess nanoparticles. Once the extracellular nanoparticle concentration gradient was removed, exocytosis of nanoparticles occurred with about 65% of the internalized fraction undergoing exocytosis in 30 min [68]. Energy depletion reduced exocytosis, indicating the ATP-dependent nature of this active sorting process. In addition, protein adsorption on nanoparticle surface has a major influence on sorting of nanoparticles into exocytic pathway as exocytosis was abrogated in serum-free culture medium. Slowing et al. have elegantly shown that when two separately cultured flasks of human umbilical vein endothelial cell (HUVECs) incubated with either FITC-labeled mesoporous silica nanoparticles or TRITC-labeled nanoparticles were pooled, individual cells contained both FITC- and TRITC-labeled nanoparticles, providing indirect evidence that internalized mesoporous silicon nanoparticles could undergo exocytosis by HUVEC, leading to reuptake and exchanging of expelled nanoparticles [69]. However, no such effect was observed in the HeLa cell line. In another experiment, about 50% of internalized D-penicillamine-coated zwitterionic quantum dots (DPA–QDs) of 4 nm radius was expelled by HeLa cells. The exocytosed fraction increased with a half-life of 21 min and reached saturation after 2 h [70]. Similar to quantum dots, nanodiamonds made of nanocarbon allotrope with negatively charged nitrogen defects as fluorescent centers have good photostability and therefore are candidates for long-term labeling of stem cells. Retention rate is an important parameter to consider before using fluorescent nanodiamonds to track a specific cell of interest. There was little ($\sim$15% or less) excretion of the endocytosed nanodiamond after 6 days of labeling from both HeLa and 489-2.1 cells, but exocytosis occurred more readily (up to 30%) for 3T3-L1 preadipocytes [71]. On the contrary, addition of silver nanoparticles at a concentration of 50 μg/ml to RBL-2H3 mast cell line triggered degranulation instantaneously [72]. Decisive *in vivo* and mechanistic experiments investigating exocytosis of nanoparticles are lacking. It remains to be seen whether exocytosis patterns seen *in vitro* would be observed *in vivo* and whether exocytosis has any major implications for nanoparticle pharmacokinetics and pharmacodynamics.

11.4   IN-DEPTH ELUCIDATION OF INTRACELLULAR TRAFFICKING EVENTS ENTAILS HIGH RESOLUTION IMAGING TECHNIQUES

The advent and application of more advanced imaging techniques will enable in-depth understanding of the intracellular trafficking events. Spinning disk confocal microscopy has increasingly become the method of choice for live cell imaging due to its quick data acquisition compared to more widely available conventional confocal instrumentation. In conjunction with short-term overexpression of fluorescently tagged Rab family proteins by transfection, the intracellular trafficking of 40 nm fluorescent polystyrene nanoparticles in HeLa cells has been observed with high spatial and temporal resolution [73]. The fraction of nanoparticles co-localized with Rab5, a marker for early endosome, decreased over time, while the fraction of nanoparticles co-localized with Rab7/9, markers for late endosome, increased correspondingly until reaching a plateau of 80%, which was consistent with the data achieved by using LAMP1 as the lysosome marker. Only miniscule co-localization of nanoparticles with Rab11 was observed, suggesting that nanoparticle trafficking into recycling vesicles was not a major sorting route in this experimental setup.

The resolution of conventional optical microscopy, including fluorescent confocal microscopy, is diffraction limited. For most studies, the resolving power of the conventional techniques cannot surpass the limit of 200–300 nm, depending on the instrumentation and fluorescent emission wavelength under study. However, the diameter of most nanoparticles employed in current research is below this diffraction-limited resolution. Conceivably, as researchers gain access to more advanced imaging techniques, such as super-resolution microscopy capable of generating subdiffraction limit resolution, our understanding of nanoparticle intracellular trafficking and fate can progress dramatically. It was reported that two 40 nm nanoparticles separated by an 85 nm center-to-center distance could be resolved by stimulated emission depletion (STED) microscopy, a scenario unresolvable by diffraction-limited conventional confocal microscopy [74]. The uptake and aggregation of silica nanoparticle with 128 nm diameter in lung carcinoma cell line A549 have also been visualized by STED microscopy [75]. Another report shows that 32 nm silica nanoparticle could gain access to nucleus and form aggregates within nucleus progressively. In contrast, 80 nm silica nanoparticles not only migrate at a slower rate intracellularly but also fail to penetrate nuclear membrane [76]. This report also demonstrated the unique strength of super-resolution technique in imaging intracellular fate of nanoparticles.

11.5 FUTURE PERSPECTIVES

Nanoparticulate systems hold great promise in enhancing the therapeutic efficacy of drugs with intracellular sites of action [77] and improving the efficacy of nucleic acid-based therapies encompassing siRNA and gene delivery [78]. The last decade has witnessed tremendous advancement of our understanding of cellular entry and intracellular trafficking of nanoparticles. This progress could guide the rational design of next generation of nanoparticles with improved therapeutic efficacy. However, one limitation in the field has been the understandable lack of consistency across research groups in the cell and animal models used. Nanoparticle–cell interaction is highly dependent on the cell line and treatment conditions used. For example, it has been shown that even closely related cell lines such as PC3, PC3-flu, and PC3-PSMA (the latter two were derived from the former) demonstrate dramatically differential interaction profiles with quantum dots. In PC3 cells, punctate cytoplasmic localization was observed, while in PC-PSMA cell line, co-localization of nanoparticles with transferrin was observed at the perinuclear recycling compartment [79]. The preferential sequestration of DNA polyplexes in perinuclear microtubule organizing compartment (MTOC) of PC3-PSMA cells (similar to that seen with quantum dots) was associated with the poor transfection efficiency. In contrast, the cytoplasmic localization of DNA polyplexes in PC3 cells resulted in relatively higher transfection efficiency [80]. Thus, caution should be exercised when reaching broad conclusions from a specific experimental setup and extrapolating observations from an experiment done with one cancer cell type to another cancer cell type or worse, to noncancerous cells and professional phagocytes.

From a therapeutic perspective, heterogeneity in tumor cell–nanoparticle interaction within a given population of tumor cells could pose significant problems. Differences in uptake of cytotoxin-armed nanoparticles within a given population could result in differential cytotoxicity in the population and impose a selection pressure that could result in the enrichment of drug-resistant clones. Thus, in the context of clinical applications, future nanoparticle design may have to consider nanoparticles' interaction with the entire tumor cell population as well as a whole gamut of other tumor-associated cell types (tumor-associated macrophages, endothelial cells, mesenchymal stem cells, to name a few) *in vivo*. This would require a "systems biology" approach to better understand the complex interactions that nanocarriers likely have with biological systems.

REFERENCES

1. Hillaireau, H., Couvreur, P. (2009). Nanocarriers' entry into the cell: relevance to drug delivery. *Cellular and Molecular Life Sciences 66*, 2873–2896.

2. Le Roy, C., Wrana, J. L. (2005). Clathrin- and non-clathrin-mediated endocytic regulation of cell signalling. *Nature Reviews Molecular Cell Biology 6*, 112–126.

3. Kirchhausen, T. (2000). CLATHRIN. *Annual Review of Biochemistry 69*, 699–727.

4. Henne, W. M., Boucrot, E., Meinecke, M., Evergren, E., Vallis, Y., Mittal, R., Mcmahon, H. T. (2010). FCHo proteins are nucleators of clathrin-mediated endocytosis. *Science 328*, 1281–1284.

5. Sweitzer, S. M., Hinshaw, J. E. (1998). Dynamin undergoes a GTP-dependent conformational change causing vesiculation. *Cell 93*, 1021–1029.

6. Mcmahon, H. T., Boucrot, E. (2011). Molecular mechanism and physiological functions of clathrin-mediated endocytosis. *Nature Reviews Molecular Cell Biology 12*, 517–533.

7. Ehrlich, M., Boll, W., Van Oijen, A., Hariharan, R., Chandran, K., Nibert, M. L., Kirchhausen, T. (2004). Endocytosis by random initiation and stabilization of clathrin-coated pits. *Cell 118*, 591–605.

8. Cureton, D. K., Massol, R. H., Saffarian, S., Kirchhausen, T. L., Whelan, S. P. J. (2009). Vesicular stomatitis virus enters cells through vesicles incompletely coated with clathrin that depend upon actin for internalization. *PLoS Pathogenesis 5*, e1000394.

9. Traub, L. M. (2009). Tickets to ride: selecting cargo for clathrin-regulated internalization. *Nature Reviews Molecular Cell Biology 10*, 583–596.

10. Mayor, S. Pagano, R. E. (2007). Pathways of clathrin-independent endocytosis. *Nature Reviews Molecular Cell Biology 8*, 603–612.

11. Nabi, I. R., Le, P. U. (2003). Caveolae/raft-dependent endocytosis. *The Journal of Cell Biology 161*, 673–677.

12. Conner, S. D., Schmid, S. L. (2003). Regulated portals of entry into the cell. *Nature 422*, 37–44.

13. Bareford, L. M., Swaan, P. W. (2007). Endocytic mechanisms for targeted drug delivery. *Advanced Drug Delivery Reviews 59*, 748–758.

14. Pelkmans, L., Helenius, A. (2002). Endocytosis via caveolae. *Traffic 3*, 311–320.

15. Swanson, J. A., Watts, C. (1995). Macropinocytosis. *Trends in Cell Biology 5*, 424–428.

16. Falcone, S., Cocucci, E., Podini, P., Kirchhausen, T., Clementi, E. Meldolesi, J. (2006). Macropinocytosis: regulated coordination of endocytic and exocytic membrane traffic events. *Journal of Cell Science 119*, 4758–4769.

17. Lim, J. P., Gleeson, P. A. (2011). Macropinocytosis: an endocytic pathway for internalising large gulps. *Immunology and Cell Biology 89*, 836–843.

18. Mercer, J., Helenius, A. (2009). Virus entry by macropinocytosis. *Nature Cell Biology 11*, 510–520.

19. Kaplan, I. M., Wadia, J. S., Dowdy, S. F. (2005). Cationic TAT peptide transduction domain enters cells by macropinocytosis. *Journal of Controlled Release 102*, 247–253.

20. Nakase, I., Niwa, M., Takeuchi, T., Sonomura, K., Kawabata, N., Koike, Y., Takehashi, M., Tanaka, S., Ueda, K., Simpson, J. C., Jones, A. T., Sugiura, Y., Futaki, S. (2004). Cellular uptake of arginine-rich peptides: roles for macropinocytosis and actin rearrangement. *Molecular Therapy 10*, 1011–1022.

21. Groves, E., Dart, A. E., Covarelli, V., Caron, E. (2008). Molecular mechanisms of phagocytic uptake in mammalian cells. *Cellular and Molecular Life Sciences 65*, 1957–1976.

22. Jutras, I., Desjardins, M. (2005). Phagocytosis: at the crossroads of innate and adaptive immunity. *Annual Review of Cell and Developmental Biology 21*, 511–527.

23. Brannon-Peppas, L., Blanchette, J. O. (2004). Nanoparticle and targeted systems for cancer therapy. *Advanced Drug Delivery Reviews 56*, 1649–1659.

24. Erwig, L. P., Henson, P. M. (2007). Clearance of apoptotic cells by phagocytes. *Cell Death Differentiation 15*, 243–250.

25. Nam, H. Y., Kwon, S. M., Chung, H., Lee, S.-Y., Kwon, S.-H., Jeon, H., Kim, Y., Park, J. H., Kim, J., Her, S., Oh, Y.-K., Kwon, I. C., Kim, K., Jeong, S. Y. (2009). Cellular uptake mechanism and intracellular fate of hydrophobically modified glycol chitosan nanoparticles. *Journal of Controlled Release 135*, 259–267.

26. Chithrani, B. D., Ghazani, A. A., Chan, W. C. W. (2006). Determining the size and shape dependence of gold nanoparticle uptake into mammalian cells. *Nano Letters 6*, 662–668.

27. dos Santos, T., Varela, J., Lynch, I., Salvati, A., Dawson, K. A. (2011). Quantitative assessment of the comparative nanoparticle-uptake efficiency of a range of cell lines. *Small 7*, 3341–3349.

28. Gao, H., Shi, W., Freund, L. B. (2005). Mechanics of receptor-mediated endocytosis. *Proceedings of the National Academy of Sciences of the United States of America 102*, 9469–9474.

29. Nakai, T., Kanamori, T., Sando, S., Aoyama, Y. (2003). Remarkably size-regulated cell invasion by artificial viruses. Saccharide-dependent self-aggregation of glycoviruses and its consequences in glycoviral gene delivery. *Journal of the American Chemical Society 125*, 8465–8475.

30. Jiang, W., Kimbetty, Y. S., Rutka, J. T., Chanwarren, C. W. (2008). Nanoparticle-mediated cellular response is size-dependent. *Nature Nanotechnology 3*, 145–150.

31. Jin, H., Heller, D. A., Sharma, R., Strano, M. S. (2009). Size-dependent cellular uptake and expulsion of single-walled carbon nanotubes: single particle tracking and a generic uptake model for nanoparticles. *ACS Nano 3*, 149–158.

32. Becker, M. L., Fagan, J. A., Gallant, N. D., Bauer, B. J., Bajpai, V., Hobbie, E. K., Lacerda, S. H., Migler, K. B., Jakupciak, J. P. (2007). Length-dependent uptake of DNA-wrapped single-walled carbon nanotubes. *Advanced Materials 19*, 939–945.

33. Gratton, S. E. A., Ropp, P. A., Pohlhaus, P. D., Luft, J. C., Madden, V. J., Napier, M. E., Desimone, J. M. (2008). The effect of particle design on cellular internalization pathways. *Proceedings of the National Academy of Sciences of the United States of America 105*, 11613–11618.

34. Huang, X., Teng, X., Chen, D., Tang, F., He, J. (2010). The effect of the shape of mesoporous silica nanoparticles on cellular uptake and cell function. *Biomaterials 31*, 438–448.

35. Meng, H., Yang, S., Li, Z., Xia, T., Chen, J., Ji, Z., Zhang, H., Wang, X., Lin, S., Huang, C., Zhou, Z. H., Zink, J. I., Nel, A. E. (2011). Aspect ratio determines the quantity of mesoporous silica nanoparticle uptake by a small GTPase-dependent macropinocytosis mechanism. *ACS Nano 5*, 4434–4447.

36. Khan, J. A., Pillai, B., Das, T. K., Singh, Y., Maiti, S. (2007). Molecular effects of uptake of gold nanoparticles in HeLa cells. *ChemBioChem 8*, 1237–1240.

37. Verma, A., Stellacci, F. (2010). Effect of surface properties on nanoparticle–cell interactions. *Small 6*, 12–21.

38. Schaeublin, N. M., Braydich-Stolle, L. K., Schrand, A. M., Miller, J. M., Hutchison, J., Schlager, J. J., Hussain, S. M. (2011). Surface charge of gold nanoparticles mediates mechanism of toxicity. *Nanoscale 3*, 410–420.

39. Harush-Frenkel, O., Rozentur, E., Benita, S., Altschuler, Y. (2008). Surface charge of nanoparticles determines their endocytic and transcytotic pathway in polarized MDCK cells. *Biomacromolecules 9*, 435–443.

40. He, C., Hu, Y., Yin, L., Tang, C., Yin, C. (2010). Effects of particle size and surface charge on cellular uptake and biodistribution of polymeric nanoparticles. *Biomaterials 31*, 3657–3666.

41. Arvizo, R. R., Miranda, O. R., Thompson, M. A., Pabelick, C. M., Bhattacharya, R., Robertson, J. D., Rotello, V. M., Prakash, Y. S., Mukherjee, P. (2010). Effect of nanoparticle surface charge at the plasma membrane and beyond. *Nano Letters 10*, 2543–2548.

42. Poon, Z., Chang, D., Zhao, X., Hammond, P. T. (2011). Layer-by-layer nanoparticles with a pH-sheddable layer for in vivo targeting of tumor hypoxia. *ACS Nano 5*, 4284–4292.

43. Verma, A., Uzun, O., Hu, Y., Hu, Y., Han, H.-S., Watson, N., Chen, S., Irvine, D. J., Stellacci, F. (2008). Surface-structure-regulated cell-membrane penetration by monolayer-protected nanoparticles. *Nature Materials 7*, 588–595.

44. Yoon, D. J., Liu, C. T., Quinlan, D. S., Nafisi, P. M., Kamei, D. T. (2011). Intracellular trafficking considerations in the development of natural ligand-drug molecular conjugates for cancer. *Annals of Biomedical Engineering 39*, 1235–1251.

45. Doherty, G. J., Mcmahon, H. T. (2009). Mechanisms of endocytosis. *Annual Review of Biochemistry 78*, 857–902.

46. Grant, B. D., Donaldson, J. G. (2009). Pathways and mechanisms of endocytic recycling. *Nature Reviews Molecular Cell Biology 10*, 597–608.

47. Qian, Z. M., Li, H., Sun, H., Ho, K. (2002). Targeted drug delivery via the transferrin receptor-mediated endocytosis pathway. *Pharmacological Reviews 54*, 561–587.

48. Choi, C. H. J., Alabi, C. A., Webster, P., Davis, M. E. (2010). Mechanism of active targeting in solid tumors with transferrin-containing gold nanoparticles. *Proceedings of the National Academy of Sciences of the United States of America 107*, 1235–1240.

49. Davis, M. E., Zuckerman, J. E., Choi, C. H. J., Seligson, D., Tolcher, A., Alabi, C. A., Yen, Y., Heidel, J. D., Ribas, A. (2010). Evidence of RNAi in humans from systemically administered siRNA via targeted nanoparticles. *Nature 464*, 1067–1070.

50. Ferris, D. P., Lu, J., Gothard, C., Yanes, R., Thomas, C. R., Olsen, J.-C., Stoddart, J. F., Tamanoi, F., Zink, J. I. (2011). Synthesis of biomolecule-modified mesoporous silica nanoparticles for targeted hydrophobic drug delivery to cancer cells. *Small 7*, 1816–1826.

51. Wang, J., Tian, S., Petros, R. A., Napier, M. E., Desimone, J. M. (2010). The complex role of multivalency in nanoparticles targeting the transferrin receptor for cancer therapies. *Journal of the American Chemical Society 132*, 11306–11313.

52. Yang, X., Koh, C. G., Liu, S., Pan, X., Santhanam, R., Yu, B., Peng, Y., Pang, J., Golan, S., Talmon, Y., Jin, Y., Muthusamy, N., Byrd, J. C., Chan, K. K., Lee, L. J., Marcucci, G., Lee, R. J. (2008). Transferrin receptor-targeted lipid nanoparticles for delivery of an antisense oligodeoxyribonucleotide against Bcl-2. *Molecular Pharmaceutics 6*, 221–230.

53. Felber, A. E., Castagner, B., Elsabahy, M., Deleavey, G. F., Damha, M. J., Leroux, J.-C. (2011). siRNA nanocarriers based on methacrylic acid copolymers. *Journal of Controlled Release 152*, 159–167.

54. Sabharanjak, S., Mayor, S. (2004). Folate receptor endocytosis and trafficking. *Advanced Drug Delivery Reviews 56*, 1099–1109.

55. Yang, J., Chen, H., Vlahov, I. R., Cheng, J.-X., Low, P. S. (2006). Evaluation of disulfide reduction during receptor-mediated endocytosis by using FRET imaging. *Proceedings of the National Academy of Sciences of the United States of America 103*, 13872–13877.

56. Park, K. M., Lee, D. W., Sarkar, B., Jung, H., Kim, J., Ko, Y. H., Lee, K. E., Jeon, H., Kim, K. (2010). Reduction-sensitive, robust vesicles with a non-covalently modifiable surface as a multifunctional drug-delivery platform. *Small 6*, 1430–1441.

57. Pun, S. H., Davis, M. E. (2002). Development of a nonviral gene delivery vehicle for systemic application. *Bioconjugate Chemistry 13*, 630–639.

58. Kreuter, J. R., Shamenkov, D., Petrov, V., Ramge, P., Cychutek, K., Koch-Brandt, C., Alyautdin, R. (2002). Apolipoprotein-mediated transport of nanoparticle-bound drugs across the blood-brain barrier. *Journal of Drug Targeting 10*, 317–325.

59. Oba, M., Fukushima, S., Kanayama, N., Aoyagi, K., Nishiyama, N., Koyama, H., Kataoka, K. (2007). Cyclic RGD peptide-conjugated polyplex micelles as a targetable gene delivery system directed to cells possessing alphavbeta3 and alphavbeta5 integrins. *Bioconjugate Chemistry 18*, 1415–1423.

60. Oba, M., Aoyagi, K., Miyata, K., Matsumoto, Y., Itaka, K., Nishiyama, N., Yamasaki, Y., Koyama, H., Kataoka, K. (2008). Polyplex micelles with cyclic RGD peptide ligands and disulfide cross-links directing to the enhanced transfection via controlled intracellular trafficking. *Molecular Pharmaceutics 5*, 1080–1092.

61. Nativo, P., Prior, I. A., Brust, M. (2008). Uptake and intracellular fate of surface-modified gold nanoparticles. *ACS Nano 2*, 1639–1644.

62. Panyam, J., Zhou, W.-Z., Prabha, S., Sahoo, S. K., Labhasetwar, V. (2002). Rapid endo-lysosomal escape of poly(DL-lactide-*co*-glycolide) nanoparticles: implications for drug and gene delivery. *The FASEB Journal 16*, 1217–1226.

63. Patil, Y., Panyam, J. (2009). Polymeric nanoparticles for siRNA delivery and gene silencing. *International Journal of Pharmaceutics 367*, 195–203.

64. Kang, B., Mackey, M. A., El-Sayed, M. A. (2010). Nuclear targeting of gold nanoparticles in cancer cells induces DNA damage, causing cytokinesis arrest and apoptosis. *Journal of the American Chemical Society 132*, 1517–1519.

65. Weyland, M., Manero, F., Paillard, A., Gree, D., Viault, G., Jarnet, D., Menei, P., Juin, P., Chourpa, I., Benoit, J. P., Gree, R., Garcion, E. (2011). Mitochondrial targeting by use of lipid nanocapsules loaded with SV30, an analogue of the small-molecule Bcl-2 inhibitor HA14-1. *Journal of Controlled Release 151*, 74–82.

66. Boddapati, S. V., D'souza, G. G. M., Erdogan, S., Torchilin, V. P., Weissig, V. (2008). Organelle-targeted nanocarriers: specific delivery of liposomal ceramide to mitochondria enhances its cytotoxicity in vitro and in vivo. *Nano Letters 8*, 2559–2563.

67. Agemy, L., Friedmann-Morvinski, D., Kotamraju, V. R., Roth, L., Sugahara, K. N., Girard, O. M., Mattrey, R. F., Verma, I. M., Ruoslahti, E. (2011). Targeted nanoparticle enhanced proapoptotic peptide as potential therapy for glioblastoma. *Proceedings of the National Academy of Sciences of the United States of America 108*, 17450–17455.

68. Panyam, J., Labhasetwar, V. (2003). Dynamics of endocytosis and exocytosis of poly (D,L-lactide-*co*-glycolide) nanoparticles in vascular smooth muscle cells. *Pharmaceutical Research 20*, 212–220.

69. Slowing, I. I., Vivero-Escoto, J. L., Zhao, Y., Kandel, K., Peeraphatdit, C., Trewyn, B. G., Lin, V. S. Y. (2011). Exocytosis of mesoporous silica nanoparticles from mammalian cells: from asymmetric cell-to-cell transfer to protein harvesting. *Small 7*, 1526–1532.

70. Jiang, X., Roĉker, C., Hafner, M., Brandholt, S., Doŕlich, R. M., Nienhaus, G. U. (2010). Endo- and exocytosis of zwitterionic quantum dot nanoparticles by live HeLa cells. *ACS Nano 4*, 6787–6797.

71. Fang, C. Y., Vaijayanthimala, V., Cheng, C. A., Yeh, S. H., Chang, C. F., Li, C. L., Chang, H. C. (2011). The exocytosis of fluorescent nanodiamond and its use as a long-term cell tracker. *Small 7*, 3363–3370.

72. Yang, W., Lee, S., Lee, J., Bae, Y., Kim, D. (2010). Silver nanoparticle-induced degranulation observed with quantitative phase microscopy. *Journal of Biomedical Optics 15*, 045005.

73. Sandin, P., Fitzpatrick, L. W., Simpson, J. C., Dawson, K. A. (2012). High-speed imaging of Rab family small GTPases reveals rare events in nanoparticle trafficking in living cells. *ACS Nano 6*, 1513–1521.

74. Willig, K. I., Keller, J., Bossi, M., Hell, S. W. (2006). STED microscopy resolves nanoparticle assemblies. *New Journal of Physics 8*, 106.

75. Schübbe, S., Cavelius, C., Schumann, C., Koch, M., Kraegeloh, A. (2010). STED microscopy to monitor agglomeration of silica particles inside A549 cells. *Advanced Engineering Materials 12*, 417–422.

76. Schubbe, S., Schumann, C., Cavelius, C., Koch, M., Müller, T., Kraegeloh, A. (2011). Size-dependent localization and quantitative evaluation of the intracellular migration of silica nanoparticles in Caco-2 cells. *Chemistry of Materials 24*, 914–923.

77. Patil, Y., Sadhukha, T., Ma, L., Panyam, J. (2009). Nanoparticle-mediated simultaneous and targeted delivery of paclitaxel and tariquidar overcomes tumor drug resistance. *Journal of Controlled Release 136*, 21–29.

78. Patil, Y. B., Swaminathan, S. K., Sadhukha, T., Ma, L., Panyam, J. (2010). The use of nanoparticle-mediated targeted gene silencing and drug delivery to overcome tumor drug resistance. *Biomaterials 31*, 358–365.

79. Barua, S., Rege, K. (2009). Cancer-cell-phenotype-dependent differential intracellular trafficking of unconjugated quantum dots. *Small 5*, 370–376.

80. Barua, S., Rege, K. (2010). The influence of mediators of intracellular trafficking on transgene expression efficacy of polymer-plasmid DNA complexes. *Biomaterials 31*, 5894–5902.

12

TOXICOLOGICAL ASSESSMENT OF NANOMEDICINE

HAYLEY NEHOFF AND SEBASTIEN TAURIN

Department of Pharmacology and Toxicology, Otago School of Medical Sciences, University of Otago, Dunedin, New Zealand

KHALED GREISH

Department of Pharmacology and Toxicology, Otago School of Medical Sciences, University of Otago, Dunedin, New Zealand; Department of Oncology, Faculty of Medicine, Suez Canal University, Egypt

12.1 INTRODUCTION

The rapid evolution of nanotechnology has led to the realization of nanomedicine, defined as the use of engineered nanostructures for molecular level monitoring, repair, construction, and control of human biological systems [1]. Projections indicate that the nano pharmaceutical market will reach US$180 billion by 2015 [2]. Nanomedicine has originally focused on drug delivery and the treatment of cancer; however, new applications are emerging such as vaccine adjuvants, medical imaging, and gene therapy, the latter being applicable to a wide range of therapies ranging from Parkinson's disease treatment to osteogenesis [3]. The potential of nanotechnology to improve medicine is not limited to the use as a vehicle for medicines but also applicable for streamline discovery, design, development, and production of medications. These advances may reduce the time it takes to move a medicine from the bench to the clinic (currently 8–14 years) and the cost (current average of US$1.8 billion) [4]. On the other hand, the abundant use of and exposure to nanomaterials naturally raises

Nanoparticulate Drug Delivery Systems: Strategies, Technologies, and Applications, First Edition.
Edited by Yoon Yeo.

the concern of potential toxicity. However, limited data are available regarding toxicological evaluation of these new nanoconstructs, despite the accelerating rate of research papers describing new nanomaterials. The situation is similar regarding the investments in the emerging nanotechnology. For example, only 6.6% of the investment of the U.S. government in nanotechnologies is being spent on investigating the health and environmental consequences of their use [5].

In the United States, nanomaterials may, at the discretion of regulatory bodies, be deemed to fall under regulations pertaining to laws such as the toxic substances control act, while in the European Union they may fall under the regulation of the Registration, Evaluation, and Authorization of Chemicals Act [6]. Nanomedicine approval by regulatory bodies is subject to the standard preclinical, clinical, and postrelease regulations of traditional drugs. The majority of current nanomaterial regulation outlines are based on regulations developed for standard materials [7]. Early experimental studies to assess the toxicity of nanoparticles (NPs) rely on knowledge obtained from the studies on the nuisance of metal fumes, exposure to asbestos, synthetic fibers, silica, and dust. However, the increasing diversity of engineered nanomaterials such as metals, dendrimers, chitosan, silica, carbon nanotubes, and others used for medical applications have resulted in the need for improved methodologies to evaluate the safety of these new nanoconstructs [8]. Standard tests to evaluate toxicity are not always applicable to nanomedicine. For example, carbon nanotubes (CNTs) have been shown to cause a false cytotoxic effect in the 3-(4,5-dimethylthiazol-2-yl)-2,5-diphenyltetrazolium bromide (MTT) assay, and titanium NPs (TiNPs) bind lactic dehydrogenase (LDH) to interfere with LDH assays [9–11]. A glaring example of the shortcomings of classical toxicological evaluation is dose measurements. Traditional mg/kg dose measurements still need to be investigated for comparison with other medicines and poisons. However, for the purposes of having meaningful measurements of the toxicity of nanomedicine, surface area should be considered as the most important determinant of reactivity and hence possible toxic interaction with biological systems. A battery of other characteristics such as shape, size, charge, and porosity are as important as surface area. Currently there is no satisfactory standard or consensus on how to define nanomedicine exposure and assess their toxicity. As shown in Table 12.1, most of the toxic attributes of a nanomedicine are governed by the physicochemical parameters of the nanoconstructs. Complete understanding of these parameters is essential for the prediction of a specific nanomedicines safety. Section 12.2 is a summary of the current knowledge on safety of commonly used nanoconstructs in medicine, followed by our recommendations for conducting safety studies for novel nanomedicines.

12.2 SAFETY OF COMMONLY USED NANOCONSTRUCTS IN NANOMEDICINE

12.2.1 Liposomes

Liposomes are widely used for drug and gene delivery [46,47]. Several types have been approved by the Food and Drug Administration (FDA) for clinical use [48] such

TABLE 12.1 Description of the Administration Routes, Characteristics Affecting the Biodistribution and Molecular Effects of Various Types of Nanoparticles

	Liposome	Polymeric Micelle	Metal Nanospheres and Nanorods	Carbon	Dendrimers	Silica Nanoparticles
Administration route	Parenteral [12] Subcutaneous [13] Dermal [14] Wound infiltration [15]	Parenteral [16] Oral [17]	Parenteral [18,19] Oral [18] Inhalation [18]	Parenteral [20]	Parenteral [21,22] Topical [21] Oral [21] Transdermal [21] Ocular [21]	Parenteral [22]
Characteristics affecting biodistribution	Size [23] Charge [24]	Size [25] Core [25] Surface [25] Charge [25]	Size [26] Charge [27] Surface [28]	Agglomeration [29] Size [30]	Size [22] Charge [22] Surface [22]	Size [31] Shape [31] Charge [32]
Molecular effects	Opsonization [33] Astrocytosis [34] Edema [34] Inflammation [34]	Chronic liver toxicity [35]	Oxidative stress [36] DNA damage [36] DNA binding [37] Inflammation [38] Apoptosis [39]	Inflammation [40]	Hemotoxicity [22] Hepatotoxicity [41]	Oxidative stress [42,43] Inflammation [43–45]

as Doxil, Danuxome, Myocet, Depocyt, and MEPACT [49–53]. Several studies have demonstrated that liposomes accumulate in organs such as the liver and spleen and are consumed by macrophages making these tissues susceptible to toxicity (Figure 12.1) [54]. Excessive liposome deposition in macrophages may affect their phagocytic capacity or alter their gene expression. Several *in vitro* experiments have shown that liposomes may downregulate the production of nitric oxide and tumor necrosis factor (TNF)-α by macrophages and may cause macrophage cell death [55]. However, there is no scientific evidence that therapeutic doses of various liposomes can induce clinically relevant immunosuppressive effects [56].

Depending on their composition, liposomes have also been observed to stimulate the innate immune responses [56]. Doxil, a polyethylene glycol (PEG)-grafted liposomal carrier for doxorubicin, has been shown to promote allergic reactions, despite the view of PEG as immunologically inert [57]. In the central nervous system (CNS) of rats, liposome administration has been observed to cause focal astrogliosis, extracellular edema, and inflammation [34]. The dependence of liposome toxicity on

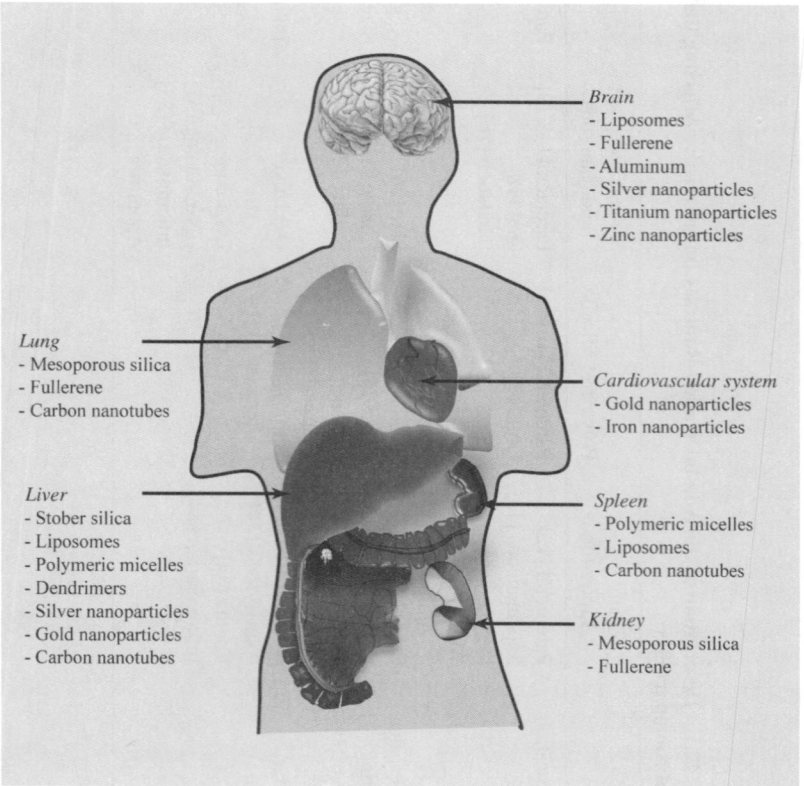

FIGURE 12.1 Schematic representation of the toxicity associated with specific nanomedicines on human organs.

their composition has been documented both *in vitro* [58] and *in vivo* [59]. Lecithin–cholesterol–dicetyl phosphate or lecithin–cholesterol–stearylamine administered intracerebrally to mice produced neuronal edema and/or necrosis, generalized epileptic seizures and death due to respiratory failure immediately after injection, whereas lecithin–cholesterol–phosphatidic acid liposomes produced relatively little effect [59].

12.2.2   Polymer Conjugates and Polymeric Micelles

Biodegradable polymers including poly(D,L-lactic acid) and poly(D,L-lactic-*co*-glycolic acid) have been commonly used to structure NPs and encapsulate a high load of poorly soluble material [60]. Since the approval of the first polymeric conjugates, PEG conjugated adenosine deaminase (ADA) [61] and styrene maleic acid conjugated neocarzinostatin polymer (SMANCS), for cancer treatment in the early 1990s, the field of polymer-based nanomedicine has grown exponentially [35,36]. Synthetic polymers such as *N*-(2-hydroxypropyl) methacrylamide (HPMA) copolymers, poly (ethyleneimine) (PEI), divinyl ether and maleic anhydride (DIVEMA), poly(vinylpyrrolidone) (PVP), PEG, and linear polyamidoamines as well as pseudosynthetic polymers such as poly(amino acids), and natural polymers including dextran, dextrin, hyaluronic acid, and chitosans have been used at a clinical level as a polymeric platform for drug delivery [62].

A major drawback of the use of polymeric carriers is related to their rapid uptake and accumulation by the reticuloendothelial system (RES) located in the liver, spleen, and lung [63] (Figure 12.1), which reduce the dose that can reach the tumors and raise concerns of chronic toxicity in these organs [64]. An example is HPMA copolymers that accumulate in RES organs in a size-dependent manner [65]. Furthermore, polymers can influence the biological responses. For example, SMANCS and DIVEMA caused increased secretion of interferon-γ and -β and increased the activity of natural killer and cytotoxic T-lymphocyte cells [66–69]. Another aspect limiting the use of certain polymers is their degradation profile, essentially HPMA and PEG copolymers which are nonbiodegradable, a parameter essential to consider for nanomedicine development [70].

12.2.3   Dendrimers

Dendrimers have gained a lot of interest over the last few years as drug carriers because of their well-defined polymeric structure, low polydispersity index (PDI), and functionalized surface diversity. In 1985, Tomalia et al. described the synthesis of polyamidoamine (PAMAM) dendrimers [71], and subsequently an array of dendrimers has been synthesized utilizing Tomalia's principles. More recently, on the basis of their ability to disrupt protein aggregates, dendrimers have been studied for the treatment of prion diseases and neurodegenerative diseases and as an antiviral agent for the treatment of AIDS [72].

To date, only two dendrimers have reached clinical development. These are the contrast agent Gadomer®, which is in Phase II clinical trial for the delineation of microvascular obstruction regions [73], and the intravaginal topical virucidal agent VivaGel® in Phase I clinical trial [74]. The complexity of obtaining chemically pure dendrimers is a major challenge for clinical development. Moreover, PAMAM dendrimers showed a distinct *in vivo* toxicity profile depending on the functional groups on their surface. Amine-terminated PAMAM dendrimers showed higher toxicity than carboxyl- or hydroxyl-terminated PAMAM dendrimers. Amine-terminated dendrimers triggered fatal hemotoxicity manifested as disseminated intravascular coagulation-like condition (DIC) and hemolysis.

12.2.4   Carbon Nanotubes

Carbon-based nanomaterials are used either as hollow spheres, ellipsoids or cylinders. Spheres and ellipsoids are referred to as fullerenes while cylinders are categorized as nanotubes. Fullerenes, especially C60, have exhibited some promise for drug delivery and for antiviral treatment as photosensitizers [75]. Pristine CNT and C60 are highly hydrophobic, chemically inert, and insoluble. Various biological and bioactive molecules such as proteins, enzymes, vaccines, DNA, and siRNA have been attached to CNTs and used for different biomedical applications [76]. More recently, CNT's photothermal properties have been exploited for their effect on cancer ablation [77]. CNTs exist in different forms such as single wall nanotubes (SWNT) and multiwalled nanotubes (MWNT), which can both be chemically modified and/or functionalized with bioactive molecules to enhance their solubility and drug targeting [78].

   However, there is a growing concern on the use of fullerene and CNTs on the basis of *in vitro* and *in vivo* animal studies (Figure 12.1). For example, incubation of keratinocytes and bronchial epithelial cells with SWNTs has been shown to cause oxidative stress and generation of reactive oxygen species (ROS) [79]. Other reports described the oxidative stress and production of ROS following treatment by CNTs on macrophages and other cell lines, either be due to metal impurities [80] or as a result of the experimental conditions [81].

   Acute and subacute toxicity studies of fullerenes by intraperitoneal injection in rats determined an approximate LD_{50} of 600 mg/kg with kidney impairment often observed in deceased animals [82]. There was no mortality in an acute oral and skin toxicity test [82]. Lipid peroxidation in the brain of largemouth bass was observed after 48 h exposure to 0.5 ppm of soluble fullerene C60 [83]; although these effects were later found to be associated with impurities from the solvent used for the preparation of soluble fullerene [84], pure soluble fullerene showed antioxidant activity [84]. Soluble fullerene modified with two cationic chains (fulleropyrrolidine derivatives) triggered hemolysis and appeared cytotoxic to a variety of cell lines at concentration between 20 and 60 μM [85]. Several studies have looked at the potential effects of fullerene on embryonic development. For example, intraperitoneal injection of 50 mg/kg CNTs to pregnant rats (11 days gestation) with fullerene C60 solubilized with polyvinyl pyrrolidone was shown to cross the placental barrier and distribute into embryonic tissues, causing the death of the embryo as a consequence of several anomalies and malformations [86].

CNTs bear remarkable structural similarities to the carcinogenic asbestos fibers and have been shown to induce asbestos-like inflammation and foreign body granulomas in mice [87]. Pulmonary exposure of rodents to CNTs has been shown to induce inflammation and granuloma formation but evidence that these granulomas progress to mesotheliomas is yet to be found [88,89]. Pulmonary exposure to fullerene in rodents has also shown to induce lung inflammation, fibrosis, and DNA damage [90,91]. Exposure of adult male rats to 200 μg of CNTs has been observed to cause decreased sperm production [92]. One major concern for the application of fullerenes and CNTs in biomedicine is the lack of consistent data regarding the effect of functionalized particles and impurities associated with their synthesis on the biological environment. All of these parameters fundamentally contribute to the difference in biological responses observed across the different studies.

12.2.5   Silica NPs

Mesoporous silica NPs (MSNs) have a well-defined structure that can be function- alized with a wide range of organic functional groups [93]. In recent years, MSNs have been investigated for biomedical applications such as biosensors, enzyme immobilization, detection of cancer cells *in vivo*, targeted cancer therapy, DNA delivery, and drug delivery [94–99]. Recently, the FDA approved the first human clinical trial of Cornell dots, multimodal silica NPs that enclose several dye molecules to be used for the detection of cancer cells [100].

There are a few studies investigating toxic effects of MSNs. Studies determine that the cytotoxicity of MSNs depends on the shape and size and functionalized surface type [101]. Size-dependent cytotoxicity is well documented with smaller particles generally being more cytotoxic. For example, MSNs with an identical geometry can trigger a size-dependent necrosis in human endothelial cells *in vitro* [102]. Although subcutaneously injected MSNs did not show toxicity, intraperitoneal and intravenous injection of nonfunctionalized MSNs with sizes varying from 150 to 4000 nm, have been shown to be lethal in mice [103]. The effect is partially based on the hydrodynamic sizes of SiO_2 postprotein exposure in relation with the physicochemical characteristics of SiO_2 NPs and the vasculature which further promoted multiple organ failures [104]. This toxicity appears to result from the particles themselves and not from contaminants or degradation products [103]. MSN was also shown to increase hemolysis at concentrations above 100 μg/ml, an effect influenced by the surface area and the geometry of the MSNs [105]. Furthermore, silica NPs were shown to enhance thrombus formation following intratracheal instillation or intravenous injection in a hamster model of thrombosis [106]. Intravenous injection of MSNs has been found to induce liver injury such as inflammation, silicotic nodular-like lesions, fibrosis, Kupffer cell activation, and the development of oxidative stress-associated liver damage at high doses (50 mg/kg) in mice [45]. Altered biliary function and glomerular function have also been observed in mice following 20 mg/kg MSNs intravenous injection [107]. MSNs have been shown to cross the blood–brain barrier (BBB) and accumulate in the striatum, where they deplete dopamine and induce oxidative stress and inflammation, and to damage the murine placenta and enter the fetal brain [43,108].

12.2.6 Metal-Based NPs

Metal-based NPs such as aluminum, gold, silver, iron, zinc, and titanium are currently being developed as nanomedicines; however, their unique properties make it difficult to predict their toxicity. Several studies were conducted *in vitro* and *in vivo* to demonstrate the influence of intrinsic parameters such as size and shape on their adverse toxicological effects. Figure 12.2 summarizes the physicochemical properties of the different nanomedicines that may influence biocompatibility and the possible pathways involved in their toxicity.

Aluminum (Al) is currently utilized for medical applications such as antacids, buffered aspirin, and vaccine adjuvants [109]. Al has no known biological function but has been demonstrated to cause significant toxicity such as dialysis encephalopathy, osteomalacia, and microcytic anemia [110]. The role of Al in dialysis encephalopathy is of particular concern and has been recognized in humans as early as 1979 when its ability to pass through the BBB and accumulate in the CNS to cause dementia was first reported in patients on chronic hemodialysis [111]. Of the various forms of dementia thought to be related to Al exposure, Alzheimer's disease has since been the most intensely studied [112]. Despite this research as well as observations of increased Al in the brains of Alzheimer patients and its ability to enhance amyloid beta penetration [113,114], there is no concrete evidence for the causality and this topic is still controversial [115].

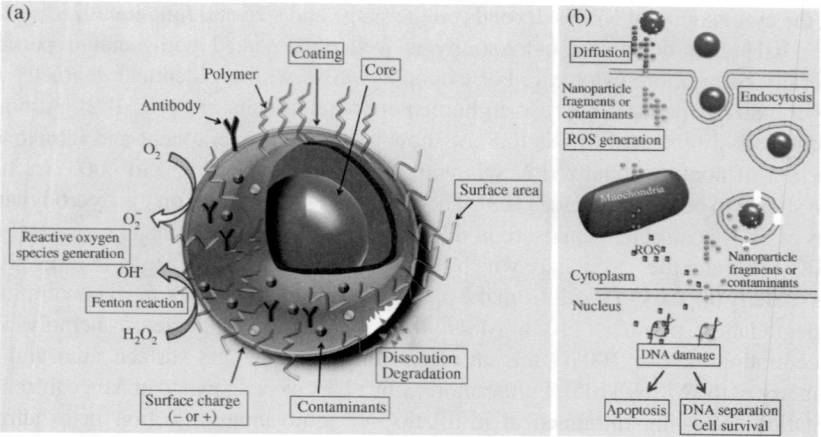

FIGURE 12.2 Physicochemical properties of nanomedicines that may influence biocompatibility. (a) Illustration of the importance of the nanomaterial composition, surface coating, solubility, degradation, dissolution, and contaminants with biological tissue. (b) Possible mechanistic pathway for toxicity induced by exposure to nanomedicines. Nanomedicines may degrade rapidly outside the cells or following endocytosis. The resulting fragments or contaminants may target mitochondria and promote oxidative stress due to the increased production of ROS. The ROS and the nanoparticles fragments or contaminants trigger downstream signaling responses that promote genotoxicity. ROS, reactive oxygen species.

Gold NPs (Au NPs) have been extensively studied for use in nanomedicine. Colloidal gold has been used since the 1920s for the treatment of tuberculosis before the emergence of antibiotics and has recently found some applications for the treatment of rheumatic disease [116]. More recently, Au NPs have been investigated for photothermal therapy, imaging, gene therapy, and drug delivery [117–120]. Biodistribution studies of Au NPs indicate that they preferentially accumulate in the liver, spleen, kidney, and lungs in a size-dependent manner with smaller particles (<50 nm) showing dispersion throughout all tissues [121]. Several studies have examined the effect of Au NP size on toxicity and have demonstrated that nanoclusters of 1.4 nm exhibit a higher cytotoxicity compared to nanoclusters of 15 nm [122]. Animal studies have also demonstrated that Au NPs can cause damage to the liver, cardiovascular system, and immune system [123,124] in a size-dependent manner. Additional studies have determined the influence of the charge, shape, and surfactant used for synthesis of the Au NPs on the potential cytotoxicity [122].

Metal-based NPs have also been demonstrated to induce oxidative stress following administration. Iron (Fe) NPs are being developed for use in magnetic resonance imaging, photothermal therapy, tissue repair, and drug delivery. The Fe NP Feraheme was approved in 2009 for the treatment of iron deficiency [125,126]. Oxidative stress is consistently observed following Fe NP treatment [125,127,128] and is highly concerning due to its potential implication in a variety of diseases including neurodegenerative diseases, cardiovascular diseases, and cancer [129–131]. This oxidative stress is also likely to be the cause of the Fe NP-induced genotoxicity [132,133]. The enhanced permeability observed *in vitro* at 50 µg/ml and the enhanced microvasculature permeability observed *in vivo* at 0.8 mg/kg [134,135] indicate that Fe NPs may play a role in the development of cardiovascular diseases, which is supported by the observation of Fe NPs in human atherosclerotic plaques [130,136]. Silver (Ag) NPs have been shown to cross the BBB and trigger oxidative stress in the CNS and impair short-term memory [137,138]. Titanium (Ti) NPs and zinc (Zn) NPs have also been shown to be associated with CNS toxicity and impairment of the spatial recognition and memory [139–141].

12.3   EVALUATION OF NANOMATERIAL TOXICITY

A classical scheme for toxicity assessment is represented in Figure 12.3. In Section 12.3.1, we briefly summarize the principles of toxicological assessment of nanomaterials. Toxicity assessments are classified into acute, subacute, subchronic, and chronic studies. Absorption, distribution, metabolism, excretion, and pharmacokinetics (ADME/PK), carcinogenic, and teratogenic studies are also required for complete evaluation of nanomaterial toxicity.

12.3.1   Characterization

Thorough and systematic physicochemical characterization of materials is essential prior to studying their toxic effects *in vivo*. One of the main reasons is to ensure the

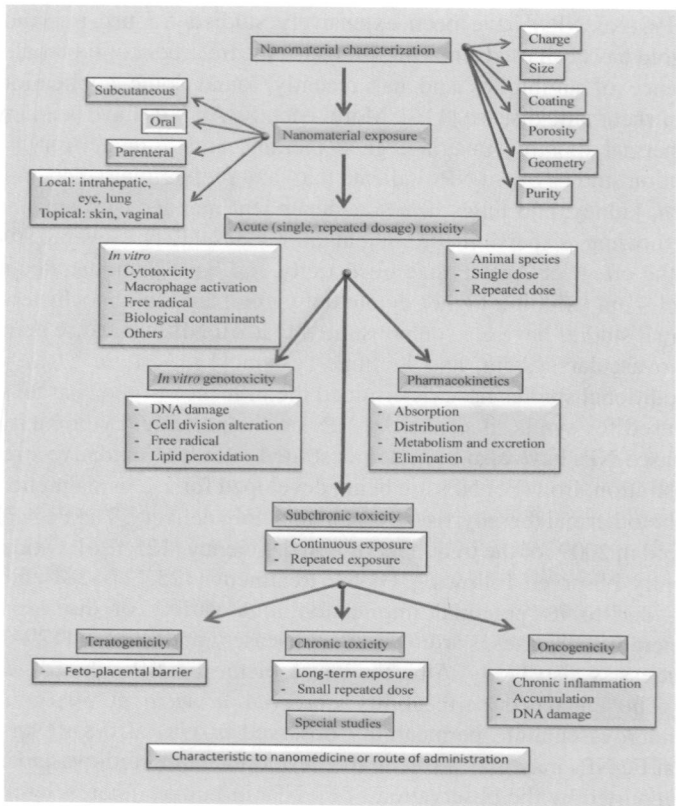

FIGURE 12.3 Proposed paradigm for toxicity testing of nanomedicines.

accuracy and reproducibility of the observed toxic effects of the studied NPs. Toxicity of NPs is significantly influenced by their physicochemical properties such as size, shape, surface charge, charge density, composition, density of structure, presence of pores, and surface activating sites. Hence, documentation of the characteristics of the NP under evaluation becomes crucial in order to correlate with the observed biological effects *in vivo*. Characterization data are important to facilitate comparison of toxicity results of a given NP with other nanoconstructs. Instrumentation techniques have advanced considerably in the past decades to sufficiently characterize the physicochemical properties of NPs, including dynamic light scattering techniques for size determination, zeta potential measurements for relative particle charge, electron microscopy (scanning and transmission) to qualitatively analyze the size and shape, nuclear magnetic resonance to accurately determine chemical structures, X-ray photoelectron spectroscopy to measure chemical compositions and chemical and electronic states, mass spectrometry for mass and elemental analysis and spectroscopy techniques that measure absorption, emission, or scattering of either wavelength or frequency for the determination of the size and distribution. An important note of caution while characterizing nanomaterials is to evaluate these properties under physiologically relevant conditions.

12.3.2   Absorption, Distribution, Metabolism, and Excretion (ADME)

Understanding the pharmacokinetics of NPs is an integral part of assessing their toxicity as it can facilitate an in-depth understanding of nanomaterial behavior in a biological system and provide valuable information which serves as a prelude to Phase I clinical study in humans. ADME studies are performed in at least one rodent species and one nonrodent species. In humans, ADME studies (usually a Phase I clinical trial) are performed in 8–10 healthy male subjects, unless the drug in question is a cytotoxic cancer treatment which is studied in patients instead of healthy volunteers [142]. The route of administration should be the proposed route for clinical application and usually a single dose is sufficient [142].

Absorption studies investigate the systemic exposure relative to the route of administration. *In vitro* absorption studies can be used to estimate *in vivo* absorption such as the Caco-2 cell monolayer to estimate intestinal absorption or a human skin equivalent model for dermal absorption [143,144]. *In vivo* oral absorption is usually investigated in a rodent model, where the plasma concentration is measured at certain time points following oral administration [145]. Microscopy methods such as scanning electron microscopy and multiphoton microscopy can be used to determine dermal NP absorption *in vivo* [146]. The *in vivo* whole body distribution of a radiolabeled nanomedicine may be quantitatively evaluated using whole body autoradiography or inductively coupled plasma mass spectroscopy [147,148]. At specified intervals, plasma samples may be taken to determine multiple pharmacokinetic parameters such as $T_{1/2}$, C_{max}, AUC, T_{max} and V_d [142]. It is critical that the radiolabel/fluorophore does not dissociate from the nanomedicine during metabolism and does not interfere with the activity of the nanomedicine, its metabolism, and/or stability [142]. The hepatic microsome test investigates potential metabolism of nanomedicines by the enzymes of the smooth endoplasmic reticulum (SER), for example, CYP450 enzymes and glucuronosyltransferases. Studies utilizing isolated hepatocytes are also conducted as these retain SER, cytosolic, and mitochondrial enzymes which may also metabolize the nanomedicine [149,150]. Excreted metabolites may be identified to determine the likely metabolic pathways *in vivo* such as hydroxylation, glucuronidation, sulfation, and methylation [142]. Following the dosing, urine, fecal, and bile samples are obtained at specified intervals to study the excretion mechanism of the nanomedicine (for more information consult [142]).

12.3.3   Acute and Chronic Toxicity

Acute and subacute toxicity studies (Table 12.2) may be measured based on a single or repeated dose administration via various methods of administration such as intravenous, oral (one dose per day), dermal, or inhalation. Typically, acute toxicity studies require five animals/sex/dose to be monitored over a period of ∼14 days with

TABLE 12.2 *In Vitro* and *In Vivo* **Assays to Evaluate the Potential Toxic Effect of Nanomedicines**

Toxicity	*In Vitro*	*In Vivo*
Genotoxicity	Ames test	Micronuclei test
	Comet test	Unscheduled DNA test
Carcinogenicity	Ames test	Two-year rodent cancer bioassay
	Comet test	
	Syrian hamster embryo assay	
	Immortalized aneuploid mouse cell assay	
Reproductive toxicity	Mouse morula assay	One generation rodent reproductive toxicity study
	Mouse blastocyst assay	Two generation rodent reproductive toxicity study
	Sperm viability assay	Prenatal development toxicity study
Acute and subacute toxicity	—	Rodent acute and subacute toxicity studies
		Dog acute and subacute toxicity studies
		Primate acute and subacute toxicity studies
Chronic and subchronic toxicity	—	Rodent chronic and subchronic toxicity studies
		Primate chronic and subchronic toxicity studies

outcomes of interest being weight change, response to the dose, function of organ systems, mortality, clinical pathology, and gross necropsy [151]. For subacute studies, 10 rodents/sex/dose are monitored over 4–5 weeks. Subchronic and chronic studies measure the effects of chronic dosing and exposure over the majority of the lifespan of the animal. Subchronic studies span ∼13 weeks while chronic studies span 18–30 months. Subchronic and chronic studies measure the same outcomes as acute studies with the addition of ocular toxicity, cardiovascular function, neurotoxicity, and immunotoxicity [151].

12.3.4 Genotoxicity

Genotoxicity is a vital test (see Table 12.2) for any new medication as it is an important indicator of the potential for the medication to induce teratogenic or carcinogenic effects. The most common *in vitro* tests are the Ames test and the comet test. The Ames involves histidine-dependent cell lines of *Salmonella typhimurium*, which are unable to grow on histidine-deficient media. If the bacteria are able to grow on the media following incubation with the nanomedicine in significantly higher numbers than the control, it is likely that the material is mutagenic [152]. The comet test involves mammalian cells, usually lymphocytes, being exposed to the material and then lysed, their genetic material extracted and then run on an electrophoresis gel. Broken DNA unwinds and so is unable to move through the gels with the same speed as intact DNA, resulting in a dense ball of intact DNA with a spread of DNA following in a pattern much similar to the tail of a comet [153].

The *in vivo* tests commonly performed are the micronuclei test and the unscheduled DNA synthesis test. The micronuclei test is reliable and well validated. The premise is that, as an erythroblast develops into an erythrocyte, the nucleus of the cell is excised but the micronuclei (a signal of incorrect chromosome distribution during anaphase) may be left behind; therefore, the presence of a significantly larger number of micronuclei than the negative control signals genotoxicity of the material [154]. The unscheduled DNA synthesis test measures DNA synthesis in mammalian liver cells that are not undergoing S phase scheduled DNA synthesis. DNA synthesis that is not scheduled is assumed to be a result of DNA repair following damage by the compound being investigated [155].

12.3.5 Carcinogenicity

Carcinogenicity is a serious concern for any medication. It may occur via two main mechanisms known as genotoxic and nongenotoxic carcinogenicity with genotoxic carcinogens being up to 50 times more potent than nongenotoxic carcinogens [156]. Nongenotoxic carcinogens have their effect via disruption of cellular metabolic processes and signaling pathways. Nongenotoxic carcinogens may, for example, alter the production of hormones that regulate the cellular mitotic rate (17β-estradiol), act as immunosuppressant (cyclosporin) or induce tissue-specific toxicity and inflammatory responses (arsenic) [157].

In vitro carcinogenicity tests (see Table 12.2) generally involve much of the same genotoxicity tests as above but also involve cell transformation assays using Syrian hamster embryo cell lines and immortalized aneuploid mouse cell lines [158]. The Syrian hamster embryo cell test is often used and signals carcinogenicity of a substance via morphological transformation of cell cultures [158]. The immortalized aneuploid mouse cell lines signals carcinogenicity via a progression from immortality to tumorigenicity [158]. The two-year cancer bioassay in rodents is considered the gold standard of carcinogenicity tests as it monitors the rodent for the greater portion of their lifespan. This test usually requires at least three different dose groups with at least 50 animals/sex/dose. The dose is administered daily for 24 months during which time tumor development, time of tumor onset, tumor location, dimensions, appearance, and progression are closely monitored [159].

12.3.6 Reproductive and Developmental Toxicity

Reproductive toxicity studies are necessary to gauge the possible effects of nano-medicine on reproductive ability and teratogenicity. Although some teratogenic effects are a consequence of genotoxicity, some are nongenotoxic such as retinoic acid. *In vitro* studies to evaluate possible reproductive and developmental toxicity include studies of sperm viability and murine embryos (Table 12.2). The evaluation of various parameters of sperm health including motility [160], apoptosis and necrosis, mitochondrial function, and LDH leakage [161] following incubation with the nanomedicines may also be informative. The effects of the nanomedicine in question on the development and proliferation of mouse morulas and blastocysts may also be studied *in vitro* to identify the effects of the nanomedicines on the viability and health of the embryo [162]. A one-generation *in vivo* study involves both males and females [163]. Sufficient numbers of animals are used in order to give 20 litters (assuming sterility is not an issue). The compound of interest is administered to the animals, males 2 weeks prior to, during, and after mating and females for 54 days to allow for 14 days premating, 14 days mating, 22 days gestation, and 4 days lactation. The parameters are pregnancy duration, stillbirths, live births, and litter number, size, sex, and abnormalities. Gross necropsy and histopathological examinations are also performed.

12.4   CONCLUSIONS

Nanomedicine has great potential to improve medical competences. Inconsistent classification, inadequate regulation, limited knowledge of its safety, and flaws in toxicity screening need to be addressed for this technology to provide the maximum benefit to mankind in general and in particular to cancer patients.

Acknowledgments

This work has been supported by Departmental fund no. PL. 108403.01.S. LM to KG and HRC Emerging Researcher First Grant (PL.108360.01.P.LM) to Sebastien Taurin from department of pharmacology and toxicology, Otago. Khaled Greish thanks Ms Rebecca Cookson for proofreading the manuscript.

REFERENCES

1. Webster, P., (2005). World Nanobiotechnology Market, Frost & Sullivan. *Nanomedicine*, *1*, 140–142.
2. Hobson, D. W. (2009). Commercialization of nanotechnology. *Wiley Interdisciplinary Reviews. Nanomedicine and Nanobiotechnology 1* (2), 189–202.
3. Krebs, M. D., Salter, E., Chen, E., Sutter, K. A., Alsberg, E. (2010). Calcium phosphate-DNA nanoparticle gene delivery from alginate hydrogels induces *in vivo* osteogenesis. *Journal of Biomedical Materials Research Part A 92* (3), 1131–1138.
4. Paul, S. M., Mytelka, D. S., Dunwiddie, C. T., Persinger, C. C., Munos, B. H., Lindborg, S. R., Schacht, A. L. (2010). How to improve R&D productivity: the pharmaceutical industry's grand challenge. *Nature Reviews Drug Discovery 9* (3), 203–214.

5. Roco, M. C. The long view of nanotechnology development: The National Nano-technology Initiative at 10 years. 1 ed. Nanotechnology research directions for societal needs in 2020, ed. M.C.M. Roco, C. A. Hersam, M. C. 2011.

6. Gwinn, M. R., Tran. L. (2010). Risk management of nanomaterials. *Wiley Interdisciplinary Reviews. Nanomedicine and Nanobiotechnology 2* (2), 130–137.

7. Chan, V. S. (2006). Nanomedicine: an unresolved regulatory issue. *Regulatory Toxicology and Pharmacology 46* (3), 218–224.

8. Fadeel, B., Garcia-Bennett, A. E. (2010). Better safe than sorry: understanding the toxicological properties of inorganic nanoparticles manufactured for biomedical applications. *Advanced Drug Delivery Reviews 62* (3), 362–374.

9. Worle-Knirsch, J. M., Pulskamp, K., Krug, H. F. (2006). Oops they did it again! Carbon nanotubes hoax scientists in viability assays. *Nano Letters 6* (6), 1261–1268.

10. Casey, A., Herzog, E., Davoren, M., Lyng, F. M., Byrne, H. J., Chambers, G. (2007). Spectroscopic analysis confirms the interactions between single walled carbon nanotubes and various dyes commonly used to assess cytotoxicity. *Carbon 45* (7), 1425–1432.

11. Zaqout, M. S., Sumizawa, T., Igisu, H., Wilson, D., Myojo, T., Ueno, S. (2011). Binding of titanium dioxide nanoparticles to lactate dehydrogenase. *Environmental Health and Preventive Medicine 17* (4), 341–345.

12. Cortes, J., Di Cosimo, S., Climent, M. A., Cortes-Funes, H., Lluch, A., Gascon, P., Mayordomo, J. I., Gil, M., Benavides, M., Cirera, L., Ojeda, B., Rodriguez, C. A., Trigo, J. M., Vazquez, J., Regueiro, P., Dorado, J. F., Baselga, J. (2009). Nonpegylated liposomal doxorubicin (TLC-D99), paclitaxel, and trastuzumab in HER-2-overexpressing breast cancer: a multicenter phase I/II study. *Clinical Cancer Research 15* (1), 307–314.

13. Buse, J., El-Aneed, A. (2010). Properties, engineering and applications of lipid-based nanoparticle drug-delivery systems: current research and advances. *Nanomedicine (Lond) 5* (8), 1237–1260.

14. Verma, D. D., Verma, S., Blume, G., Fahr, A. (2003). Particle size of liposomes influences dermal delivery of substances into skin. *International Journal of Pharmaceutics 258* (1–2), 141–151.

15. Golf, M., Daniels, S. E., Onel, E. (2011). A phase 3, randomized, placebo-controlled trial of DepoFoam(R) bupivacaine (extended-release bupivacaine local analgesic) in bunionectomy. *Advances in Therapy 28* (9), 776–788.

16. Gaucher, G., Marchessault, R. H., Leroux, J. C. (2010). Polyester-based micelles and nanoparticles for the parenteral delivery of taxanes. *Journal of Controlled Release 143* (1), 2–12.

17. Gaucher, G., Satturwar, P., Jones, M. C., Furtos, A., Leroux, J. C. (2010). Polymeric micelles for oral drug delivery. *European Journal of Pharmaceutics and Biopharmaceutics 76* (2), 147–158.

18. Arvizo, R., Bhattacharya, R., Mukherjee, P. (2010). Gold nanoparticles: opportunities and challenges in nanomedicine. *Expert Opinion on Drug Delivery 7* (6), 753–763.

19. De Jong, W. H., Hagens, W. I., Krystek, P., Burger, M. C., Sips, A. J., Geertsma, R. E. (2008). Particle size-dependent organ distribution of gold nanoparticles after intravenous administration. *Biomaterials 29* (12), 1912–1919.

20. Liu, Z., Tabakman, S., Welsher, K., Dai, H. (2009). Carbon nanotubes in biology and medicine: *in vitro* and *in vivo* detection. *Imaging and Drug Delivery. Nano Research 2* (2), 85–120.

21. Jain, N. K., Asthana, A. (2007). Dendritic systems in drug delivery applications. *Expert Opinion on Drug Delivery 4* (5), 495–512.

22. Greish, K., Thiagarajan, G., Herd, H., Price, R., Bauer, H., Hubbard, D., Burckle, A., Sadekar, S., Yu, T., Anwar, A., Ray, A., Ghandehari, H. (2011). Size and surface charge significantly influence the toxicity of silica and dendritic nanoparticles. *Nanotoxicology. 6*, 713–723.

23. Liu, D., Mori, A., Huang, L. (1992). Role of liposome size and RES blockade in controlling biodistribution and tumor uptake of GM1-containing liposomes. *Biochimica et Biophysica Acta 1104* (1), 95–101.

24. Henriksen-Lacey, M., Christensen, D., Bramwell, V. W., Lindenstrom, T., Agger, E. M., Andersen, P., Perrie, Y. (2010). Liposomal cationic charge and antigen adsorption are important properties for the efficient deposition of antigen at the injection site and ability of the vaccine to induce a CMI response. *Journal of Controlled Release 145* (2), 102–108.

25. Alexis, F., Pridgen, E., Molnar, L. K., Farokhzad, O. C. (2008). Factors affecting the clearance and biodistribution of polymeric nanoparticles. *Molecular Pharmacology 5* (4), 505–515.

26. Sonavane, G., Tomoda, K., Makino, K. (2008). Biodistribution of colloidal gold nano-particles after intravenous administration: effect of particle size. *Colloids and Surfaces B: Biointerfaces 66* (2), 274–280.

27. Hirn, S., Semmler-Behnke, M., Schleh, C., Wenk, A., Lipka, J., Schäffler, M., Takenaka, S., Möller, W., Schmid, G., Simon, U., Kreyling, W. G. (2011). Particle size-dependent and surface charge-dependent biodistribution of gold nanoparticles after intravenous administration. *European Journal of Pharmaceutics and Biopharmaceutics 77* (3), 407–416.

28. Duffin, R., Tran, L., Brown, D., Stone, V., Donaldson, K. (2007). Proinflammogenic effects of low-toxicity and metal nanoparticles *in vivo* and *in vitro*: highlighting the role of particle surface area and surface reactivity. *Inhalation Toxicology 19* (10), 849–856.

29. Wick, P., Manser, P., Limbach, L. K., Dettlaff-Weglikowska, U., Krumeich, F., Roth, S., Stark, W. J., Bruinink, A. (2007). The degree and kind of agglomeration affect carbon nanotube cytotoxicity. *Toxicology Letters 168* (2), 121–131.

30. Sato, Y., Yokoyama, A., Shibata, K., Akimoto, Y., Ogino, S., Nodasaka, Y., Kohgo, T., Tamura, K., Akasaka, T., Uo, M., Motomiya, K., Jeyadevan, B., Ishiguro, M., Hatakeyama, R., Watari, F., Tohji, K. (2005). Influence of length on cytotoxicity of multi-walled carbon nanotubes against human acute monocytic leukemia cell line THP-1 *in vitro* and subcutaneous tissue of rats *in vivo*. *Molecular Biosystems 1* (2), 176–182.

31. Decuzzi, P., Godin, B., Tanaka, T., Lee, S. Y., Chiappini, C., Liu, X., Ferrari, M. (2010). Size and shape effects in the biodistribution of intravascularly injected particles. *Journal of Controlled Release 141* (3), 320–327.

32. Tao, Z., Toms, B. B., Goodisman, J., Asefa, T. (2009). Mesoporosity and functional group dependent endocytosis and cytotoxicity of silica nanomaterials. *Chemical Research in Toxicology 22* (11), 1869–1880.

33. Patel, H. M., Moghimi, S. M. (1998). Serum-mediated recognition of liposomes by phagocytic cells of the reticuloendothelial system—the concept of tissue specificity. *Advanced Drug Delivery Reviews 32* (1–2), 45–60.

34. Bell, H., Kimber, W. L., Li, M., Whittle, I. R. (1998). Liposomal transfection efficiency and toxicity on glioma cell lines: *in vitro* and *in vivo* studies. *Neuroreport 9* (5), 793–798.

35. Kasai, T., Matsumura, S., Iizuka, T., Shiba, K., Kanamori, T., Yudasaka, M., Iijima, S., Yokoyama, A. (2011). Carbon nanohorns accelerate bone regeneration in rat calvarial bone defect. *Nanotechnology 22* (6), 065102.

36. Trouiller, B., Reliene, R., Westbrook, A., Solaimani, P., Schiestl, R. H. (2009). Titanium dioxide nanoparticles induce DNA damage and genetic instability *in vivo* in mice. *Cancer Research 69* (22), 8784–8789.

37. Goodman, C. M., McCusker, C. D., Yilmaz, T., Rotello, V. M. (2004). Toxicity of gold nanoparticles functionalized with cationic and anionic side chains. *Bioconjugate Chemistry 15* (4), 897–900.

38. Cho, W. S., Kim, S., Han, B. S., Son, W. C., Jeong, J. (2009). Comparison of gene expression profiles in mice liver following intravenous injection of 4 and 100 nm-sized PEG-coated gold nanoparticles. *Toxicology Letters 191* (1), 96–102.

39. Wang, X., Wang, Y., Chen, Z. G., Shin, D. M. (2009). Advances of cancer therapy by nanotechnology. *Cancer Research and Treatment 41* (1), 1–11.

40. Crouzier, D., Follot, S., Gentilhomme, E., Flahaut, E., Arnaud, R., Dabouis, V., Castellarin, C., Debouzy, J. C. (2010). Carbon nanotubes induce inflammation but decrease the production of reactive oxygen species in lung. *Toxicology 272* (1–3), 39–45.

41. Jain, R. K., Stylianopoulos, T. (2010). Delivering nanomedicine to solid tumors. *Nature Reviews Clinical Oncology 7* (11), 653–664.

42. Park, E. J., Park, K. (2009). Oxidative stress and pro-inflammatory responses induced by silica nanoparticles *in vivo* and *in vitro*. *Toxicology Letters 184* (1), 18–25.

43. Wu, J., Wang, C., Sun, J., Xue, Y. (2011). Neurotoxicity of silica nanoparticles: brain localization and dopaminergic neurons damage pathways. *ACS Nano 5* (6), 4476–4489.

44. Nishimori, H., Kondoh, M., Isoda, K., Tsunoda, S., Tsutsumi, Y., Yagi, K. (2009). Silica nanoparticles as hepatotoxicants. *European Journal of Pharmaceutics and Biopharmaceutics 72* (3), 496–501.

45. Liu, T., Li, L., Fu, C., Liu, H., Chen, D., Tang, F. (2012). Pathological mechanisms of liver injury caused by continuous intraperitoneal injection of silica nanoparticles. *Biomaterials 33* (7), 2399–2407.

46. Kshirsagar, N. A., Pandya, S. K., Kirodian, G. B., Sanath, S. (2005). Liposomal drug delivery system from laboratory to clinic. *Journal of Postgraduate Medicine 51*S5–S15.

47. Sonoke, S., Ueda, T., Fujiwara, K., Kuwabara, K., Yano, J. (2011). Galactose-modified cationic liposomes as a liver-targeting delivery system for small interfering RNA. *Biological & Pharmaceutical Bulletin 34* (8), 1338–1342.

48. Boisseau, P., Loubaton, B. (2011). Nanomedicine, nanotechnology in medicine. *Comptes Rendus Physique 12* (7), 620–636.

49. FDA. (1996). FDA approves DaunoXome as first-line therapy for Kaposi's sarcoma. Food and Drug Administration. *Journal of International Association of Physicians in AIDS Care 2* (5), 50–51.

50. Gordon, A. N., Fleagle, J. T., Guthrie, D., Parkin, D. E., Gore, M. E., Lacave, A. J. (2001). Recurrent epithelial ovarian carcinoma: a randomized phase III study of pegylated liposomal doxorubicin versus topotecan. *Journal of Clinical Oncology 19* (14), 3312–3322.

51. Chhikara, B. S., Parang, K. (2010). Development of cytarabine prodrugs and delivery systems for leukemia treatment. *Expert Opinion on Drug Delivery 7* (12), 1399–1414.

52. Batist, G., Barton, J., Chaikin, P., Swenson, C., Welles, L. (2002). Myocet (liposome-encapsulated doxorubicin citrate): a new approach in breast cancer therapy. *Expert Opinion on Pharmacotherapy 3* (12), 1739–1751.

53. Ando, K., Mori, K., Corradini, N., Redini, F., Heymann, D. (2011). Mifamurtide for the treatment of nonmetastatic osteosarcoma. *Expert Opinion on Pharmacotherapy 12* (2), 285–292.

54. Yatvin, M. B., Lelkes, P. I. (1982). Clinical prospects for liposomes. *Medical Physics 9* (2), 149–175.

55. Filion, M. C., Phillips, N. C. (1997). Toxicity and immunomodulatory activity of liposomal vectors formulated with cationic lipids toward immune effector cells. *Biochimica et Biophysica Acta 1329* (2), 345–356.

56. Szebeni, J., Moghimi, S. M. (2009). Liposome triggering of innate immune responses: a perspective on benefits and adverse reactions. *Journal of Liposome Research 19* (2), 85–90.

57. Hunter, A. C., Moghimi, S. M. (2002). Therapeutic synthetic polymers: a game of Russian roulette? *Drug Discovery Today 7* (19), 998–1001.

58. Mayhew, E., Ito, M., Lazo, R. (1987). Toxicity of non-drug-containing liposomes for cultured human cells. *Experimental Cell Research 171* (1), 195–202.

59. Adams, D. H., Joyce, G., Richardson, V. J., Ryman, B. E., Wisniewski, H. M. (1977). Liposome toxicity in the mouse central nervous system. *Journal of the Neurological Sciences 31* (2), 173–179.

60. Kuijpers, S. A., Coimbra, M. J., Storm, G., Schiffelers, R. M. (2010). Liposomes targeting tumour stromal cells. *Molecular Membrane Biology 27* (7), 328–340.

61. Chun, J. D., Lee, N., Kobayashi, R. H., Chaffee, S., Hershfield, M. S., Stiehm, E. R. (1993). Suppression of an antibody to adenosine-deaminase (ADA) in an ADA-deficient patient receiving polyethylene glycol modified adenosine deaminase. *Annals of Allergy 70* (6), 462–466.

62. Duncan, R. (2003). The dawning era of polymer therapeutics. *Nature Reviews Drug Discovery 2* (5), 347–360.

63. Nishiyama, N., Kataoka, K. (2006). Current state, achievements, and future prospects of polymeric micelles as nanocarriers for drug and gene delivery. *Pharmacology & Therapeutics 112* (3), 630–648.

64. Garnett, M. C., Kallinteri, P. (2006). Nanomedicines and nanotoxicology: some physiological principles. *Occupational Medicine 56* (5), 307–311.

65. Allmeroth, M., Moderegger, D., Biesalski, B., Koynov, K., Rosch, F., Thews, O., Zentel, R. (2011). Modifying the body distribution of HPMA-based copolymers by molecular weight and aggregate formation. *Biomacromolecules 12* (7), 2841–2849.

66. Suzuki, F., Pollard, R. B., Maeda, H. (1989). Stimulation of non-specific resistance to tumors in the mouse using a poly(maleic-acid-styrene)-conjugated neocarzinostatin. *Cancer Immunology, Immunotherapy 30* (2), 97–104.

67. Breslow, D. S., Edwards, E. I., Newburg, N. R. (1973). Divinyl ether-maleic anhydride (pyran) copolymer used to demonstrate the effect of molecular weight on biological activity. *Nature 246* (5429), 160–162.

68. Suzuki, F., Matsumoto, K., Schmitt, D. A., Pollard, R. B., Maeda, H. (1993). Immunomodulating activities of orally administered SMANCS, a polymer-conjugated derivative of the proteinaceous antibiotic neocarzinostatin, in an oily formulation. *International Journal of Immunopharmacology 15* (2), 175–183.

69. Pratten, M. K., Duncan, R., Cable, H. C., Schnee, R., Ringsdorf, H., Lloyd, J. B. (1981). Pinocytic uptake of divinyl ether-maleic anhydride (pyran copolymer) and its failure to stimulate pinocytosis. *Chemico-Biological Interactions 35* (3), 319–330.

70. Duncan, R., Vicent, M. J. (2010). Do HPMA copolymer conjugates have a future as clinically useful nanomedicines? A critical overview of current status and future opportunities. *Advanced Drug Delivery Reviews 62* (2), 272–282.

71. Tomalia, D. A., Baker, H., Dewald, J., Hall, M., Kallos, G., Martin, S., Roeck, J., Ryder, J., Smith, P. (1985). A new class of polymers: starburst-dendritic macromolecules. *Polymer Journal 17* (1), 117–132.

72. Gajbhiye, V., Palanirajan, V. K., Tekade, R. K., Jain, N. K. (2009). Dendrimers as therapeutic agents: a systematic review. *The Journal of Pharmacy and Pharmacology 61* (8), 989–1003.

73. Barkhausen, J., Ebert, W., Heyer, C., Debatin, J. F., Weinmann, H. J. (2003). Detection of atherosclerotic plaque with Gadofluorine-enhanced magnetic resonance imaging. *Circulation 108* (5), 605–609.

74. O'Loughlin, J., Millwood, I. Y., McDonald, H. M., Price, C. F., Kaldor, J. M., Paull, J. R. (2010). Safety, tolerability, and pharmacokinetics of SPL7013 gel (VivaGel): a dose ranging, phase I study. *Sexually Transmitted Diseases 37* (2), 100–104.

75. Bakry, R., Vallant, R. M., Najam-ul-Haq, M., Rainer, M., Szabo, Z., Huck, C. W., Bonn, G. K. (2007). Medicinal applications of fullerenes. *International Journal of Nanomedicine 2* (4), 639–649.

76. Varkouhi, A. K., Foillard, S., Lammers, T., Schiffelers, R. M., Doris, E., Hennink, W. E., Storm, G. (2011). SiRNA delivery with functionalized carbon nanotubes. *International Journal of Pharmaceutics 416* (2), 419–425.

77. Burke, A., Ding, X., Singh, R., Kraft, R. A., Levi-Polyachenko, N., Rylander, M. N., Szot, C., Buchanan, C., Whitney, J., Fisher, J., Hatcher, H. C., D'Agostino, R., Kock, N. D., Ajayan, P. M., Carroll, D. L., Akman, S., Torti, F. M., Torti, S. V. (2009). Long-term survival following a single treatment of kidney tumors with multiwalled carbon nanotubes and near-infrared radiation. *Proceedings of the National Academy of Sciences of the United States of America. 106*, 12897–12902.

78. Gomez-Gualdron, D. A., Burgos, J. C., Yu, J., Balbuena, P. B. (2011). Carbon nanotubes: engineering biomedical applications. *Progress in Molecular Biology and Translational Science 104*, 175–245.

79. Shvedova, A. A., Castranova, V., Kisin, E. R., Schwegler-Berry, D., Murray, A. R., Gandelsman, V. Z., Maynard, A., Baron, P. (2003). Exposure to carbon nanotube material: assessment of nanotube cytotoxicity using human keratinocyte cells. *Journal of Toxicology and Environmental Health. Part A 66* (20), 1909–1926.

80. Liu, D., Yi, C., Zhang, D., Zhang, J., Yang, M. (2010). Inhibition of proliferation and differentiation of mesenchymal stem cells by carboxylated carbon nanotubes. *ACS Nano 4* (4), 2185–2195.

81. Nel, A., Xia, T., Mädler, L., Li, N. (2006). Toxic potential of materials at the nanolevel. *Science 311* (5761), 622–627.

82. Chen, H. H., Yu, C., Ueng, T. H., Chen, S., Chen, B. J., Huang, K. J., Chiang, L. Y. (1998). Acute and subacute toxicity study of water-soluble polyalkylsulfonated C60 in rats. *Toxicologic Pathology 26* (1), 143–151.

83. Oberdorster, E. (2004). Manufactured nanomaterials (fullerenes, C60) induce oxidative stress in the brain of juvenile largemouth bass. *Environmental Health Perspectives 112* (10), 1058–1062.

84. Andrievsky, G., Klochkov, V., Derevyanchenko, L. (2005). Is the C60 fullerene molecule toxic?! *Fullerenes, Nanotubes and Carbon Nanostructures 13* (4), 363–376.

85. Bosi, S., Feruglio, L., Da Ros, T., Spalluto, G., Gregoretti, B., Terdoslavich, M., Decorti, G., Passamonti, S., Moro, S., Prato, M. (2004). Hemolytic effects of water-soluble fullerene derivatives. *Journal of Medicinal Chemistry 47* (27), 6711–6715.

86. Tsuchiya, T., Oguri, I., Yamakoshi, Y. N., Miyata, N. (1996). Novel harmful effects of [60]fullerene on mouse embryos *in vitro* and *in vivo*. *FEBS Letters 393* (1), 139–145.

87. Poland, C. A., Duffin, R., Kinloch, I., Maynard, A., Wallace, W. A., Seaton, A., Stone, V., Brown, S., Macnee, W., Donaldson, K. (2008). Carbon nanotubes introduced into the abdominal cavity of mice show asbestos-like pathogenicity in a pilot study. *Nature Nanotechnology 3* (7), 423–428.

88. Morimoto, Y., Hirohashi, M., Ogami, A., Oyabu, T., Myojo, T., Todoroki, M., Yamamoto, M., Hashiba, M., Mizuguchi, Y., Lee, B. W., Kuroda, E., Shimada, M., Wang, W. N., Yamamoto, K., Fujita, K., Endoh, S., Uchida, K., Kobayashi, N., Mizuno, K., Inada, M., Tao, H., Nakazato, T., Nakanishi, J., Tanaka, I. (2011). Pulmonary toxicity of well-dispersed multi-wall carbon nanotubes following inhalation and intratracheal instillation. *Nanotoxicology. 6* (6), 587–599.

89. Varga, C., Szendi, K. (2010). Carbon nanotubes induce granulomas but not mesotheliomas. *In Vivo 24* (2), 153–156.

90. Kamata, H., Tasaka, S., Inoue, K., Miyamoto, K., Nakano, Y., Shinoda, H., Kimizuka, Y., Fujiwara, H., Ishii, M., Hasegawa, N., Takamiya, R., Fujishima, S., Takano, H., Ishizaka, A. (2011). Carbon black nanoparticles enhance bleomycin-induced lung inflammatory and fibrotic changes in mice. *Experimental Biology and Medicine 236* (3), 315–324.

91. Wessels, A., Van Berlo, D., Boots, A. W., Gerloff, K., Scherbart, A. M., Cassee, F. R., Gerlofs-Nijland, M. E., Van Schooten, F. J., Albrecht, C., Schins, R. P. (2011). Oxidative stress and DNA damage responses in rat and mouse lung to inhaled carbon nanoparticles. *Nanotoxicology 5* (1), 66–78.

92. Yoshida, S., Hiyoshi, K., Oshio, S., Takano, H., Takeda, K., Ichinose, T. (2010). Effects of fetal exposure to carbon nanoparticles on reproductive function in male offspring. *Fertility and Sterility 93* (5), 1695–1699.

93. Qhobosheane, M., Santra, S., Zhang, P., Tan, W. (2001). Biochemically functionalized silica nanoparticles. *The Analyst 126* (8), 1274–1278.

94. Zhang, Q., Zhang, L., Liu, B., Lu, X., Li, J. (2007). Assembly of quantum dots-mesoporous silicate hybrid material for protein immobilization and direct electrochemistry. *Biosensors & Bioelectronics 23* (5), 695–700.

95. Wang, Y., Caruso, F. (2004). Enzyme encapsulation in nanoporous silica spheres. *Chemical Communications (Cambridge, England)* (13) 1528–1529.

96. Benezra, M., Penate-Medina, O., Zanzonico, P. B., Schaer, D., Ow, H., Burns, A., DeStanchina, E., Longo, V., Herz, E., Iyer, S., Wolchok, J., Larson, S. M., Wiesner, U., Bradbury, M. S. (2011). Multimodal silica nanoparticles are effective cancer-targeted probes in a model of human melanoma. *The Journal of Clinical Investigation 121* (7), 2768–2780.

97. Rosenholm, J. M., Peuhu, E., Bate-Eya, L. T., Eriksson, J. E., Sahlgren, C., Linden, M. (2010). Cancer-cell-specific induction of apoptosis using mesoporous silica nanoparticles as drug-delivery vectors. *Small 6* (11), 1234–1241.

98. Popat, A., Hartono, S. B., Stahr, F., Liu, J., Qiao, S. Z., Qing Max, G. Lu. (2011). Mesoporous silica nanoparticles for bioadsorption, enzyme immobilisation, and delivery carriers. *Nanoscale 3* (7), 2801–2818.

99. Liu, Q., Zhang, J., Sun, W., Xie, Q. R., Xia, W., Gu, H. (2012). Delivering hydrophilic and hydrophobic chemotherapeutics simultaneously by magnetic mesoporous silica nanoparticles to inhibit cancer cells. *International Journal of Nanomedicine 7*, 999–1013.

100. Choi, J., Burns, A. A., Williams, R. M., Zhou, Z., Flesken-Nikitin, A. Zipfel, W. R., Wiesner, U., Nikitin, A. Y. (2007). Core-shell silica nanoparticles as fluorescent labels for nanomedicine. *Journal of Biomedical Optics 12* (6), 064007.

101. Vallhov, H., Gabrielsson, S., Stromme, M., Scheynius, A., Garcia-Bennett, A. E. (2007). Mesoporous silica particles induce size dependent effects on human dendritic cells. *Nano Letters 7* (12), 3576–3582.

102. Napierska, D., Thomassen, L. C., Rabolli, V., Lison, D., Gonzalez, L., Kirsch-Volders, M. Martens, J. A., Hoet, P. H. (2009). Size-dependent cytotoxicity of monodisperse silica nanoparticles in human endothelial cells. *Small 5* (7), 846–853.

103. Hudson, S. P., Padera, R. F., Langer, R., Kohane, D. S. (2008). The biocompatibility of mesoporous silicates. *Biomaterials 29* (30), 4045–4055.

104. Yu, T., Greish, K., McGill, L. D., Ray, A., Ghandehari, H. (2012). Influence of geometry, porosity, and surface characteristics of silica nanoparticles on acute toxicity: their vasculature effect and tolerance threshold. *ACS Nano. 6* (3), 2289–2301.

105. Yu, T., Malugin, A., Ghandehari, H. (2011). Impact of silica nanoparticle design on cellular toxicity and hemolytic activity. *ACS Nano 5* (7), 5717–5728.

106. Nemmar, A., Nemery, B., Hoet, P. H., Van Rooijen, N., Hoylaerts, M. F. (2005). Silica particles enhance peripheral thrombosis: key role of lung macrophage-neutrophil crosstalk. *American Journal of Respiratory and Critical Care Medicine 171* (8), 872–879.

107. Huang, X., Li, L., Liu, T., Hao, N., Liu, H., Chen, D., Tang, F. (2011). The shape effect of mesoporous silica nanoparticles on biodistribution, clearance, and biocompatibility *in vivo*. *ACS Nano 5* (7), 5390–5399.

108. Yamashita, K., Yoshioka, Y., Higashisaka, K., Mimura, K., Morishita, Y., Nozaki, M., Yoshida, T., Ogura, T., Nabeshi, H., Nagano, K., Abe, Y., Kamada, H., Monobe, Y., Imazawa, T., Aoshima, H., Shishido, K., Kawai, Y., Mayumi, T., Tsunoda, S., Itoh, N., Yoshikawa, T., Yanagihara, I., Saito, S., Tsutsumi, Y. (2011). Silica and titanium dioxide nanoparticles cause pregnancy complications in mice. *Nature Nanotechnology 6* (5), 321–328.

109. Exley, C., Siesjo, P., Eriksson, H. (2010). The immunobiology of aluminium adjuvants: how do they really work? *Trends in Immunology 31* (3), 103–109.

110. Becaria, A., Campbell, A., Bondy, S. C. (2002). Aluminum as a toxicant. *Toxicology and Industrial Health 18* (7), 309–320.

111. Arieff, A. I., Cooper, J. D., Armstrong, D., Lazarowitz, V. C. (1979). Dementia, renal failure, and brain aluminum. *Annals of Internal Medicine 90* (5), 741–747.

112. Bondy, S. C. (2010). The neurotoxicity of environmental aluminum is still an issue. *Neurotoxicology 31* (5), 575–581.

113. Yumoto, S., Kakimi, S., Ohsaki, A., Ishikawa, A. (2009). Demonstration of aluminum in amyloid fibers in the cores of senile plaques in the brains of patients with Alzheimer's disease. *Journal of Inorganic Biochemistry 103* (11), 1579–1584.

114. Banks, W. A., Niehoff, M. L., Drago, D., Zatta, P. (2006). Aluminum complexing enhances amyloid beta protein penetration of blood-brain barrier. *Brain Research 1116* (1), 215–221.

115. Frisardi, V., Solfrizzi, V., Capurso, C., Kehoe, P. G., Imbimbo, B. P., Santamato, A., Dellegrazie, F., Seripa, D., Pilotto, A., Capurso, A., Panza, F. (2010). Aluminum in the diet and Alzheimer's disease: from current epidemiology to possible disease-modifying treatment. *Journal of Alzheimer's Disease 20* (1), 17–30.

116. Hafejee, A., Burke, M. J. (2004). Acute pneumonitis starting 2 hours after intramuscular gold administration in a patient with rheumatoid arthritis. *Annals of the Rheumatic Diseases 63* (11), 1525–1526.

117. Elsherbini, A. A., Saber, M., Aggag, M., El-Shahawy, A., Shokier, H. A. (2011). Laser and radiofrequency-induced hyperthermia treatment via gold-coated magnetic nano-composites. *International Journal of Nanomedicine 6*, 2155–2165.

118. Reuveni, T., Motiei, M., Romman, Z., Popovtzer, A., Popovtzer, R. (2011). Targeted gold nanoparticles enable molecular CT imaging of cancer: an *in vivo* study. *International Journal of Nanomedicine 6*, 2859–2864.

119. Rink, J. S., McMahon, K. M., Chen, X., Mirkin, C. A., Thaxton, C. S., Kaufman, D. B. (2010). Transfection of pancreatic islets using polyvalent DNA-functionalized gold nanoparticles. *Surgery 148* (2), 335–345.

120. Kumari, S., Singh, R. P. (2011). Glycolic acid-*g*-chitosan-gold nanoflower nanocomposite scaffolds for drug delivery and tissue engineering. *International Journal of Biological Macromolecules. 50* (3), 878–883.

121. Sun, Y. N., Wang, C. D., Zhang, X. M., Ren, L., Tian, X. H. (2011). Shape dependence of gold nanoparticles on *in vivo* acute toxicological effects and biodistribution. *Journal of Nanoscience and Nanotechnology 11* (2), 1210–1216.

122. Aillon, K. L., Xie, Y., El-Gendy, N., Berkland, C. J., Forrest, M. L. (2009). Effects of nanomaterial physicochemical properties on *in vivo* toxicity. *Advanced Drug Delivery Reviews 61* (6), 457–466.

123. Abdelhalim, M. A., Jarrar, B. M. (2011). Gold nanoparticles induced cloudy swelling to hydropic degeneration, cytoplasmic hyaline vacuolation, polymorphism, binucleation, karyopyknosis, karyolysis, karyorrhexis and necrosis in the liver. *Lipids in Health and Disease 10*, 166.

124. Zhang, X. D., Wu, D., Shen, X., Liu, P. X., Yang, N., Zhao, B., Zhang, H., Sun, Y. M., Zhang, L. A., Fan, F. Y. (2011). Size-dependent *in vivo* toxicity of PEG-coated gold nanoparticles. *International Journal of Nanomedicine 6*, 2071–2081.

125. Naqvi, S., Samim, M., Abdin, M., Ahmed, F. J., Maitra, A., Prashant, C., Dinda, A. K. (2010). Concentration-dependent toxicity of iron oxide nanoparticles mediated by increased oxidative stress. *International Journal of Nanomedicine 5*, 983–989.

126. Rosner, M. H., Auerbach, M. (2011). Ferumoxytol for the treatment of iron deficiency. *Expert Review of Hematology 4* (4), 399–406.

127. Voinov, M. A., Sosa Pagan, J. O., Morrison, E., Smirnova, T. I., Smirnov, A. I. (2011). Surface-mediated production of hydroxyl radicals as a mechanism of iron oxide nanoparticle biotoxicity. *Journal of the American Chemical Society 133* (1), 35–41.

128. Mahmoudi, M., Laurent, S., Shokrgozar, M. A., Hosseinkhani, M. (2011). Toxicity evaluations of superparamagnetic iron oxide nanoparticles: cell "vision" versus physicochemical properties of nanoparticles. *ACS Nano 5* (9), 7263–7276.

129. Patel, V. P., Chu, C. T. (2011). Nuclear transport, oxidative stress, and neurodegeneration. *International Journal of Clinical Experimental Pathology 4* (3), 215–229.

130. Lee, S., Park, Y., Zuidema, M. Y., Hannink, M., Zhang, C. (2011). Effects of interventions on oxidative stress and inflammation of cardiovascular diseases. *World Journal of Cardiology 3* (1), 18–24.

131. Klaunig, J. E., Kamendulis, L. M., Hocevar, B. A. (2010). Oxidative stress and oxidative damage in carcinogenesis. *Toxicologic Pathology 38* (1), 96–109.

132. Maiti, S. (2011). Nanotoxicity of gold and iron nanoparticles. *Journal of Biomedical Nanotechnology 7* (1), 65.

133. Singh, N., Jenkins, G. J., Nelson, B. C., Marquis, B. J., Maffeis, T. G., Brown, A. P., Williams, P. M., Wright, C. J., Doak, S. H. (2012). The role of iron redox state in the genotoxicity of ultrafine superparamagnetic iron oxide nanoparticles. *Biomaterials 33* (1), 163–170.

134. Apopa, P. L., Qian, Y., Shao, R., Guo, N. L., Schwegler-Berry, D., Pacurari, M., Porter, D., Shi, X., Vallyathan, V., Castranova, V., Flynn, D. C. (2009). Iron oxide nanoparticles induce human microvascular endothelial cell permeability through reactive oxygen species production and microtubule remodeling. *Particle and Fibre Toxicology 6*, 1.

135. Zhu, M. T., Feng, W. Y., Wang, B., Wang, T. C., Gu, Y. Q., Wang, M., Wang, Y., Ouyang, H., Zhao, Y. L., Chai, Z. F. (2008). Comparative study of pulmonary responses to nano- and submicron-sized ferric oxide in rats. *Toxicology 247* (2–3), 102–111.

136. Kooi, M. E., Cappendijk, V. C., Cleutjens, K. B., Kessels, A. G., Kitslaar, P. J., Borgers, M., Frederik, P. M., Daemen, M. J., van Engelshoven, J. M. (2003). Accumulation of ultrasmall superparamagnetic particles of iron oxide in human atherosclerotic plaques can be detected by *in vivo* magnetic resonance imaging. *Circulation 107* (19), 2453–2458.

137. Rahman, M. F., Wang, J., Patterson, T. A., Saini, U. T., Robinson, B. L., Newport, G. D., Murdock, R. C., Schlager, J. J., Hussain, S. M., Ali, S. F. (2009). Expression of genes related to oxidative stress in the mouse brain after exposure to silver-25 nanoparticles. *Toxicology Letters 187* (1), 15–21.

138. Hritcu, L., Stefan, M., Ursu, L., Neagu, A., Mihasan, M., Tartau, L., Melnig, V. (2011). Exposure to silver nanoparticles induces oxidative stress and memory deficits in laboratory rats. *Central European Journal of Biology 6* (4), 497–509.

139. Gao, X., Yin, S., Tang, M., Chen, J., Yang, Z., Zhang, W., Chen, L., Yang, B., Li, Z., Zha, Y., Ruan, D., Wang, M. (2011). Effects of developmental exposure to TiO2 nanoparticles on synaptic plasticity in hippocampal dentate gyrus area: an *in vivo* study in anesthetized rats. *Biological Trace Element Research 143* (3), 1616–1628.

140. Han, D., Tian, Y., Zhang, T., Ren, G., Yang, Z. (2011). Nano-zinc oxide damages spatial cognition capability via over-enhanced long-term potentiation in hippocampus of Wistar rats. *International Journal of Nanomedicine 6*, 1453–1461.

141. Hu, R., Zheng, L., Zhang, T., Gao, G., Cui, Y., Cheng, Z., Cheng, J., Hong, M., Tang, M., Hong, F. (2011). Molecular mechanism of hippocampal apoptosis of mice following exposure to titanium dioxide nanoparticles. *Journal of Hazardous Materials 191* (1–3), 32–40.

142. Zhang, D. C. S., ADME studies in animals and humans: experimental design, metabolite profiling and identification, and data presentation, In *Drug Metabolism. Drug Design and Development: Basic Concepts and Practice*, ed. Zhang, D., Zhu, M.; Humphreys, W. G., Wiley, Hoboken, NJ, 2007.

143. Kobori, T., Watanabe, J., Nakao, H. (2012). Gold nanoparticles as localization markers for direct and live imaging of particle absorption through a Caco-2 cell monolayer using dark-field microscopy. *Analytical Sciences 28* (1), 61.

144. Jeong, S. H., Kim, J. H., Yi, S. M., Lee, J. P., Kim, J. H., Sohn, K. H., Park, K. L., Kim, M. K., Son, S. W. (2010). Assessment of penetration of quantum dots through *in vitro* and *in vivo* human skin using the human skin equivalent model and the tape stripping method. *Biochemical and Biophysical Research Communications 394* (3), 612–615.

145. Iqbal, J., Vigl, C., Moser, G., Gasteiger, M., Perera, G., Bernkop-Schnurch, A. (2011). Development and *in vivo* evaluation of a new oral nanoparticulate dosage form for leuprolide based on polyacrylic acid. *Drug Delivery 18* (6), 432–440.

146. Zvyagin, A. V., Zhao, X., Gierden, A., Sanchez, W., Ross, J. A., Roberts, M. S. (2008). Imaging of zinc oxide nanoparticle penetration in human skin *in vitro* and *in vivo*. *Journal of Biomedical Optics 13* (6), 064031.

147. Castellino, S., Groseclose, M. R., Wagner, D. (2011). MALDI imaging mass spectrometry: bridging biology and chemistry in drug development. *Bioanalysis 3* (21), 2427–2441.

148. Arnida, M. M., Janat-Amsbury, A., Ray, C. M., Peterson, Ghandehari, H. (2011). Geometry and surface characteristics of gold nanoparticles influence their biodistribution and uptake by macrophages. *European Journal of Pharmaceutics and Biopharmaceutics 77* (3), 417–423.

149. Winters, D. K., Cederbaum, A. I. (1992). Expression of a catalytically active human cytochrome P-4502E1 in Escherichia coli. *Biochimica et Biophysica Acta 1156* (1), 43–49.

150. Li, A. P. (2001). Screening for human ADME/Tox drug properties in drug discovery. *Drug Discovery Today 6* (7), 357–366.

151. Tyner, K., Sadrieh, N. (2011). Considerations when submitting nanotherapeutics to FDA/CDER for regulatory review. *Methods in Molecular Biology 697*, 17–31.

152. Mortelmans, K., Zeiger, E. (2000). The Ames Salmonella/microsome mutagenicity assay. *Mutation Research 455* (1–2), 29–60.

153. Cotelle, S., Ferard, J. F. (1999). Comet assay in genetic ecotoxicology: a review. *Environmental and Molecular Mutagenesis 34* (4), 246–255.

154. Weber, E., Bidwell, K., Legator, M. S. (1975). An evaluation of the micronuclei test using triethylenemelamine, trimethylphosphate, hycanthone and niridazole. *Mutation Research 28* (1), 101–106.

155. OECD, *Test No. 486: Unscheduled DNA Synthesis (UDS) Test with Mammalian Liver Cells* in vivo: OECD Publishing.

156. Parodi, S., Malacarne, D., Romano, P., Taningher, M. (1991). Are genotoxic carcinogens more potent than nongenotoxic carcinogens? *Environmental Health Perspectives 95*, 199–204.

157. Hernandez, L. G., van Steeg, H., Luijten, M., van Benthem, J. (2009). Mechanisms of non-genotoxic carcinogens and importance of a weight of evidence approach. *Mutation Research 682* (2–3), 94–109.

158. Creton, S., Aardema, M. J., Carmichael, P. L., Harvey, J. S., Martin, F. L., Newbold, R. F., O'Donovan, M. R., Pant, K., Poth, A., Sakai, A., Sasaki, K., Scott, A. D., Schechtman, L. M., Shen, R. R., Tanaka, N., Yasaei, H. (2012). Cell transformation assays for prediction of carcinogenic potential: state of the science and future research needs. *Mutagenesis 27* (1), 93–101.

159. OECD, *Test No. 451: Carcinogenicity Studies*: OECD Publishing.

160. Wiwanitkit, V., Sereemaspun, A., Rojanathanes, R. (2009). Effect of gold nanoparticles on spermatozoa: the first world report. *Fertility and Sterility 91* (1), e7–e8.

161. Brendler-Schwaab, S., Hartmann, A., Pfuhler, S., Speit, G. (2005). The *in vivo* comet assay: use and status in genotoxicity testing. *Mutagenesis 20* (4), 245–254.

162. Chan, W. H., Shiao, N. H. (2008). Cytotoxic effect of CdSe quantum dots on mouse embryonic development. *Acta Pharmacologica Sinica 29* (2), 259–266.

163. OECD, *Test No. 415: One-Generation Reproduction Toxicity Study*: OECD Publishing.

INDEX

Nanoparticulate Drug Delivery Systems: Strategies, Technologies, and Applications, First Edition.
Edited by Yoon Yeo.
© 2013 John Wiley & Sons, Inc. Published 2013 by John Wiley & Sons, Inc.